WILLIAM L BAILEY

W9-BZC-205

Risk Assessment: Logic and Measurement

Edited by

Michael C. Newman
Carl L. Strojan

Ann Arbor Press
Chelsea, Michigan

Library of Congress Cataloging-in-Publication Data

    Risk assessment : logic and measurement / edited by Michael C. Newman, Carl L.
      Strojan.
          p.   cm.
      Includes bibliographical references and index.
      ISBN 1-57504-048-4
      1. Environmental risk assessment. 2. Human ecology. 3. Environmental toxicol-
    ogy. I. Newman, Michael C. II. Strojan, Carl. L.
    GE145.R58   1998
    364.73--dc21   97-46420
                                                    CIP

ISBN 1-57504-048-4

COPYRIGHT 1998 by Sleeping Bear Press
ALL RIGHTS RESERVED

This book represents information obtained from authentic and highly regarded sources. Reprinted material is quoted with permission, and sources are indicated. A wide variety of references are listed. Every reasonable effort has been made to give reliable data and information, but the author and publisher cannot assume responsibility for the validity of all materials or for the consequences of their use.

Neither this book nor any part may be reproduced or transmitted in any form or by any means, electronic or mechanical, including photocopying, microfilming, and recording, or by any information storage or retrieval system, without permission in writing from the publisher.

ANN ARBOR PRESS
121 South Main Street, Chelsea, Michigan 48118
Ann Arbor Press is an imprint of Sleeping Bear Press

PRINTED IN THE UNITED STATES OF AMERICA
10 9 8 7 6 5 4 3 2 1

Acknowledgments

This book contains information generated during the Savannah River Ecology Laboratory (SREL) symposium, *Risk Assessment: Logic and Measurement*. This symposium was held August 19 to 21, 1996 in Aiken, South Carolina (U.S.A.) and was funded under U.S. Department of Energy (DOE) Financial Assistance Award (Number DE-FC09-96SR18546) from the DOE to the University of Georgia Research Foundation. The editors are grateful to Dr. Nat Frazer for providing support. Special thanks are extended to Ms. Edith Towns, who acted as symposium coordinator and SREL technical editor for this book. The editors thank the following individuals for their timely and thoughtful reviews of chapter manuscripts.

Darvene Adams—U.S. EPA
Peter Chapman—Zeneca Agrochemicals
Mark Crane—Royal Holloway University of London
Philip Dixon—Savannah River Ecology Laboratory
Scott Ferson—Applied Biomathematics
Richard Gilbert—Battelle Washington Office
Lowell Greenbaum—Medical College of Georgia
Tom Hakonson—Colorado State University
Barbara Hamm—Westinghouse Savannah River Company
Tom Hinton—Savannah River Ecology Laboratory
Annie M. Jarabek—U.S. EPA National Center for Environmental Assessment
Robert C. Lee—University of Washington
Craig Loehle—Argonne National Laboratory
Geoffrey Matthews—Western Washington University
Lyman McDonald—WEST, Inc.
Frank Noppert—Ministry of Transport and Waterworks, The Netherlands
Sue Norton—U.S. EPA
John Pinder—Savannah River Ecology Laboratory
Richard Pratt—Portland State University
Don Rodier—U.S. EPA
Tim Sparks—Institute of Terrestrial Ecology
Mark Sprenger—U.S. EPA
John Thomas—Washington State University
Eberhard Voit—Medical University of South Carolina
Barry Zajdlik—Zajdlik Associates

About the Editors

Michael C. Newman

Dr. Newman is presently a Full Professor in the Department of Environmental Sciences at the College of William and Mary's Virginia Institute of Marine Science. Until 1997, he was a Senior Research Scientist at the University of Georgia Savannah River Ecology Laboratory and Head of its Environmental Toxicology, Remediation, and Risk Assessment (ETRRA) Group. After receiving B.A. (biological sciences) and M.S. (zoology) degrees from the University of Connecticut, he earned M.S. and Ph.D. degrees in Environmental Sciences from Rutgers University. After postdoctoral fellowships at the University of Georgia and the University of California–San Diego, he joined the research faculty at the University of Georgia (1983). His research interests include toxicity and bioaccumulation models, toxicant effects on populations, factors modifying toxicity and bioaccumulation, quantitative methods for ecological risk assessment, statistical toxicology, and inorganic water chemistry. He has published more than 70 scientific articles on these topics. He authored the book *Quantitative Methods in Aquatic Ecotoxicology* (1995) and was senior editor of *Metal Ecotoxicology: Concepts and Applications* (1991, with A.W. McIntosh) and *Ecotoxicology: A Hierarchical Treatment* (1996, with C.H. Jagoe). He also directed the development of UNCENSOR, a program that produces univariate statistics for data sets containing "below detection limit" observations.

Carl L. Strojan

Dr. Strojan is a Research Manager at the University of Georgia Savannah River Ecology Laboratory. He received a B.A. degree in Biology from Antioch College and a Ph.D. in Ecology from Rutgers University. He also spent a year at the Universität von Tübingen in Tübingen, Germany. Following positions as a postdoctoral fellow at UCLA and a senior scientist at the Solar Energy Research Institute in Golden, Colorado, he joined the Savannah River Ecology Laboratory. He has conducted or managed environmental research in basic and applied ecology, the environmental impact of energy technologies, and contaminant fate and effects. In his present position, Dr. Strojan has been responsible for the day-to-day management of issues dealing with people, programs, and resources in areas such as ecotoxicology, environmental chemistry, and radiation ecology. He is a member of the American Association for the Advancement of Science, the Ecological Society of America, and Sigma Xi. He is the author or co-author of over 25 scientific publications.

About the Contributors

Steven M. Bartell
SENES Oak Ridge, Inc.
Center for Risk Analysis
102 Donner Drive
Oak Ridge, TN 37830

John J. Bascietto
U.S. Department of Energy
1000 Independence Ave., SW
Washington, DC 20585

J.J.M. Bedaux
Vrije Universiteit
Department of Theoretical Biology
de Boelelaan 1087
NL-1081 HV Amsterdam
The Netherlands

Kym Rouse Campbell
SENES Oak Ridge, Inc.
Center for Risk Analysis
102 Donner Drive
Oak Ridge, TN 37830

Philip M. Dixon
Savannah River Ecology Laboratory
University of Georgia
Aiken, SC 29802

Scott Ferson
Applied Biomathematics
100 North Country Road
Setauket, NY 11733

A.A.M. Gerritsen
TNO-Toxicology
Department of Environmental
   Toxicology
P.O. Box 6011
NL-2600 JA Delft
The Netherlands

A.O. Hanstveit
TNO-Toxicology
Department of Environmental
   Toxicology
P.O.Box 6011
NL-2600 JA Delft
The Netherlands

Thomas G. Hinton
Savannah River Ecology Laboratory
University of Georgia
Aiken, SC 29802

Lucinda B. Johnson
Natural Resources Research Institute
University of Minnesota
5013 Miller Trunk Hwy.
Duluth, MN 55811

Thomas B. Kirchner
Carlsbad Environmental Monitoring
   and Research Center
New Mexico State University
1400 University Drive
Carlsbad, NM 88220

S.A.L.M. Kooijman
Vrije Universiteit
Department of Theoretical Biology
de Boelelaan 1087
NL-1081 HV Amsterdam
The Netherlands

Wayne G. Landis
Institute of Environmental Toxicology
   and Chemistry
Huxley College of
   Environmental Studies
Western Washington University
Bellingham, WA 98225

Thomas F. Long
ChemRisk,
    A Division of McLaren-Hart
29225 Chagrin Boulevard
Cleveland, OH 44122

Geoffrey B. Matthews
Computer Science Department
Western Washington University
Bellingham, WA 98225

Robin A. Matthews
Institute for Watershed Studies
Huxley College of
    Environmental Studies
Western Washington University
Bellingham, WA 98225

David A. Mauriello
U.S. Environmental Protection Agency
Office of Pollution Prevention
    and Toxics
Risk Assessment Division 7403
401 M Street, SW
Washington, DC 20460

Michael C. Newman
Department of Environmental Sciences
Virginia Institute of Marine Science
College of William and Mary
Gloucester Point, VA 23062

H. Oldersma
TNO-Toxicology
Department of Environmental
    Toxicology
P.O.Box 6011
NL-2600 JA Delft
The Netherlands

Carl Richards
Natural Resources Research Institute
University of Minnesota
5013 Miller Trunk Hwy.
Duluth, MN 55811

Donald Rodier
U.S. Environmental Protection Agency
Office of Pollution Prevention and Toxics
Health and Environmental Review
    Division 7403
401 M Street, SW
Washington, DC 20460

P. Barry Ryan
Department of Environmental and
    Occupational Health
Rollins School of Public Health
Emory University
1518 Clifton Road, NE
Atlanta, GA 30322

Mary K. Schubauer-Berigan
Department of Biometry and
    Epidemiology
Medical University of South Carolina
Charleston, SC 29425-2503

Francis Stay
U.S. Environmental Protection Agency
Office of Research and Development
Mid-Continent Ecology Division
6201 Congdon Boulevard
Duluth, MN 55804

Carl L. Strojan
Savannah River Ecology Laboratory
University of Georgia
P.O. Drawer E
Aiken, SC 29802

Glenn W. Suter II
Environmental Sciences Division
Oak Ridge National Laboratory
Oak Ridge, TN 37831-6038

Eberhard O. Voit
Department of Biometry and
    Epidemiology
Medical University of South Carolina
Charleston, SC 29425-2503

Contents

Part 4
Conclusion

Part I
Introduction

1 ‖ Risk Assessment: Logic and Measurement

Michael C. Newman and Carl L. Strojan

> *Now, what I want is, Facts. Teach these boys and girls nothing but Facts. Facts alone are wanted in life. Plant nothing else, and root out everything else. You can only form the minds of reasoning animals upon Facts: nothing else will ever be of service to them.*
>
> —Dickens (1854)

THE ONE THING NEEDFUL?

Charles Dickens opens his novel, *Hard Times,* with Mr. Gradgrind advocating this dreary means of producing informed citizens. He taught accordingly from a "monotonous vault of a school room" in Coketown, a town of red brick smudged with smoke and ash. Like Scrooge, Gradgrind undergoes a profound transformation as the story progresses, realizing that Facts are not at all the one needful thing in life.

Similarly, during the last fifty years, environmental toxicologists have been preoccupied with facts—details and particulars—to the relative neglect of sound logical constructs, quantitative methods, and models to best predict risk to health and environment (Newman, 1995; 1996). Toxicity data have been compiled for countless species and chemicals under various conditions without careful evaluation of their value for assessing risk in a natural setting. For example, regardless of the impressive size of our current 96 h LC_{50} database, effective application of such information to estimate lethal consequences of contaminant releases to natural populations is greatly compromised because ecotoxicologists have inadequately addressed exposure duration. The changes in lethal impact over time are largely ignored and effects are noted only at one time endpoint. This produces inadequate information because risk assessments require consideration of duration as well as magnitude (concentration) of exposure (Norton et al., 1992; EPA, 1989a). Consequently, considerable uncertainty is introduced into risk assessments during extrapolation to other exposure durations from such data (Newman and Aplin, 1992; Newman, 1995; Newman and McCloskey, 1996). Further, important covariates are controlled in laboratory assays—so much so that the results are unrealistic and unifaceted. For example, the use of animals of uniform size in an ecotoxicity test compromises prediction of effect on a field population composed of animals of different sizes. Most databases containing NOAEC/LOAEC and associated RfD[1] data share these fundamental flaws. As another example, fact gathering for ecological risk assessments remains biased toward lower levels of ecological organization, such as the individual (Clements and Kiffney, 1994), despite the acknowledged higher relevance associated with information about effects at higher levels, e.g., effects to communities or landscapes (Taub, 1989; Cairns, 1993).

THE FOUR THINGS NEEDFUL

Fortunately, movement away from excessive and unthoughtful data collection to the neglect of other important factors in environmental toxicology began about fifteen years ago. Although bias still exists toward unthoughtful collection of facts, other important things are also being addressed. The important factors relative to present efforts in improving risk assessment can be broken into four general categories: generation of relevant data, logical framework, quantitative measurement techniques, and appropriate models (Figure 1.1). All four must eventually come together in a balanced manner to effectively assess risk to health and environment. The intent of this book is to foster this integration.

Relevant Data

Generation of data of the appropriate type and quality remains an essential activity. However, our efforts to acquire more data to the neglect of efforts in the three other categories of effort lead to inconsistency between the kinds of data being generated and the kinds of data required, e.g., a predominance of single species effects data despite a strong need for population and community effects data (Barnthouse et al., 1987). This type of data generation seems to have been characteristic of the science of ecotoxicology for decades, as evidenced by Moriarty's (1983) quote from his classic textbook that "...the [ecotoxicology] literature is both enormous and, in large part, trivial." Newman (1995, 1996) attributes this bias partially to the transition of ecotoxicology from a young science preoccupied with fact generation to a mature science with an appropriate balance between fact gathering and the generation or testing of paradigms. Ecotoxicology is now making the transition to a mature science and is beginning to slowly move away from this bias (Newman, 1995, 1996). Our capacity for accurate risk assessment can only improve as a result of such movement.

Also essential is the assessment of data quality and the organization of high-quality information in a readily accessible format. This also has improved in the last decade. Relative to human health risk assessment, the EPA has compiled RfD, slope factors, drinking water advisories, and other crucial information in IRIS (Integrated Risk Information System), a database accessible from the U.S. Government's Right-to-Know web site (http://www.rtk.net/T866). For aquatic biota, wildlife, terrestrial plants, sediments, and soil invertebrates and microbes, Oak Ridge National Laboratory has a similar web site (http://www.hsrd.ornl.gov/ecorisk). These and similar databases contribute tremendously to risk assessment activities. However, the logic associated with their generation and application remains compromised, e.g.; RfDs are based on the NOAEC/LOAEC methods.

---

[1] The reference dose (Rfd) for noncarcinogenic effects is the best estimate of the daily exposure that will result in no significant risk of an adverse effect if it is not exceeded.

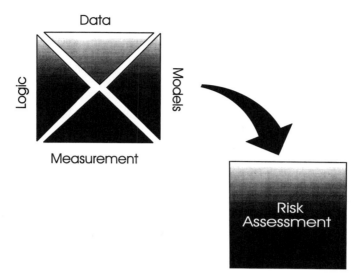

Figure 1.1. Four general categories of effort that contribute to effective risk assessment. Sound and relevant data organized in a consistent and accessible manner are essential. A logical framework for developing hypotheses, planning sampling programs, and defining the structure to the assessment activity must also be established. Quantitative measurement methods, including statistical methods, must be applied correctly to data in the context of the logical framework. For prediction of or definition of fate and effects, quantitative models are developed that draw from relevant data, the logical framework of the risk assessment, and quantitative measurement techniques.

Logical Framework

Within the last fifteen years, numerous documents have detailed a sound logical framework for risk assessment. The National Research Council (1983) produced a pivotal document from which subsequent U.S. federal documents evolved. Specifically, Environmental Protection Agency documents describing human health assessments at Superfund sites (EPA, 1989a), and ecological risk assessments (EPA, 1989b, 1996; Norton et al., 1992) contain permutations of this logical paradigm.

Despite the soundness of these documents, formalization and enrichment of the logic of risk assessment remain unfinished tasks. For example, although an EPA Risk Assessment Oversight Group (EPA, 1991) stated that "In characterizing ecological risks, it is preferable that probabilities of adverse effects...be quantified," most ecological risk assessments fail to do so. As pointed out in Chapters 1 and 7 of this book, most generate semiquantitative or qualitative statements of the assessors' confidence in conclusions that lack accurate effect probabilities. Further, assessments often hinge on a weight-of-evidence approach for inference. The weight-of-evidence concept is applied vaguely to mean that a *reasonable* person reviewing the available information *could* agree that the conclusion was plausible (modified from Apple et al. (1986) by Newman (in press)). Some of this vagueness can be eliminated by drawing upon methods of other disciplines. Risk assessors have yet to make full use of formal methods of working with such an inferential approach, i.e., formalized abductive inference (Josephson and Josephson, 1996) or Bayesian induction (Howson and Urbach,

1989). This is unfortunate, as they lend themselves to, and have been applied to, similar situations using computer programs based on artificial intelligence algorithms (Josephson and Josephson, 1996).

Recognizing the need for enrichment of the logic with which we approach risk assessments, Chapters 2 through 6 of this book provide further context and explanation for the logic applied to human and ecological risk assessments.

Quantitative Measurement Techniques

Quantitative measurement techniques are also essential to accurately describe or to predict the risks from contamination. Techniques have begun to coalesce around the logical framework of the NAS paradigm (NRC, 1983), e.g., Suter (1993). These include many diverse methods, ranging from criteria for the selection of measurement endpoints to calculations of intake, exposure, and risk (e.g., EPA, 1989a); however, further refinement is required to effectively apply many of these techniques. Several of the chapters in this book provide useful insights and suggestions for improvement. Chapter 7 describes quantitative methods for estimating risk to humans and aquatic biota from radionuclides, suggesting that those used for humans are more advanced than those applied for ecological entities. Suter's chapter (Chapter 8) outlines various types of extrapolations done in ecological assessments and provides guidance for their application. Kooijman et al. (Chapter 9) demonstrate an ecologically meaningful way to apply ecotoxicity test data to predict effect. Chapter 10 describes and then contrasts the various measurements for ecological communities that can be focused on assessing effect at this poorly-handled level of ecological organization. Broadening the scale further, Chapter 11 provides discussion of measurement techniques for assessing effects at the landscape level. This has become necessary because many risk assessments must now address the integrated effects of many sources within a landscape feature such as a watershed.

Chapters 12 to 14 explore three different and important considerations in accurately predicting risk. Dixon (Chapter 12) compares the traditional hypothesis testing approach of measuring significant differences to tests of equivalence. Detailed examples are provided for the application of equivalence methods. These methods could greatly enhance the effectiveness of measurement at several stages of risk assessment, including comparison of background contaminant concentrations to those at the site of concern. Kirchner (Chapter 13) discusses essential points for accurate aggregation and integration of data. This is done using insightful examples, such as estimating intake based on different estimates of contaminant concentrations in food. Ferson and Long (Chapter 14) then address uncertainty reduction in risk calculations via deconvolution methods, e.g., Monte Carlo simulations. They discuss important, but often overlooked, features of combining random variables in probabilistic risk assessment calculations. Directly relevant examples are provided, such as toxicant dose estimation.

Appropriate Models

Quantitative prediction of contaminant fate and effects requires some level of model formulation. Models ranging from simple to complex are quickly coming to play key roles in risk assessment (Bartell et al., 1992; McCarthy and Mackay, 1993). Indeed, many of the chapters in this book are structured around the logic, data requirements, and measurement techniques for effectively generating models for risk assessments. Mauriello et al. (Chapter 4) define the logic of and the role of simulation models in ecological risk assessment. Campbell and Bartell (Chapter 5) provide more detail on applicable ecological models for addressing questions at the levels of the individual, population, community, ecosystem, and landscape. Placing more emphasis on human risk, Voit and Schubauer-Berigan (Chapter 6) evaluate canonical models as a common approach to both human and ecological assessments. Kooijman et al. (Chapter 9) also discuss a theory-rich, dynamic model that draws on available ecotoxicity data to predict effects on populations.

This book attempts to enrich the detail surrounding risk assessment logic, measurement methods, and models. Both human and ecological risk assessments were incorporated in the hopes that practitioners of each could benefit from the other. These improvements are required in order for the regulatory needs of the United States to be met in an intelligent and effective way. The current framework for risk assessment, although generally sound, is inadequately supported in these areas.

ACKNOWLEDGMENT

This work was supported by Financial Assistance Award Number DE-FC09-96SR18546 from the U.S. Department of Energy to the University of Georgia Research Foundation.

REFERENCES

Apple, G.J., W.G. Hunter, and S. Bisgaard. Scientific data and environmental regulation, in *Statistics and the Law*, M.H. DeGroot, S.E. Fienberg, and J.B. Kadane, Eds., John Wiley & Sons, Inc., New York, NY, 1986.

Barnthouse, L.W., G.W. Suter, II, A.E. Rosen, and J.J. Beauchamp. Estimating responses of fish populations to toxic contaminants. *Environ. Toxicol. Chem.* 6, pp. 811–824, 1987.

Bartell, S.M., R.H. Gardner, and R.V. O'Neill. *Ecological Risk Estimation*. Lewis Publishers, Boca Raton, FL, 1992, p. 283.

Cairns, Jr., J. Will there ever be a field of landscape toxicology? *Environ. Toxicol. Chem.* 12, pp. 609–610, 1993.

Clements, W.H. and P.M. Kiffney. Assessing contaminant effects at higher levels of biological organization. *Environ. Toxicol. Chem.* 13, pp. 357–359, 1994.

EPA. *Risk Assessment Guidance for Superfund. Volume 1, Human Health Evaluation Manual (Part A)*, EPA 540-1-89-002, December 1989a.

EPA. *Ecological Assessment of Hazardous Waste Sites*, EPA 600/3-89/013, March 1989b.

EPA. *Summary Report on Issues in Ecological Risk Assessment*, EPA/625/3-91/018, February 1991.

EPA. *Proposed Guidelines for Ecological Risk Assessment*, EPA/630/R-95/002B, August 1996.

Howson, C. and P. Urbach. *Scientific Reasoning. The Bayesian Approach*. Open Court Publishing Co., La Salle, IL, 1989, p. 312.

Josephson, J.R. and S.G. Josephson. *Abductive Inference. Computation, Philosophy, Technology*. Cambridge University Press, Cambridge, UK, 1996, p. 306.

McCarthy, L.S. and D. Mackay. Enhancing ecotoxicological modeling and assessment. *Environ. Sci. & Technol.* 27, pp. 1719–1728, 1993.

Moriarty, F. *Ecotoxicology. The Study of Pollutants in Ecosystems.* Academic Press, Inc., London, 1983, p. 233.

National Research Council (NRC). *Risk Assessment in the Federal Government: Managing the Process.* National Academy Press, Washington, DC, 1983, p. 191.

Newman, M.C. *Quantitative Methods in Aquatic Ecotoxicology.* Lewis Publishers, Boca Raton, FL, 1995, p. 426.

Newman, M.C. Ecotoxicology as a science, in *Ecotoxicology. A Hierarchical Treatment*, Newman, M.C. and C.H. Jagoe, Eds., Lewis Publishers, Boca Raton, FL, 1996.

Newman, M.C. *Fundamentals of Ecotoxicology*, Ann Arbor Press, Inc., in press.

Newman, M.C. and M.S. Aplin. Enhancing toxicity data interpretation and prediction of ecological risk with survival time modeling: An illustration using sodium chloride toxicity to mosquitofish (*Gambusia holbrooki*). *Aquat. Toxicol. (Amst)* 23, pp. 85–96, 1992.

Newman, M.C. and J.T. McCloskey. Time-to-event analyses of ecotoxicology data. *Ecotoxicology* 5, pp. 187–196, 1996.

Norton, S.B., D.J. Rodier, J.H. Gentile, W.H. Van der Schalie, W.P. Wood, and M.W. Slimak. A framework for ecological risk assessment at the EPA. *Environ. Toxicol. Chem.* 11, pp. 1663–1672, 1992.

Suter, G.W., II. *Ecological Risk Assessment.* Lewis Publishers, Boca Raton, FL, 1993, p. 538.

Taub, F.B. Standardized aquatic microcosms. *Environ. Sci. & Technol.* 23, pp. 1064–1066, 1989.

Part 2
Logical Framework

2 ||| A Framework for Ecological Risk Assessment: Beyond the Quotient Method

John J. Bascietto

A FRAMEWORK FOR ECOLOGICAL RISK ASSESSMENT

The Regulatory Paradigm

In the spring of 1984, seven hundred Atlantic brant geese (*Branta bernicla*) were killed by an application of a pesticide to a Long Island, New York golf course. Six years later on July 12, 1990, after a protracted and contentious regulatory battle, the federal registration[1] of the use of diazinon pesticide on turfgrasses (golf courses and sod farms) in the United States was canceled by the U.S. Environmental Protection Agency (EPA).

This chapter will attempt to set a context for subsequent chapters in this symposium on "Risk Assessment: Logic and Measurement." The Long Island Brant kill and eventual cancellation action taken against diazinon are recounted here because these events set in motion a long-term process that eventually resulted in the publication by EPA of the first set of ecological risk assessment guidelines (EPA, 1996). These events in no small way helped EPA and others to decide to move away from a traditional reliance on a qualitative ecological risk assessment and descriptive ecology, to a more quantitative approach.

It is important to realize that in the mid-1980s, EPA's decision to perform a special review of diazinon was indeed extraordinary, because it marked one of the few times that EPA launched a serious regulatory action of this nature solely on the basis of potential ecological risk. Unlike the 1972 DDT case, no permanent adverse effects of diazinon on wildlife populations were ever demonstrated or determined through wildlife monitoring studies. The diazinon action was based primarily on a qualitative assessment of risk—and on an expectation of future harm (i.e., "risk" of injury) based on inferences about the future of waterfowl populations that could be exposed to what appeared to be acutely toxic levels of the pesticide in their environment. In the words of then U.S. EPA Administrator William K. Reilly, "diazinon use on golf courses and sod farms causes an unreasonable risk to birds commonly and with considerable frequency." EPA had determined that the benefits Soci-

---

[1] All pesticides used in the United States must be registered by EPA pursuant to the Federal Insecticide, Fungicide, and Rodenticide Act (FIFRA).

ety gained by allowing the use of diazinon on turfgrasses did not outweigh the waterfowl injuries which could materialize, and thus met the statutory standard: "when used in accordance with widespread and commonly recognized practice, generally causes unreasonable adverse effects on the environment...."[2]

In general, ecological risk assessments performed by the EPA are based on the model for human health risk assessment reported by the National Research Council in 1983 (NRC, 1983). By that time ecological risk assessments, albeit mostly qualitative and mostly based on ecotoxicity data, had gone on in one form or another in three fundamental areas: priority-setting; the development of standards or guidelines for environmental quality; and in risk management decision-making. EPA, however, has had to expand the NRC model considerably because of the many significant difficulties in translating single-organism and single-species effects into the systems effects we associate with the term "ecological risk." The Pesticides and Toxic Substances Programs are, for example, concerned about the potential impacts of pesticides and toxic chemicals on organisms, including aquatic and terrestrial communities. The Water Program is required by the Clean Water Act to restore and maintain the biological integrity of the Nation's waters and specifically to ensure the protection and propagation of a balanced population of fish, shellfish, and wildlife. The Office of Solid Waste and Emergency Response (which includes the Superfund Program) has responsibility for assessment of effects from hazardous waste and for remediation of uncontrolled hazardous waste sites.

In 1991, EPA announced the phaseout of another, more highly toxic anticholinesterase agent, the granular form of the carbamate pesticide carbofuran, which had a much broader agricultural use than diazinon, having registrations on many important food crops, such as corn, soybeans, and rice. The carbofuran case also took six years to decide.

The lack of formal ecological risk assessment guidelines in both pesticide cases led to bitter disagreements and protracted regulatory posturing between EPA and the agricultural chemical industry. Both sides had developed widely divergent conclusions about the significance of the ecological "risk" associated with these two pesticides. Even so, by the late 1980s, it was ecological risk assessment methodology itself, not the level of risk to be managed, that was the center of controversy.

Qualitative Approaches and the Quotient Method

Prior to the pesticide cases of the mid-1980s and early 1990s, many regulatory agencies relied heavily on qualitative assessments to perform "impact" studies (descriptive ecology), and semiquantitative "risk" assessments that used ecotoxicity data. The former merely describe current and/or past harm (or imply future harm), but have virtually no predictive power, while the latter have only marginal if any predictive power. The main components of these semiquantitative assessments include estimates of the acute and chronic toxicities of the hazardous substances under study; characterizations of the sizes and compositions of populations likely to be exposed to the chemicals being assessed; the use of some form of "safety factor" (e.g., a multiple of a toxicity value estimate to account for uncertainties);

---

[2] See "Remand Decision" In re: Ciba-Geigy Corporation, et al. v. U.S. EPA, FIFRA Docket Nos. 562. et al.

and a simple exposure model relating the toxicity observed in laboratory studies to the organisms in the field.

It is not uncommon for a semiquantitative approach to be taken in the initial stages of a tiered risk assessment procedure. There, increasingly complex laboratory and field studies can yield ever larger data sets which are used to define toxic potencies of environmental substances. These data are also used to predict how contaminants will behave in the environment, including whether their environmental concentrations will exceed previously determined toxic thresholds, called "quotients." When toxic thresholds are indeed exceeded, adverse effects on biological organisms and populations are inferred, because it is assumed that the toxic materials are present in the environment in sufficient quantities to elicit toxic responses from exposed environmental organisms. For example, rapid and simple tests—such as "LD_{50}" and "LC_{50}" studies—could give rapid estimates of acute toxicity, and are used for the initial tiers, with the more costly chronic studies reserved for materials that exceed preestablished threshold values. When available, measured environmental concentrations are to be used to estimate exposure. More frequently, however, acute toxicity values are compared to simple exposure models in deciding to move up to the advanced tiers (i.e., chronic toxicity tests and/or reproductive impairment studies). Alternatively, instead of using discrete values for the exposure concentration and the toxicity terms, risk inferences can be drawn from distributions of exposure and distributions of toxicity data, although this practice is still rare (Bartell et al., 1992; Suter, 1993; Weigert and Bartell, 1994).

Many of these assessments rely on single-species toxicity tests, but even as early as the early 1970s small-scale simple laboratory ecosystems (microcosms) had been employed to study the toxic effects of chemicals on small aquatic organisms. By the early 1980s, EPA began to use larger mesocosms (usually replications of ¼-acre artificial ponds), to estimate ecological effects of chemicals on multiple species in complex systems. By the mid-1980s, EPA made a limited number of risk management decisions affecting pesticide registration applications, partially on the basis of results from aquatic mesocosm and terrestrial community field studies designed to establish threshold ecological effects concentrations. The focus, however, continued to be on the threshold value which, if exceeded, triggered risk management measures to either reduce or eliminate the threshold exposure in the environment.

This long-standing risk management approach probably explains the popularity of the most widely used ecological risk assessment method, the quotient method (QM), whereby a ratio between the expected environmental concentration (EEC) of a hazardous substance and some threshold of toxicity is used to estimate the degree of harm, which is synonymous with "risk" in this context. The threshold is an "all or nothing" approach, and should essentially be considered a semiquantitative approach. It has little predictive power and adds virtually nothing to our knowledge about how the environment works. This is particularly true with respect to effects at higher levels of biological organization. (It should be noted, however, that the method was never intended for these purposes).

The quotient method offers a quick, inexpensive means of screening out high- or low-risk situations, possibly making further and more resource-intensive evaluations unnecessary. In this context, the QM provides a consistent and easily understood approach for decisions that have to be made on a wide variety of chemical stressors. The limitations of QM have been extensively reviewed (see Smith and Cairns, 1993; Suter, 1993). They include: little predictive capability; possibly inappropriate use of an endpoint; and having to make adjustments to the toxicity measurement, usually in the form of "uncertainty" or "safety" factors, which may be needed to compensate for lack of knowledge about the behavior of the toxicant in the system being assessed. The QM ratio establishes finite limits of exposure tolerance, which implicitly assumes that adverse effects will always occur. However, should an LC_{50} be used as the endpoint of an acute dietary test, 50% mortality may not be sufficiently protective of the ecological resource at risk, particularly if the magnitude of the response is as important as the endpoint itself. A 50% effect on reproduction, for example, may not be protective.

Inherent with the quotient method is the assumption that the exposure in the system being assessed will last as long or longer than the test duration for which the effects were measured, and that there is no change in exposure concentration over time. QM cannot easily elucidate secondary or food chain effects without the use of other methods, such as ecosystem models, and field validations will usually be required to determine the margin by which a quotient overestimates or underestimates significant ecological effects. After an extensive review of dozens of ecological assessment case studies, EPA concluded that when the quotient method is used in risk characterization, at least a qualitative description of key study uncertainties and limitations should be provided (EPA, 1994).

MOVING BEYOND THE QUOTIENT METHOD

By the mid-1980s, EPA's efforts to regulate acutely toxic pesticides using risk assessments that relied heavily on the quotient method met with a significant challenge from the regulated industry, which contended that EPA's threshold values were always set too low and *over*predicted risks. Even though field studies and pesticide incident reports on the most acutely toxic materials indicated that wildlife were indeed dying in large numbers due to the applications of these pesticides for horticultural and agricultural purposes, regulators and industry representatives disagreed about the significance of the acute mortality. Pesticide registrants correctly contended that field studies were needed to validate the quotient method assessments. Such validations faced a number of difficulties, including: establishment of appropriate control or reference sites spatially or temporally isolated from the potential experimental effects; separation of effects of the hazardous substances from the effects of intrinsic environmental factors; violation of assumptions of rigorous statistical analysis techniques; and determination of appropriate endpoints with which to assess "significant" ecological effects, risk, or impact.

Characterizing Risk: Semiquantitative Approaches

Earlier it was suggested that the diazinon case was a watershed event in EPA's efforts to develop ecological risk assessment guidelines. It is truly a landmark in environmental case law because it definitively established the precedent that discrete ecological risks could have limits set upon them. The diazinon case (and the subsequent carbofuran case) also represented the first regulatory and judicial tests of the use of the weight-of-evidence approach to ecological risk assessment.

From 1984 through 1988 EPA's wildlife biologists, working with the Canadian Wildlife Service, conducted a special review of diazinon. It was during this review that EPA refined the weight-of-evidence approach in ecological risk assessment, which laid out several lines of risk evidence available on the environmental behavior and effects of the compound: ecotoxicological data indicating high acute toxicity in sensitive waterfowl species; environmental monitoring and modeling data indicating that after approved applications of diazinon, residues lingered on Brant and Widgeon (*Mareca americana*) food items at levels exceeding acute thresholds for time periods that allowed for significant dietary exposure; the field kill history of diazinon involving sensitive avian species; and experimental field studies of acute waterfowl mortality after approved use. EPA reasoned that the weight-of-evidence approach is powerful because each new line of evidence increases confidence in the overall conclusions of the assessment.

Although the weight-of-evidence approach to the diazinon risk assessment was reviewed and approved by the EPA's Pesticide Scientific Advisory Panel in 1986, the regulatory process of cancellation had only just begun for diazinon. In 1987 (about three years after the Long Island Brant kill), EPA's Assistant Administrator for Pesticides and Toxic Substances issued a notice of intent to cancel the use of diazinon on golf courses and sod farms because of what the EPA termed "unreasonable" adverse effects on the environment. The manufacturer requested a hearing before an EPA administrative law judge (ALJ) who found that diazinon's registration should *not* be canceled but merely restricted because, in the ALJ's opinion, the benefits outweighed the harm. EPA's Administrator overruled this finding in 1988, holding that diazinon generally caused adverse effects out of proportion to the benefits received by society, interpreting FIFRA's standard of "generally causes unreasonable adverse effects" to mean "considering the overall picture" (EPA, 1988).

A cancellation notice for diazinon was issued by the EPA on April 5, 1988 (EPA, 1988), four years after the Long Island brant kill. EPA's action was appealed to the Fifth U.S. Circuit Court,[3] with the manufacturer contending that even though its product killed birds, insufficient adverse environmental effects resulted. Furthermore, the manufacturer argued that EPA had failed to prove that registered uses of diazinon killed birds "generally," which the company argued meant "more often than not." In 1989, the Fifth Circuit rejected both the registrant's and the EPA Administrator's interpretation of FIFRA's standard, instead saying that "generally causes unreasonable adverse effects" means "commonly" or "with considerable frequency," though not necessarily 51% of the time. Because FIFRA §22(b) defines unreasonable adverse effects to include not only the harmful consequences but also "unreasonable risk," the court recognized that diazinon use need not generally cause actual bird kills to warrant cancellation, but that cancellation could be justified if it generally causes an unreasonable risk of bird kills. Nonetheless, the court remanded the decision back to the EPA Administrator for clarification of whether the risk occurred commonly or with considerable frequency, and directed the Administrator to sufficiently tailor any cancellation orders to reflect distinctions in risk frequency among different kinds of uses. Ultimately, the EPA canceled the golf course and turfgrass uses (EPA, 1990).

What's the point of the story? What should strike us about this case is the apparently relaxed standard used to draw conclusions about the risk of diazinon's use. Regardless of the confidence in the correctness of the risk management decision (which, it's fair to say, is

---

[3] Ciba-Geigy Corp. v. EPA, 874 F. 2d 277, 279 [5th Cir. 1989]

fairly high), at no time during the Special Review, the peer review, the cancellation hearings, or the appellate process was quantitative evidence addressing the projected state of waterfowl populations exposed to diazinon, or birds in general, offered or requested. No quantitative standard or index of "unreasonable" ecological risk—or even "significant" biological effect—was developed. No prediction was made, nor expectation expressed by the Agency, regarding how frequently we might expect such effects to actually materialize.

The golf course incident that resulted in the death of the 700 geese turned out to be a misapplication of the pesticide. The ecotoxicological database for diazinon would have been the same had merely 70 or only 7 of the ill-fated flock been killed on that golf course. The case offers only marginal new knowledge about how diazinon works on bird populations, or in the ecosystem in general.

Clearly, the court did validate the use of the concept of "risk" to interpret statutory standards. However, the process proved extremely difficult and divisive. There was, both in and out of EPA, a certain lack of peer support for the risk management decision, which was rooted in the qualitative nature of the risk assessment; in addition, the expense of the long regulatory and judicial proceedings did not escape EPA's attention. It became clear that EPA needed to develop formal and rigorous ways to estimate ecological risk: methods that would both hold up to legal scrutiny as well as scientific peer review.

EPA (1992) notes that risk characterization may be qualitative or quantitative, and that very often the risk assessor must rely on his or her professional (scientific) judgment. Opinions on this statement vary, with some feeling that use of the terms "risk" assessment, as compared to "impact" or "hazard" assessment, requires that predictive or probabilistic statements be made about future events, necessarily implying a quantitative approach. Others feel, and EPA seems to agree, that either qualitative or quantitative statements of risk can be made, as long as the discussion involves causal relationships (EPA, 1993).

Granular Carbofuran Special Review

The granular carbofuran case[4] has been cited in many publications and EPA reports as a precedent-setting assessment of ecological risk. Again, it is a case involving acute avian mortality, and is in essence somewhat qualitative (descriptive). However, it employed several lines of rigorously collected quantitative ecotoxicity and environmental monitoring data, and makes predictive statements about the magnitude of future risks. The assessment inferred that up to two million bird deaths per year could occur from the continued use of granular carbofuran, but did not employ standard inferential techniques and did not quantify the uncertainties of the methods it did use.

Carbofuran is very highly toxic to a variety of avian species in laboratory toxicity studies. Granular formulations are also very highly toxic (a single granule can kill a small songbird) and have been implicated in many bird kill incidents, even excluding known misuses and mortality observed during field studies of the granular products.

---

[4] The special review of granular carbofuran was also discussed in an EPA publication of case studies in ecological risk assessment (EPA, 1993). However, the 1993 discussion was based on secondary and tertiary sources. The case discussed here is based on the original case report (Bascietto et al., 1987).

The risk assessment considered both exposure estimates and actual measured environmental concentrations of granular carbofuran to determine if these exposures exceeded avian acute toxicity values determined in the laboratory studies. By closely examining and testing the label-directed farming practices used to apply the material "at planting," EPA could both measure and estimate the number of granules that would remain exposed on the soil surface and available for consumption by wildlife. (EPA hypothesized that birds directly ingest exposed granules as dietary grit and observed that birds obtained doses from contaminated soil invertebrate food items such as earthworms.)

The Agency found that environmental levels did indeed far exceed the acute toxicity thresholds. The assessment also demonstrated that measured levels of carbofuran and carbofuran granules in agricultural fields following "at-planting" applications and the carbofuran residue levels found in dead and dying wildlife and on wildlife food items taken from treated fields also exceeded threshold toxicity values. These facts constituted the lines of evidence used in the risk characterization.

The risk assessment also relied heavily on reports of birds killed after label-directed uses of carbofuran. The evidence, in fact, was so overwhelming that peer reviewers later concluded that any uncertainties in the assessment of risk were more than compensated for by the sheer weight of the bird kill evidence (EPA, 1993). Kills occurred with migratory as well as nonmigratory species, and with endangered species. Reports indicated that birds were killed over a wide geographic area, were not confined to any particular crop, and could occur following applications other than the "at-planting" regimen.

In order to better understand the environmental conditions under which bird kills occurred, the risk assessment considered bird mortality data and exposure data gathered during studies the EPA required of the 10% active ingredient (a.i.)—known as "10G"—and 15% a.i. or "15G" granular material in a variety of field settings, attempting to replicate the label-directed uses. Field tests consistently showed that label-directed uses of both 10G and 15G resulted in avian mortality regardless of the application rate and despite using commonly practiced techniques for soil incorporation of the granules.

The risk assessment concluded that bird mortality following the label-directed use of 10G and 15G was a "frequent and regular occurrence," and that it was likely that the magnitude of the risk of acute mortality in fields (which the assessment concluded would always contain an excess of toxic granules) was more likely to depend on the extent of usage of those fields by birds, rather than on the pesticide application rate or technique.[5] The risk assessment concluded that

"...carbofuran's toxicity, and the probability of avian exposure to granules routinely combine to provide a virtually certain hazard" (Bascietto et al., 1987).

The assessment also detailed concerns for bird populations exposed to the granular product's use over several million acres of agricultural fields. By establishing an average

---

[5] In later risk assessments of granular carbofuran, EPA supplemented the risk methodology by including a new index of acute toxicity: the "LD_{50} per square foot." This measure is a modification of the quotient method which EPA used to relate the relative amount of granules exposed on the soil surface to an acute toxicity threshold. An Agency peer review panel concluded that the "LD_{50} per square foot" index is a unique measure for surface applied granular pesticides, and was probably not generally useful for other types of pesticide risk assessments (EPA, 1993).

"bird mortality per acre" rate (approximately 30 birds per 100 acres treated) as a basis for projecting risk,[6] the assessment inferred that as many as two million birds per year (based on total granular usage) could be lost to the combined use of 10G and 15G.

Concerns for adverse "impacts" upon populations of declining bird species (as measured by an annual Breeding Bird Survey) and for members of endangered species were described but not quantified, nor was causation established with respect to the use or misuse of carbofuran. The assessment made no actual estimate of any bird population effect.

Finally, the assessment focused on secondary poisonings of birds of prey and scavenging birds as a particular concern. Of course, such species are attracted to dead and dying smaller birds and small mammals that may be affected by granular carbofuran in newly planted fields. Dead or incapacitated prey make for an easy catch for avian predators and scavengers. The latter are themselves poisoned after ingesting carbofuran or its metabolites present in postabsorptive tissues of prey individuals, or by unassimilated chemical present in the gut of a small animal.[7] Secondary poisonings were, in fact, observed in several field studies, and monitored in several bird kill incidents.

EPA risk assessors expressed a particular concern for secondary poisonings because the species affected generally produce only a few young per year, and are also slow to reach maturity. Thus, the assessment contended, the death of each individual of these species can have "important consequences for a population." As an illustration, a simple computer program was used to simulate the deleterious effects of low levels of pesticide-induced acute mortality on a small breeding population of bald eagles (*Haliaeetus leucocephalus*) in southeastern Virginia, in which several secondary mortalities were observed after applications to cornfields.[8]

Documented population declines of several raptor species, which were known to heavily use the very agricultural fields being treated with granular carbofuran, were invoked. While the risk assessment did not quantitatively tie these population reductions to the pesticide, it did infer that a certain number of secondary poisonings of these species by granular carbofuran

---

[6] Bird mortality rate per acre was determined by calculating an average for the number of birds killed in each field study divided by the acres of fields tested in that study. Individual field test mortality rates varied within at least a ten-fold range. However, EPA considered that the observed rates of bird mortality were likely to be lower than actual mortality rates in any particular study. The mortality data used in the assessment are uncorrected for investigators' inability to account for all dead birds that were known to be in any particular field, and losses of bird carcasses to scavengers.

[7] Cases of secondary poisonings due to unassimilated chemical in the gut of a small prey item were thought to occur frequently. Since carbofuran was observed to be an extremely quick-acting toxicant, especially in small birds, small animals could die prior to digesting and metabolizing all of the granules.

[8] The Special Review of granular carbofuran may never have been initiated had it not been for the serendipitous discovery of a dead adult bald eagle from this population. The bird, which was later found to contain significant carbofuran residue, was found below an eagle's nest at the base of a tree. Young eaglets in the nest, and the remains of a pigeon (hypothesized to have been fed to the young) were also found to have been contaminated. The remains of the adult bird were well hidden under thick brush; and had it not been for a passerby prospecting for Civil War artifacts with a metal detector, may never have been discovered. The bird's metallic leg band had given it away.

were inevitable, and because these could not be reliably monitored, were suggested to be all the more risky because the level would be uncontrollable. The secondary poisonings, the assessment suggested, could therefore be "an important additive factor in the decline of these species."

TOWARD GUIDELINES FOR ECOLOGICAL RISK ASSESSMENT

Newman (1995), in a recent review of quantitative aquatic toxicology, argued that the discipline of ecotoxicology is characterized by a strong need for prediction at all levels of organization. He offered, however, that the discipline lacks sufficient knowledge for making such predictions, particularly at the higher levels (i.e., populations, community, and ecosystem), and lamented the bias toward more "tractable,"[9] lower-level studies, because that approach is less apt to instruct us on ecological relevance, which for Newman "is highest for effects at the higher levels." This dilemma is very aptly illustrated by the diazinon and carbofuran pesticide cases. A semiquantitative, weight-of-evidence approach enabled EPA to reasonably infer that a future state of ecological risk due to granular carbofuran's agricultural use could be unacceptable, but the method essentially did not permit an evaluation of the strength of the inference.

Both pesticide cases rest upon an essentially descriptive approach to risk characterization, and were perhaps well suited to the EPA's purpose. The carbofuran approach, like the diazinon case before it, was ultimately successful not because of the showing of risk *per se*, but because FIFRA's demand—that registrants bear the burden of showing that the continued registration of the pesticide did not pose "unreasonable risks" of harm—could not be met by the manufacturers. A description of the hazard may be all that is required to meet the law's demand under unequivocal circumstances of harm. The success of these cases rested almost entirely on the undeniable evidence of the frequent and seemingly uncontrollable injury that these materials actually caused, combined with a strong expectation that the injuries would continue unabated. The uncertainties inherent in the assessments were drowned out by a sea of information concerning actual harm.

The risk management decisions (i.e., cancellation of use) ultimately rested on weighing the facts of the injury case against the lesser hazards and similar costs of reasonably effective and readily available alternatives. Critics of EPA's pesticide assessments, however, demanded a more rigorous approach that included hypothesis testing, quantification of uncertainties, and a projection of the probability and magnitude of acute mortalities that could result in population reductions in birds. While Newman likely did not contemplate the diazinon and carbofuran cases in his 1995 review, his admonishment regarding the lackluster predictive capabilities of ecotoxicology would have rung very true for EPA's pesticide staff had it been made ten years earlier.

In 1992, EPA issued its seminal report, "Framework for Ecological Risk Assessment" (EPA, 1992). The Agency's stated purpose in issuing the framework report was to offer a simple, flexible structure for conducting and evaluating ecological risks. It was also hoped that it would help foster a consistent Agency approach for conducting and evaluating ecological risk assessments, identify key issues, and provide operational definitions for the technical terms used in these assessments.

---

[9] Here Newman is referring to the ability to extract information from studies performed at higher levels of organization.

The framework consists of three major phases: problem formulation; analysis (characterization of exposure and ecological effects); and risk characterization. Problem formulation is a planning and scoping process that establishes the goals, breadth, and focus of the risk assessment. It includes a preliminary characterization of exposure and effects, as well as examination of scientific data and data needs, policy and regulatory issues, and site-specific factors to define the feasibility, scope, and objectives for the risk assessment. Its end products are assessment endpoints, a conceptual model, and an analysis plan.

A conceptual model identifies the environmental values to be protected (the assessment endpoints), the relationships among "assessment" and "measurement" endpoints, the data needed, and the methodologies that will be used to analyze the data. For chemical stressors, the exposure scenario usually involves consideration of sources, environmental transport, partitioning of the chemical among various environmental media, chemical/biological transformation or speciation processes, and identification of potential routes of exposure (e.g., ingestion). For nonchemical stressors, such as temperature changes or physical disturbance, the exposure scenario describes the ecological components exposed and the general temporal and spatial patterns of their co-occurrence with the stressor. Although many hypotheses may be generated during problem formulation, only those that are considered most likely to contribute to risk are selected for further evaluation in the analysis phase.

The analysis phase develops profiles of environmental exposure and the ecological effects of the stressor. The purpose of exposure characterization is to predict or measure the spatial and temporal distribution of a stressor and its co-occurrence or contact with the ecological components of concern. The purpose of ecological effects characterization is to identify and quantify the adverse effects elicited by a stressor and, to the extent possible, to evaluate cause-and-effect relationships. The types of effects data that are evaluated depend largely on the nature of the stressor and the ecological component being evaluated. Effects may range from mortality and reproductive impairment in individuals and populations to disruptions in community and ecosystem function such as primary productivity. The evaluation process relies on professional judgment. Data are preferred that minimize the need for extrapolation, and that are relevant to the measurement and assessment endpoints. Data from both field observations and laboratory tests can be used to evaluate ecological effects.

Characteristics of the ecosystem can greatly modify the ultimate nature and distribution of the stressor. Chemical stressors can be modified through biotransformation by microbial communities or through other environmental fate processes, such as photolysis, hydrolysis, and sorption. The bioavailability of chemical stressors also can be affected by the environment, which in turn influences the exposure of ecological receptors. Physical stressors can be also modified by the ecosystem. For example, siltation in streams depends on sediment volume, flow regime, and physical stream characteristics.

Risk characterization integrates the exposure and effects profiles. Risks can be expressed quantitatively or qualitatively. This component includes a summary of the assumptions used, the scientific uncertainties, and the strengths and weaknesses of the analyses. It is important that risks are described in terms of the assessment endpoints; the ecological significance of the effects; recovery potential; and the overall confidence in the assessment. The purpose is to provide a complete picture of the analysis and results. The weight-of-evidence discussion of risk characterization provides the risk manager with insight about the confidence of the conclusions reached in the risk assessment by comparing the positive and negative aspects of the data, including uncertainties identified throughout the process.

Considerations most useful in a weight-of-evidence discussion include: the relevance of the evidence to the assessment endpoint; the relevance of the evidence to the conceptual model; the sufficiency and quality of the data and experimental designs used in key studies; the strength of cause-and-effect relationships; and the relative uncertainties of each line of evidence.

CONCLUSION

The stated purpose of this chapter was to attempt to set a context for the subsequent chapters produced from the August, 1996 Savannah River Ecology Laboratory symposium on "Risk Assessment: Logic and Measurement." The subsequent chapters in this volume, it is hoped, serve as a point of departure for students and practitioners of ecological risk assessment who want to reach beyond the quotient method, semiquantitative, and descriptive methods, toward a more rigorous and systematic approach to quantitative ecological risk assessment and risk management.

REFERENCES

Bartell, S.M., R.H. Gardner, and R.V. O'Neill. *Ecological Risk Estimation*. Lewis Publishers, Boca Raton, FL, 1992, p. 252.

Bascietto, J., D. Urban, and M. Slimak. (1987). "Transmittal of ecological effects branch hazard assessment for granular carbofuran and alternatives," memorandum to I. Sunzenauer, Special Review Branch, OPP/OPTS. Prepared for the Special Review of Granular Carbofuran, Office of Pesticides and Toxic Substances, U.S. Environmental Protection Agency, June 4, 1987.

EPA. (Environmental Protection Agency). Notice of Cancellation for Use of Diazinon Pesticide on Golf Courses and Sod Farms. *Federal Register* 53, pp. 11119, 1988.

EPA. (1990). "Remand Decision," In re: Ciba Geigy Corporation, et al.; FIFRA Docket Nos. 562, et al., W.K. Reilly, Administrator; Attachment to July 16, 1990 Letter from B. Hammiel, Hearing Clerk, U.S. EPA, to K.W. Weinstein, Esq., T. Putsavage, and S. Wolfson, Esq.

EPA. *Framework for Ecological Risk Assessment*. EPA 630/R-92/001. Risk Assessment Forum, Washington, DC, 1992.

EPA. *A Review of Ecological Assessment Case Studies from a Risk Assessment Perspective*. EPA/630/R-92/005. Risk Assessment Forum, Washington, DC, 1993.

EPA. *A Review of Ecological Assessment Case Studies from a Risk Assessment Perspective*. Vol. II. EPA/630/R-94/003. Risk Assessment Forum, Washington, DC, 1994.

EPA. Proposed Guidelines for Ecological Risk Assessment. *Federal Register* 61, pp. 47552, 1996.

National Research Council (NRC). *Risk Assessment in the Federal Government: Managing the Process*. National Research Council, National Academy Press, Washington, DC, 1983, p. 191.

Newman, M.C. *Quantitative Methods in Aquatic Toxicology*. Lewis Publishers, Boca Raton, FL, 1995, p. 426.

Smith, E.P. and J. Cairns, Jr. Extrapolation methods for setting ecological standards for water quality: statistical and ecological concerns. *Ecotoxicology* 2, pp. 203–219, 1993.

Suter, G.W., II. A critique of ecosystem health concepts and indexes. *Environ. Toxicol. Chem.* 12, pp. 1533–1539, 1993.

Weigert, R.G. and S.M. Bartell (1994). "Issue paper on risk integration methods." Ecological risk assessment issue papers. Risk Assessment Forum, Environmental Protection Agency, Washington, DC, pp. 9-1 to 9-66. EPA/630/R-94/009.

3 | Historical Perspective on the Role of Exposure Assessment in Human Risk Assessment

P. Barry Ryan

INTRODUCTION

Although the discipline of quantitative risk assessment is a relatively new one, its historical roots go back at least 50 years (NRC, 1994). Toxicologists and industrial hygienists had at that time begun to understand the relationship between exposure to hazardous substances in the workplace (and elsewhere) and adverse health effects. The concepts of Threshold Limit Values, No Observable Effects Levels, and Lowest Observable Effects mark the beginning of quantitative risk assessment as applied to environmental effects. Other concepts grew out of work associated with the nuclear power industry. Reactor failure and safety were modeled extensively in the 1950s and 1960s, with many of the quantitative methods later referred to as "risk assessment" developed and implemented by engineers. Engineering concepts such as design redundancy, failure management, and fault tolerance were eventually transferred to the new science.

The Risk Assessment Paradigm

In the 1980s, risk assessment began to become an accepted "science." The *Journal of Risk Analysis* was founded, with its primary role being the quantitative assessment of risk in engineering and the environment. The early 1980s was a watershed period for risk assessment. The National Academy of Sciences convened a panel to quantify the risk assessment process. The output of this panel became the major reference for the risk assessment professional (NRC, 1983). This so-called "Red Book" describes a quantitative method of assessing risk. This risk assessment paradigm divided the risk assessment process into four components: Hazard Identification, Dose-Response Assessment, Exposure Assessment, and Risk Characterization. Each is described briefly below.

Hazard Identification

Hazard identification is the association of an activity, location, or pollutant with a hazard. Certain activities, such as skydiving, swimming with sharks, and bungee jumping, can be considered intrinsically hazardous. Other activities may not be considered intrinsically dangerous, but still may have an associated risk. In excess of 40,000 people are killed each year in automobile accidents, but given the number of miles driven per year, one may view this as a relatively safe activity.

In terms of environmental hazards, hazard identification is usually performed by exposing animals (or plants) to environmental contaminants and observing some type of adverse outcome. Such outcomes may be loss of aesthetic appeal (e.g., browning of leaves, etc.), altered enzymatic activity, disease, or even death of the exposed species. A chemical species in the environment may be identified as a hazard when any of these observations are noted. Most hazards are identified through animal testing, with emphasis on carcinogenicity of the substance, often administered at high concentrations.

Once a material has been identified as a hazard, the risk assessment process begins. Two separate approaches are needed: one needs to understand the relationship between dose of the pollutant and the adverse outcome, and one needs to assess the exposure to the pollutant under actual conditions. Let us examine these two approaches.

Dose-Response Assessment

To paraphrase Paracelsus, the dose makes the poison. In order to examine the risk associated with a hazard, it is necessary to determine the relationship between the amount of contaminant entering the target species and the adverse outcome associated with that entry. Small amounts of contaminant may result in little or no damage. Larger amounts may result in small effects, while still larger amounts may result in death of the organism. It is important to establish the magnitude of "small," "larger," and "still larger."

For most chemical species of interest to the risk assessor, this dose-response relationship is established in the laboratory. Typically, laboratory animals are fed or otherwise given known concentrations of contaminants and the outcome is observed. From this, the quantitation of the scale of dose and response can be made. Extrapolation is then made to the human species or to another target species. This extrapolation process is potentially problematic, but such difficulties will be ignored in this discussion.

Models are often used to aid in this process (NRC, 1994). The risk assessor may extrapolate from high dose to low dose effects, postulate a threshold effect, or otherwise make use of mathematical elements in establishing the dose-response relationship. This is always done with an eye toward establishing what "real-world" environmental concentrations will have as their effect on biological systems.

Exposure Assessment

Having established that a given contaminant can cause harm (the hazard identification step) and having quantified the relationship between the amount of contaminant received by the target organism (the dose-response relationship), we are still not certain of the risk. It may be the case that a certain chemical species is indeed a hazard and that a relationship exists between the dose and the outcome, but if no human subject or other organism comes into contact with this material, there is still no risk. A concrete example is smallpox. This disease was once the scourge of mankind, causing untold pain, suffering, and death throughout history. Clearly a hazard exists and we know about dose and response—a small "dose" of smallpox can kill. But no cases of smallpox have been reported worldwide since the 1970s. That is because the disease has been eradicated, thus eliminating the *exposure* to this hazard.

Similar arguments can be made in the environment. A hazardous waste site may be filled with contaminants identified as hazards and have established dose-response relationships. Still, risk associated with this site can be minimized by controlling access to the area, thus eliminating the exposure of organisms to the hazard. The role of the risk assessment process here is to establish the magnitude, duration, and frequency of exposures experienced by biological systems (Sexton and Ryan, 1988).

Risk Characterization

The final step is the characterization of the risk. If sufficient exposure to a hazardous substance is noted, steps may be taken to control the risk experienced by the population as a whole, or to susceptible subpopulations. Such decisions are usually the purview of policy makers and reflect a final step—the risk management process. Factors influencing such management decisions may include numbers of affected individuals, economic considerations, and the specifics of the populations themselves. Policy makers may include insight developed by scientists and physicians in their decisions or may choose to use other information as their principal guidelines.

The Role of Exposure Assessment

The discussion of risk assessment usually assumes an understanding of the exposure experienced by an individual and the associated dose. Further, the risk assessor assumes a relationship between exposure, dose, and effect in order to calculate risk. Often, however, complete information is not available and assumptions must be made. This is especially true about the exposure component. The emphasis of this chapter is on the need for exposure assessment in the risk assessment process.

To illustrate this more concisely, consider Figure 3.1 (Ott, 1985). In order to make intelligent decisions about the risk individuals have, information must be available for each of the components of this diagram. Typically, information is available concerning sources. This information may be in terms of tons of sulfur dioxide emitted to the air per year from a given power plant, the level of trichloroethylene in a drinking water supply, the concentration of dioxin in cow's milk, etc. Further, information is likely to be available on the health effects of exposure to such pollutants. Animal studies may suggest a relationship between exposure to sulfur dioxide at a certain level and pulmonary problems. Similar data may be available for trichloroethylene exposure from industrial sources or for dioxin in laboratory investigations.

Often, however, little is known about actual exposures experienced by human subjects or other target organisms. However, it is incumbent upon regulators to protect the population, even if given insufficient information to make informed choices. Often in such cases, regulators are forced to rely on "exposure scenarios." An exposure scenario is a plausible set of circumstances, more or less based on reality, that affords estimation of exposure. As an example, a risk assessor may be asked to calculate the risk associated with a given waste site. He or she will gather information about the site, including materials and quantities

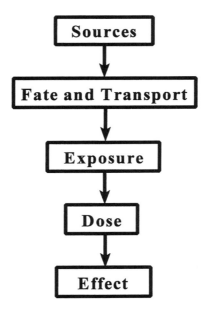

Figure 3.1. The relationship between sources of pollutants and effects (Ott, 1985).

present. The risk assessor will then adopt a plausible scenario for exposure, e.g., a child visiting the site once a week for ten years. Exposure can then be estimated using the assumptions of the given scenario. Risk is then calculated with this information.

The essential question associated with such analyses is the accuracy of the scenario. What assumptions are included? What do we know about variability in the exposure? Is the scenario at all indicative of what individuals in the surrounding communities experience? From these questions, it can be seen that exposure information is often the weak link in the chain leading to appropriate risk management decisions. If the information on exposure is poor, then so is the information on risk.

The rest of this chapter will focus on the historical development of exposure assessment as it applies to risk assessment. Examples will be drawn that illustrate the development and maturing of the science of exposure assessment during the last 25 years. Further, studies are selected to improve understanding of the exposure link in the risk assessment chain. It should be understood that if risk assessment is viewed as being in its infancy, exposure assessment must be viewed as embryonic. The need for accurate assessment of total pollutant exposure has only been realized since the development of quantitative risk assessment procedures. We will thus focus here on the historical development of exposure assessment by selecting for examination a few landmark exposure assessment investigations.

Definitions

Much scientific argument and discourse has centered on the fundamental definitions used in exposure assessment. A good deal of the uncertainty in such definitions comes from the blending of several fields in exposure assessment: toxicology, industrial hygiene, air pollution science, soil science, etc. Each of these disciplines has its own nomenclature for generic concepts such "concentration," "exposure," and "dose." It is thus worthwhile to take time here to develop the definitions. For a more complete discussion see NRC (1991). In this chapter, the definitions used by Sexton and Ryan (1988) will be used.

Concentration

The concentration of a pollutant is the amount of that material contained in a specific amount of the medium in question. For air pollutants, this is typically measured as either mass of pollutant per unit volume, e.g., µg (pollutant)/m^3 (air), or as a mixing ratio, e.g., ppmv (parts-per-million by volume). In water, the units may be µg (pollutant)/L (water) or a similar unit; in soil, dust, food, etc., the units typically used are mass of contaminant per mass of medium, e.g., g (pollutant)/g (medium). Concentration is a physical property of the medium in question. It represents the *potential* for exposure but does not represent exposure itself.

Exposure

Exposure is one step closer to indicating health effects. Exposure requires the simultaneous presence of a pollutant concentration and a biological receptor in the same location. There can be substantial concentrations of pollutants in various media in the environment, but if no biological system ever comes in contact with such concentrations, no exposure (and no biological effects) can occur. Thus exposure and concentration are different; exposure describes the interaction between the environment and the biological system in question.

Dose

Dose must be distinguished from both concentration and exposure. Concentration represents a physical characteristic of the environment that may potentially lead to an exposure. Likewise, exposure represents a possibility that the biological system may interact with the environment to result in adverse health impact. Dose requires the crossing of some boundary by the pollutant and subsequent entry into the biological system. Dose is the amount of pollutant that actually crosses the biological system body boundary and reaches target tissues.

An example may clarify the relationship between concentration, exposure, and dose. Consider a tank car recently drained of industrial solvent. Within the tank car, the remaining industrial solvent is in equilibrium with the air in the tank, resulting in a specific *concentration* of solvent in the air. No *exposure* occurs because no biological system experiences this concentration. Next, two workers enter the tank to clean it. The first of these workers enters wearing an airline respirator supplying outside air for breathing. He is now *exposed* to the solvent vapors as he is in the environment, but with respect to the inhalation route of exposure, he receives no *dose,* as his air supply is from elsewhere. The other worker, a rather unfortunate fellow, enters with no respiratory protection. He receives *exposure* as does the other worker, but because he is breathing the contaminated air he also receives a *dose*; solvent vapor crosses the boundary of his body and enters. He is then subject to health effects associated with inhalation exposure to the solvent vapors. To complicate the story a bit more, assuming neither worker is wearing other protective gear, *both* are subject to exposure and dose to solvent vapors when the dermal absorption route of exposure is considered. If this route is important, then both may suffer adverse health consequences.

Modifications of the above definitions are often necessary in specific contexts. For example, ingestion of soil may result in a dose of heavy metals, such as lead, in that the soil has crossed the boundary of the body through the ingestion itself. But it may also be true that all of the material is not biologically available because, for example, the chemical form may not be soluble under physiological conditions. In certain contexts one might then choose to refer to the "biologically available" dose. Other terms are also used occasionally. Such fine points of definition will not be addressed here, but the reader should be aware of them.

Exposure Route

The route of exposure generally is limited to one of three: inhalation, ingestion, or absorption. Each route has different characteristics associated with it that may influence the type of pollutants causing greatest concern and thus influencing risk. For the inhalation route, pollutants must be gaseous or otherwise suspended in air, e.g., particulate matter or dust. The body barrier crossed in going from exposure to dose is usually the lung epithelium, but for very soluble materials direct transfer to the bloodstream higher in the nasalpharyngeal region is also possible. It is important to note that investigation of total exposure to any pollutant must include all of these routes; exposure to a specific chemical species might well come from ingestion of contaminated water, dermal contact with such water, and inhalation of contaminant volatilized from this water system.

In the ingestion route, a pollutant is taken into the body through direct or indirect ingestion. Direct ingestion includes eating and drinking, while indirect ingestion includes mucociliary clearance of particulate matter from the lung, with subsequent swallowing. The boundary region for taking exposure to dose for this route is the digestive system epithelial layer. Local characteristics of the gut, including thickness of this layer, pH, and residence time, influence the exposure-dose relationship. In general, and for most pollutants, the digestive system epithelial layer represents a greater barrier to pollutant transfer than does the lung epithelium.

The third route of exposure is through dermal contact and subsequent absorption. Although upon absorption pollutants generally have direct access to the bloodstream, the skin—the body boundary for this exposure to dose—is generally much thicker and less subject to easy transfer. Local conditions, including type of contaminant, the matrix in which it is present, moisture content of the matrix and skin, etc., influence the exposure-dose relationship. On the other hand, if dermatitis is the outcome of interest, dermal exposure is of singular importance.

Exposure Pathway

In contrast to the exposure route, the exposure pathway describes the mechanism by which the pollutant reaches the human receptor. An example again may clarify. Consider exposure to a relatively soluble solvent such as chloroform. In an occupational setting, exposure might occur through either of two routes: inhalation or dermal absorption. The *pathway* for inhalation might be volatilization of the pure liquid followed by diffusion of the vapor to the human receptor. For dermal exposure, the pathway might include direct contact with the neat liquid and also condensation of the vapor on the skin surface. For non-occupational settings, two routes are also possible: inhalation and ingestion. These occur because drinking water is often contaminated with chloroform, either through industrial runoff, or more commonly by the production of chloroform during the chlorination process. Water containing chloroform may be ingested and exposure affected through the digestive epithelial layer. As an alternative pathway, the inhalation route may be operative if the drinking water is used for showering and bathing. During these activities, chloroform in the water may volatilize due to elevated temperatures and increased surface-to-volume ratio of the medium to release the material to the air, where it is then inhaled. Thus, the drinking water *pathway,* coupled with showering, leads to exposure via the inhalation *route.*

Ecological Risk Assessment

As the name suggests, ecological risk assessment takes a much broader view than does human risk assessment. Ecological risk assessment looks at the big picture of the impact of environmental contamination on all living things. Suter (1993), in the introduction to his monograph, describes numerous reasons why human risk assessment is an insufficient study of the larger problem. He points out seven major reasons why nonhuman systems may be more sensitive and thus give impetus to perform ecological risk assessment in an effort to understand the total impact on the environment. First, certain routes of exposure are not evident in humans, e.g., respiring water. Second, some nonhuman species may be more sensitive to certain types of environmental contaminants and thus may act as sentinels for environmental exposure. Third, large ecosystems may respond through mechanisms not available to the human species, e.g., lake eutrophication. Fourth, certain species may be more highly exposed to particular contaminants due to diet, habitat, or other restrictions. Fifth, many species have higher metabolic rates than humans and thus inhale, ingest, or absorb more chemicals per body weight than similarly exposed human subjects. Sixth, pesticides and herbicides designed to kill "pests" may result in long-term impact on nontarget species, including humans. And, seventh, the whole ecosystem is strongly coupled. Adverse effects on one segment of the ecosystem are likely to influence other parts.

Suter's arguments are compelling. There is a strong impetus among scientists and regulators to look only in the human risk arena. This should be tempered. The work of others more qualified to address these issues is included in this book. I will defer to their knowledge and discuss human risk assessment only.

HUMAN ENVIRONMENTAL EXPOSURE ASSESSMENT

The goal of human environmental exposure assessment is to improve understanding of total exposure to environmental contamination experienced by human populations through all media. As we will see below, many studies have been restricted to a single medium, with further focus on only one route and perhaps only one pathway of exposure. Such restrictions were necessary for various reasons, including logistic feasibility, belief that only a single medium-route-pathway combination was important in exposure, or lack of knowledge that other mechanisms of exposure existed. Research into so-called "nontraditional" pathways of exposure changed this substantially. Prominent in this area was the realization that indoor air quality was important in determining total exposure to airborne pollutants. This realization opened up the concept of "multimedia" exposure assessment.

The examples given below indicate the development of the exposure concept in risk analysis. The evolution of this thinking process is both interesting and enlightening. It is hoped that the discussion will show the process by which current researchers came to the conclusion that exposure is a far more complicated concept than first imagined.

Issues in Study Design

As we explore the history of exposure assessment, it is important to keep in mind the purpose behind various study designs. Chief among these are the goals of the investigation itself. What is it that the researchers sought to investigate? Secondly, did the researchers plan on making generalizations to populations based upon the data they collected? These two questions are of paramount importance in critiquing studies and in understanding the particular choices for design. Further, understanding the study design does lead to a better appreciation of the growth of the field.

Goals of Investigations

Studies of exposure and exposure-related factors, both current or historical, often look substantially different to those unfamiliar with the field. This is because the selected populations under investigation differ widely. Why, for instance, would one researcher only be interested in studying a small group of individuals living in a specific location while another investigates effects on the entire population of the United States, even when looking at the same pollutant and health effects? To understand this aspect of exposure investigation, one must look at the goals of the investigation. Typically, exposure assessment investigations choose to look at one of three types of population groups: highly exposed individuals, sensitive subpopulations, and the general population. There are compelling reasons for studying each. One may define risk as an exposure divided by some critical level. Risk may be increased either by raising the exposure level (highly exposed individual) or by reducing the critical level (as in sensitive subpopulations). However, understanding the general population mean risk can only be assessed using a population-based investigation.

Highly Exposed Populations

Certain studies are geared toward only the most highly exposed individuals. Such individuals often are exposed in an occupational setting, but this will not always be the case. The principal compelling reason for studying such individuals is their expected exposures. These exposures are postulated to be the highest among all members of the population as a whole and, therefore, one may expect the risks to be highest for these individuals. It is hypothesized that if no effects are seen in such individuals, others will not see effects either.

A shortcoming of dealing with highly exposed populations, especially occupationally exposed individuals, is that somehow they are not representative or "the same" as other individuals. By particular virtue of their high exposures, it is less likely that they would respond adversely to exposure to the pollutant in question. Had they responded in such a manner, they would have removed themselves from the environment earlier. Further, occupationally exposed workers are likely to be healthier on the average than the populations as a whole. This "healthy worker" effect is well known in the industrial hygiene and epidemiological literature. Still, highly exposed populations are studied quite often. Exposures are easier to measure, and health outcomes more likely to be seen.

Sensitive Populations

Sensitive subpopulations also represent a good choice for exposure investigations. They represent the other side of the coin from the highly exposed individuals. They are the individuals most likely to manifest adverse health outcomes when exposed to the lowest level of pollutant. The extreme case of such studies is controlled animal exposure studies on animals bred for specific susceptibility to adverse outcomes such as cancer. In human populations, sensitive individuals include the elderly, the very young, pregnant women, and individuals with specific conditions, such as asthma, immune deficiency diseases, and other biological shortcomings.

The advantage of working with such populations is clear: effects are likely to be manifested in such groups at the lowest levels of exposure. Similar to those with high exposure, if these individuals manifest no symptoms, then risk is likely to be low to the population at large.

General Population

The most general type of study is designed to afford extrapolation of the results to the general population. Usually, this is done by selecting participants in the study in a statistical fashion designed to ensure adequate representation of all segments of the population. Extrapolation back to the population as a whole is at the center of such designs.

Studies involving the general population have a great deal of power to assess the risk associated with exposures. In a good design, all components of the population are monitored, including the segments discussed above. Research from such studies can be generalized to the whole population, affording regulators further information to aid in the development of strategies to control pollutants in the most cost-effective way.

Unfortunately, such studies often require very large numbers to be truly statistically representative of all components of the population. Logistical problems, cost, and other considerations often preclude these types of studies except in very small populations. Though highly desirable, such studies are done only infrequently.

Population Generalizability

Within the framework discussed above, one still must account for generalizability or lack of generalizability to the population under investigation. Even if one is looking only at highly exposed or sensitive individuals, the statistical design of the study can be done to ensure that at least the population under investigation is well-understood. In exposure assessment studies, two general types of designs have been implemented. These include the anecdotal studies and the population-based studies. Both have their place and will be discussed presently.

Anecdotal Studies

Anecdotal studies are those in which a convenience or deliberate sample of individuals is taken. Examples include advertising for volunteers, choosing one's friends and acquaintances for participation in a study, or selecting only those for whom complaints regarding pollutant exposure are on file. Most historical exposure assessment studies have been of this type. Indeed, even those purported to be population-based (see below) suffer somewhat from what might be called the "anecdotal study problem," in that individuals may choose to participate or not for reasons somehow associated with pollutant exposure.

Anecdotal investigations are quite useful in that they can show the existence of a problem, illustrate large effects, and in many ways teach us the important factors influencing the problem. Thus they are appropriate early in the development of the science. They are not, however, suitable to draw conclusions for the population as a whole, which is often the ultimate goal of exposure assessment investigations.

Population-Based Investigations

Population-based investigations use a sample of participants drawn from the general population using techniques designed to ensure that the sample is representative of the population as a whole. These techniques may include a census of the entire population, or may choose any of several ways of selecting representative subpopulations. The principal strength of such an investigation is clear: population generalizability. Results from such studies are generalizable—within sampling error—to the whole population. Exposures measured on the sample can be used to infer exposures for the whole population.

The drawbacks to such studies include reluctance of the study population to participate, expense of data collection, and, depending upon experimental design, complexity in statistical analysis of the data at the conclusion of the study. Still, because of the power of such investigations, exposure assessment investigators have adopted this type of design more and more often, especially over the last ten years. Regulatory agencies, for example, typically require population-based investigations because of their need to obtain data applicable to the general population.

The Epidemiological Component

It is fair to say that environmental exposure assessment is as much an outgrowth of epidemiology as it is a component of the risk assessment paradigm. Even the earliest of epidemiological studies tried to establish a causal or correlative link between disease and some outside agent. In epidemiology, illness and disease in populations are studied with an eye toward determining cause-and-effect relationships and, potentially, mitigating or removing the cause.

Early epidemiological studies often displayed only weak relationships between causes and effects. Typical studies would deem an individual "exposed" if he or she displayed certain characteristics, e.g., lived in a certain town or had a certain occupation, and would then measure an effect such as disease outcome thought to be associated with the exposure. This often resulted in "no effect" studies; the hypothesized cause (exposure as measured by the weak surrogate) was not found to be statistically associated with the effect being measured. However, a weak surrogate for a true exposure can result in a misclassification bias toward the null result (Shy et al., 1978; Ozkaynak et al., 1986). Thus epidemiologists determined that better exposure assessment measurements—e.g., continuous measurement of pollutant exposure, resulting in less misclassification of such exposures—could lead to a better understanding of the epidemiological results obtained. Thus, epidemiologists as well as risk assessors began to push for better exposure assessment investigations.

The Contribution to Risk Assessment

Given the discussion above, the contribution of exposure assessment to risk assessment becomes clear. Exposure assessment is an essential part of the risk assessment paradigm. Knowing that an environmental contaminant has some human health risk associated with it is not sufficient to perform the risk analysis. One must know if populations or individuals are exposed to the contaminant in any of their activities. This requires high-quality exposure data. Simple dichotomous surrogates of exposure often lead to no-effect studies. More detail is needed to show the relationship between exposure and effects.

EXAMPLES OF HUMAN EXPOSURE ASSESSMENT STUDIES

Our attention now turns to the historical development of exposure assessment as it links to risk assessment and epidemiology. Discussion will commence with early investigations linked with toxicology, industrial hygiene, and epidemiological studies. In these early studies, exposure measurements were taken only to support the basic toxicology, industrial hygiene, or epidemiology that was the central focus of the investigation. As the science of exposure assessment progresses, an understanding of the need for better exposure assessment to support epidemiological investigations of disease will come to the fore. As we progress into more recent investigations, the need for high-quality exposure assessment investigations in support of government regulations will become evident. Modern exposure assessment studies stand on their own as examples of a quality scientific endeavor gathering needed data rather than in support of other branches of inquiry. In this we see the maturing of the science. Throughout this section, the reader is asked to keep in mind the risk assessment paradigm. The gaps in knowledge between the source of pollutants and the dose received by the biological system must be spanned in order to perform quality risk assessments.

Early Exposure Assessment Studies

Industrial hygienists have been performing exposure assessment studies after a fashion since the early part of the century. In that the primary focus of the industrial hygiene community was (and is) occupational exposures primarily to airborne pollutants, there remains a bias toward studying exposure through this medium, and by the inhalation route, to the present day. Outside the occupational realm, this was even more true. Discussion will commence with a series of epidemiological investigations of the health effects of air pollution.

Berlin, New Hampshire Studies

Studies related to air pollution epidemiology led to the beginnings of environmental exposure assessments. Perhaps the first such investigation was begun by researchers at Harvard University (Ferris and Anderson, 1962; Ferris, 1964; Ferris et al., 1973; Ferris et al., 1976). These investigations were carried out between 1958 and the early 1970s in Berlin, New Hampshire, a small New England pulp mill town. The studies were prospective epidemiological investigations: a population was followed for an extended period of time and information on it was collected. In particular, pulmonary function tests and respiratory symptoms were noted. Visits occurred every six years, during which pulmonary function tests, including spirometry, and respiratory symptoms, e.g., persistent cough, wheeze, were noted. Further, air pollution measurements were taken. In particular, total suspended particulate matter (TSP) and sulfur dioxide (SO_2) measurements (using a somewhat primitive method) were taken as a surrogate for total air pollution. Their studies found that both pulmonary function and respiratory symptoms improved between the first two visits, a time in which both TSP and SO_2 concentrations decreased. At the third visit, no such improvement was noted. This time period was characterized by further decrease in the TSP levels, but a concomitant increase in SO_2 to levels higher than at the initial visit.

This long-term investigation piqued the interest of epidemiologists everywhere. The relationship between concentration and effect was only beginning to be established. This work showed that decreases in air pollution levels could result in improved lung health. Further, the study established that a decrease in one pollutant, e.g., TSP, may not result in improved health if another pollutant, in this case SO_2, increased.

Looking through the perfect vision of hindsight, it is a wonder that any results were obtained. TSP is a very imperfect measure of exposure, especially when measurements occurred only in a single location and only during the time period when the biological measurements were being taken. Much of exposure to what is called TSP is not available as a dose, as the particles are much too large to penetrate deeply into the lung. Further, the primitive method used to assess SO_2 exposure was very imprecise and subject to numerous interferences. Finally, the "snapshot" of exposure obtained by taking short-duration environmental measurements may not be representative of exposures experienced throughout the year or throughout one's lifetime. Given the potential for misclassification bias, this study becomes a tribute to the researchers in their concerted effort to understand this problem.

The Berlin, New Hampshire studies indicate the state-of-the-art in exposure assessment in the 1960s and early 1970s. Poor surrogates for actual exposure were used; insensitive measurement techniques, flawed study designs (see below), and difficult outcome variables were the standard of the day. Yet these studies "showed the way" to the more sophisticated work that followed.

CHESS Studies

A very controversial study was carried out under EPA's aegis in the early 1970s. This investigation, known as the CHESS studies, attempted to correlate the measured concentration of SO_2 with a variety of respiratory symptoms (Finklea et al., 1975). The results suggested a positive relationship between such symptoms and the measured concentration of SO_2. This investigation was subjected to severe scrutiny and criticism on methodological grounds. Other researchers who performed analyses using subsets of the CHESS data were unable to show such effects. Further, quality assurance problems with the data brought the whole study under suspicion.

Despite the methodological problems associated with this investigation, there are several reasons to select it for discussion. First, it again showed a relationship between a measure of environmental concentration (and a weak measure of exposure) and adverse health outcomes. Given the difficulty of getting positive results in epidemiological investigations, this is itself noteworthy. Secondly, the study was cross-sectional. It compared results at a fixed time for individuals whose "exposure" differed, as measured by the surrogate concentration of sulfur dioxide. Third, it was, again, in a community setting—locations similar to those experienced by typical individuals in the greater population. Fourth, the severe criticism brought to bear on this investigation permanently changed the way other studies were done. No longer were such studies reviewed only by other scientists; they were also reviewed by industrial experts eager to show that their factories were not responsible for adverse health effects in the population, by regulators trying to understand the impact of the economic and industrial system on human health, and by the public, which through Congressional hearings on the studies, started to gain an awareness of the source-exposure-dose-health effect relationship.

With the CHESS studies, awareness of the potential effects of air pollution on human health came to the scientific and medical community. Laboratory investigations exploring dose-response relationships in animals, evaluation of source emission characteristics for air pollutants, and classic epidemiological investigations increased in number. Still lagging behind, however, was a solid understanding of the relationship between emissions from a stack and health effects. Why is it, it might have been asked, that some individuals living relatively close to a source of pollution had few effects, while those living farther away may have had more? Further, how can these effects be mitigated? This mitigation question was answered soon after these investigations, but perhaps with insufficient data. It became evident from the Berlin, New Hampshire studies and the CHESS studies that being near a source was generally bad for people. Because most sources were stationary, there was a dilemma—neither the sources nor the people "exposed" could be moved. Soon a solution presented itself. Through the study of air pollution meteorology, it was found that if an emission source such as a stack from a power plant were made taller, exposure to those near the plant could be reduced. If the stack was made tall enough, less pollution was found in the local area. The impact of pollution associated with a fixed site could be mitigated; taller stacks resulted in less local exposure and greater dilution of the pollutant "plume." Tall stacks were the order of the day in the late 1970s. Stacks in excess of 1000 feet were common for big power plants and other large, stationary point sources. The effects of these tall stacks—long-range transport of air pollution—were not realized until later.

Harvard "Six-Cities" Study

Perhaps the most famous of the air pollution epidemiological investigations was the Harvard Air Pollution and Lung Health Study, better known as the Six-Cities Study (Ferris et al., 1979; Ware et al., 1981). These studies also represented a significant leap in sophistication of data collection and analysis. Although still focusing attention on a single route of exposure (inhalation) and a single pathway (air) the data collected were far more detailed, both from the epidemiological side and from the air monitoring side. In this study, for the first time, the air monitoring program was more sophisticated than the biological monitoring.

The last statement was not meant to belittle the health effects monitoring. The data collected of this type were more detailed, using more sophisticated techniques, than any previous study. Detailed, digitized spirometry measurements were made, affording accurate assessment of new health-related parameters on a much larger population than ever before. More detailed questionnaire information was obtained which, in turn, led to the development of more sophisticated measures of pulmonary effects. Further, the study was designed with exposure in mind. The six cities for the study were chosen to have varying degrees of overall pollutant burden: two cities with low pollutant burden, two cities with intermediate pollutant burden, and two cities with high pollutant burden. Different cities were monitored at different times during the year, so that temporal effects in health outcomes could be evaluated. These studies were deemed by all to push the state-of-the-art to a new level.

But it was the air pollution measurements that truly "pushed the envelope," making this a far more sophisticated exposure assessment study than any previously completed. Multiple pollutants were measured, including real-time measurements of ozone, oxides of nitrogen, and sulfur dioxide. Particulate measurements were carried out using both TSP and a new more biologically valuable method that afforded separation of particulate matter into size fractions. Chamber studies had shown that particulate matter greater than a certain size had little impact on health because it was never delivered to the lung epithelium, but rather, was lost higher in the respiratory system. This study monitored separately the fine particles that are capable of penetrating deeply into the lung. Such sophisticated measurements led to a better understanding of pollutant concentrations and their variability in the eastern U.S.

The results of the study were also striking. Relatively small effects could be seen. Risk ratios of less than 2.0 could be determined with statistical confidence. Day-to-day, seasonal, and year-to-year differences in air pollution values were now available. Long-term trends—generally downward—in air pollutant concentrations were made evident by the eight years of measurements taken.

The studies were not without their flaws, however. The principal researchers were not yet satisfied with the results. A large amount of variability in health outcomes still remained unaccounted for. Although substantial differences between cities in both air pollution concentrations could be seen, the lack of variability in exposure measures within a city made understanding the variability of health outcomes at this level impossible to work out. These researchers closely examined this situation and devised several substudies to help pinpoint causes and effects. The outcome of these substudies led to the next level of sophistication in exposure assessment investigations.

Harvard Indoor-Outdoor Study

The Harvard researchers involved with the Six-Cities Studies had wrestled extensively with the problem of using fixed-site monitors to estimate exposures throughout a whole city. Such monitors were typically located in a central area, with siting designed to reflect the city as a whole. Yet small-scale investigations suggested that such a central site monitor may not adequately reflect the entire city. Differing impact of point sources on one area versus another, prevailing wind directions, and mobile source emission differences were part of the consideration.

At this time there was also under consideration the impact of one's own indoor environment. In the late 1970s the indoor environment was considered pristine. Air pollution was seen as an outdoor problem that could be mitigated by going—or staying—inside. An appreciation of indoor air pollution was still several years in the future.

To assess the variability within a city and to give a preliminary examination of indoor air quality, a small substudy of the Harvard Six-Cities Study (Spengler et al., 1979; Dockery et al., 1981; Spengler et al., 1981) was suggested and then implemented. In this investigation, ten individuals from each of the six cities who were already participating in the larger investigation were solicited to participate in a smaller study. In this study, nitrogen dioxide (NO_2), SO_2, respirable particulate matter (RSP), and the sulfate component of RSP (RSO_4) were monitored both outside and inside the individuals' homes. Homes were monitored using relatively unsophisticated sampling apparatus for approximately one week.

This study again proved to be a watershed event in exposure assessment. The results from the outdoor air monitoring were remarkable, but perhaps predictable. Pollutants were not uniformly distributed around each city but were different in each area. Some pollutants varied only by a small amount, and thus could be approximated by the central-site monitoring concentrations, while others differed substantially from site to site. These results alone suggested that the use of a central-site monitoring protocol for outdoor pollutants may lead to misclassification bias. The assumption that all individuals within a city experienced the same pollutant exposure was seriously called into question.

Even more remarkable were the results of the indoor air monitoring. SO_2 and RSO_4 showed the expected effect. The ratio of indoor concentration of these pollutants to the outdoor (the I/O ratio) was found to be somewhat less than one; the indoor environment offered a degree of protection from these pollutants. On the other hand, the I/O ratio for NO_2 and RSP varied considerably, with values ranging from a low of 0.4 to sometimes in excess of 5.0. Examination of questionnaire data suggested that the presence of sources in the indoor environment—combustion appliances, e.g., gas ranges for NO_2, and cigarette smoking for RSP—resulted in I/O ratios in excess of 1.0. Such homes did not offer protection from air pollutants. On the contrary, such homes were associated with more pollutant exposure than outdoor air concentrations would suggest. These observations, coupled with the observation that individuals were spending in excess of 70% of their time in their own home, cast a very bad light on using central site concentration data as a surrogate for exposure. Far superior information on exposure could be gathered by monitoring the individual's home.

These results changed the entire direction of exposure assessment research and resulted in a disconnection of exposure assessment and epidemiology. To this point, exposure assessment was always carried out in support of epidemiological research. Few small studies, and essentially no large studies, had been carried out that had as their goal an improved understanding of exposure to any pollutant for its own sake. This would change with the results of the Harvard Six-Cities Studies, and particularly the Indoor/Outdoor Study.

Houston Study on the Effects of Personal Exposures to Air Pollutants on Asthmatics

A study conducted in Houston, Texas, during May–October 1981 (Stock et al., 1985) is typical of many studies conducted in the early 1980s. With knowledge of the Harvard results, many researchers were now seeking better estimates of exposure. While designed to investigate the impact of air pollutants on asthmatic children, and thus considered to support an epidemiological investigation, this study is indicative of the thinking in environmental exposure assessment during this time period. The concept of better exposure assessment is in the forefront. In this work, the researchers selected three different methods of assessing exposure to a series of pollutants. Note that the change has now been made to looking at more than one pollutant. Although the route is still inhalation exclusively, the pathway is now more complicated. As had been done in previous studies, researchers in this investigation elected to monitor pollutants at a central site. However, they also developed a mobile monitoring platform that allowed them to go from home to home and monitor pollutants both inside and outside participants' homes. Thus they had adopted both the major air monitoring protocols developed in the Six-Cities Study. With this, the researchers were now able to examine two pathways of exposure: ambient or city-wide air, and local air as measured by both indoor and outdoor monitors.

The third component of their air monitoring is indicative of state-of-the-art procedures in the early 1980s. These researchers elected to monitor personal air using small, portable monitors that the participants carried with them as they went about their daily activities. This was a major step forward. Monitoring of surrogates, such as central-site ambient air or indoor air, was not required. The concentrations measured were those actually experienced by the participants and should, therefore, be far more closely associated with health outcomes.

The results of their investigation of airborne exposures are typical of many other studies of the early 1980s. As an example, they showed that central site ozone concentrations were weakly correlated with outdoor air concentrations at participants' homes. Further, they were essentially uncorrelated with either indoor concentrations or personal concentrations. Neither of the latter two were correlated with concentration *outside the participant's own home*. The results of this study, and other similar studies, led to a major change in exposure assessment investigations. Far less emphasis was placed on central site monitoring, while more emphasis was placed on in-home and personal monitoring.

The results of this study further suggest reasons for the dearth of observations relating pollutant concentrations and health effects in many other studies. Using an extremely poor (nearly uncorrelated) surrogate of exposure—central site pollutant monitoring data—results in only the largest of investigations suggesting any relationship at all.

The EPA TEAM Studies

The studies discussed above remained focused on single-route and, often, single-pathway exposures. But as early as 1979, researchers at EPA (Behar et al., 1979) advocated the use of multimedia exposure assessment in an effort to "...control *critical sources* of pollutants that cause major problems or threats to *critical receptors*." Even then, scientists were considering the possibility of multiple pathways and even multiple routes to exposure. From these roots sprang the concept of Total Exposure. Total Exposure means the exposure experienced by an individual through all possible pathways and routes. One must consider inhalation, ingestion, and dermal absorption if one is to determine total pollutant exposure burden. One must consider nontraditional pathways to exposure, such as the volatilization-inhalation pathway for chloroform discussed earlier. Though not technologically feasible in the late 1970s, the concept was being discussed.

The EPA Total Exposure Assessment Methodology (TEAM) Studies (Wallace, 1987) were the first to attempt Total Exposure Assessment. These studies were carried out from 1979 through 1985. To quote from the Introduction of Wallace's monograph on TEAM studies:

"...The goals of this study were: (1) to develop methods to measure total exposure (exposure through air, food, and water) and the resulting body burden of toxic and carcinogenic chemicals, and (2) to apply these methods within a probability-based sampling framework to estimate the exposures and body burdens of urban populations in several U.S. cities...."

The early pilot investigations carried out in July and December 1980 tested protocols for measuring concentrations of four groups of chemicals in environmental samples (air, water, food, house dust) and biological samples (blood, breath, urine, and hair) for a series of volatile organic compounds (VOCs), pesticides and PCBs, metals, and polynuclear aromatic hydrocarbons. Because of sampling and analytical limitations present at the time, only VOCs could be adequately assessed. Working methods and protocols were developed for the determination of these compounds in air (ambient, personal, and exhaled breath) and drinking water. No such compounds were found in food; thus food was not measured. The studies were carried out with over 600 individuals from numerous locations around the country participating.

The EPA TEAM investigations also represented a new leap forward. For the first time, multiple media were the focus of the investigation. Further, although not successful due to technological constraints, multiple pollutant classes were considered. Third, the study shifted from the anecdotal base of volunteers, as in the Harvard investigations and the Texas-type studies, to one statistically selected to be representative of the populations of the cities under investigation.

The results of the investigation were almost shocking, even to those scientists familiar with the exposure assessment literature. For example, sampling was done both indoors and outdoors in locations as diverse as Bayonne, New Jersey, and Devil's Lake, North Dakota. The New Jersey location is near a large concentration of petrochemical industries, while the North Dakota location is far removed from large industrial areas. As the pollutants under investigation were VOCs, commonly associated with the petrochemical industry, it was expected that these two locations would be substantially different in exposure to individuals, with Bayonne far exceeding Devil's Lake. This was indeed the case with respect to outdoor air concentrations. But, very surprisingly, indoor concentrations differed little in the two locations. Further, personal exposures (as measured by a personal monitor that the individuals carried with them) were also similar and exceeded even the indoor concentrations for many compounds. I/O ratios often exceeded 10 in *both* locations and occasionally exceeded 1000 for some compounds. Clearly, measurement of central site VOC concentrations was not likely to represent exposures to these compounds. Indoor sources, e.g., gasoline, solvents, paints, and varnishes stored inside, were responsible for the majority of indoor pollutant concentrations and for personal exposures experienced by individuals.

The TEAM studies strongly indicated that, for some chemical species, exposure estimates of any value could only be made using personal monitoring protocols. Further, they suggested the feasibility of multimedia assessments and suggested that such assessments may be required if accurate exposures were needed to correlate with health outcomes.

The Southern California Gas and Harvard/GRI Nitrogen Dioxide Studies

Two related studies—the Southern California Gas Residential Indoor Air Quality Characterization Study (Wilson et al., 1986), carried out in the Los Angeles area and the Harvard/ GRI Nitrogen Dioxide Studies (Ryan et al., 1988a; Ryan et al., 1988b; Spengler et al., 1994), carried out in the greater Boston area—represent state-of-the-art studies from the mid 1980s. Both of the investigations focused on a single pollutant (NO_2) and a single route (inhalation) of exposure. In both cases, multiple pathways (indoor air, outdoor air, and personal air) were considered. Both were large (400–600 participants) studies with a statistical selection criterion (although methodological difficulties in the Southern California Gas Study preclude generalizability). The similarities continued in the ancillary pollutant measurements taken. Both studies took air exchange measurements with an eye toward understanding the relationship between outdoor air concentrations and indoor air concentrations. Both were exclusively exposure assessment investigations. No health effects measurements were taken. Both took measurements in differing seasons in an effort to determine seasonal variability components. The Harvard/GRI Study took additional water vapor concentration measurements.

The studies differed in that the Harvard/GRI study had as an essential component a follow-on personal monitoring investigation in which participants carried personal monitoring badges with them as they went about their daily activities. A similar study was carried out by Harvard researchers—co-funded by GRI and Southern California Gas—in the Los Angeles area in an effort separate from the original Southern California Gas investigation.

The results of this investigation supported the earlier results of the Harvard Indoor-Outdoor Study. The presence of combustion sources, e.g., gas ranges or unvented gas appliances, were associated with elevated NO_2 concentrations. Other factors influencing indoor concentrations included ambient concentrations and degree of weatherization. Further, exposures, as measured by the personal samplers, closely tracked indoor concentrations. Elevated personal exposures were associated with combustion sources present in the home.

The value of these investigations over the anecdotal Harvard studies is the generalizability to the populations as a whole. Accurate exposure estimates for nitrogen dioxide can be made for these two populations with known sampling error.

National Human Exposure Assessment Survey (NHEXAS) Investigations

The current state-of-the-art in exposure assessment investigations is the National Human Exposure Assessment Survey (NHEXAS) (Sexton et al., 1995a; Sexton et al., 1995b). It consists of three different studies, one in the state of Arizona (Lebowitz, et al., 1995), a second in the Upper Great Lakes Region (Pellizzari et al., 1995), and a third in Maryland, designed to investigate the relationship between short-term measurements and long-term exposures (Buck et al., 1995). The three studies are related and are based on the presumption that exposure occurs simultaneously through multiple media and by various pathways. Further, for any given pollutant, it is typically true that more than one route of exposure is important. Thus all components of NHEXAS are true multimedia, multipollutant exposure assessment investigations. They are also true exposure assessment investigations. Although some biological measurements are taken in each component, exposure, not health outcome, is the essential factor being measured.

The Arizona study and the Upper Great Lakes study are similar. Both rely on population-based statistical designs to ensure generalizability to their respective populations. Both have a large number of participants and involve a nested design in which many participants are subjected to screening tests, while smaller numbers are evaluated for exposures that are more difficult, intrusive, or costly to measure. Both studies are examining exposures to four classes of compounds: metals, pesticides, PAHs, and VOCs. Several environmental media are monitored. These include air (indoor, outdoor, and personal), drinking water, house dust, soil, and food. Biological samples (blood and urine, and for the Upper Great Lakes study, hair) are taken with an eye toward evaluating the impact of the exposures measured by other techniques. The essential purposes of these studies are two: to assess exposure in the population, and to determine whether large populations can be monitored effectively in this fashion.

The third study differs from the other two in several important ways. First, the study—being carried out in metropolitan Baltimore, Maryland—is designed to assess the variability in exposures experienced through multiple media over a year-long period. Secondly, the number of participants, though still statistically selected, is much smaller, although the size of the study, due to the need for repeated visits to each person, is comparable to those of the other two investigations. The study is similar in its collection protocols, with all media monitored in the other studies monitored in this one as well. In addition, dermal wipe samples are taken in the Baltimore study. The Baltimore investigation does not, however, monitor for VOCs.

These ongoing studies are expected to yield much new data on simultaneous exposures to various compounds through various media and multiple pathways. These data on large populations will be especially useful. The data from the Baltimore study will give insight into commonly used single-measurement sampling strategies and may indicate how many monitoring periods must be included in future designs.

CONCLUSION

In this chapter the development of the science of exposure assessment has been followed from its beginnings in toxicology and epidemiology through to the current state-of-the-art. The change in perception of exposure assessment's role is evident. In the beginning, scientists believed they understood rather completely what influenced exposure of most pollutants. Today, that belief is strongly tempered by the knowledge that exposure can come from many places and it is likely that some not yet considered may be important.

The state-of-the-art NHEXAS studies indicate what is now believed to be the appropriate exposure assessment strategy. All studies should be statistically based and multimedia in design. Exposure data are essential to understanding the full risk associated with a hazard. Ten years from now, it may be true that the NHEXAS investigations will be thought of as naive. This would be consistent with the history presented here. It can only be hoped that all such studies being undertaken now, and in the future, improve our understanding of the environment, its impact upon human health, and our role in nature.

REFERENCES

Behar, J.V., E.A. Schick, R.E. Stanley, and G.B. Morgan. Integrated exposure assessment monitoring. *Environ. Sci. & Technol.* 13, pp. 34–39, 1979.

Buck, R.J., K.A. Hammerstrom, and P.B. Ryan. Estimating long-term exposures from short-term measurements. *J. Exp. Anal. Environ. Epidemiol.* 5, pp. 359–374, 1995.

Dockery, D.W., J.D. Spengler, M.P. Reed, and J.H. Ware. Relationship among personal, indoor, and outdoor NO_2 measurements. *Environ. Int.* 5, pp. 101–107, 1981.

Ferris, B.G., Jr. and D.O. Anderson. The prevalence of chronic respiratory disease in a New Hampshire town. *Am. Rev. Respir. Dis.* 86, pp. 165, 1962.

Ferris, B.G., Jr. Epidemiological studies related to air pollution: a comparison of Berlin, New Hampshire and Chilliwack, British Columbia. *Proc. R. Soc. Med.* 57, pp. 979, 1964.

Ferris, B.G., Jr., I.T.T. Higgins, M.W. Higgins, and J.M. Peters. Chronic nonspecific respiratory disease in Berlin, New Hampshire, 1961–1967. A follow-up study. *Am. Rev. Respir. Dis.* 107, pp. 110–122, 1973.

Ferris, B.G., Jr., H. Chen, S. Puleo, and R.L.H. Murphy. Chronic nonspecific respiratory disease in Berlin, New Hampshire, 1967–1973. A further follow-up study. *Am. Rev. Respir. Dis.* 113, pp. 475–485, 1976.

Ferris, B.G., Jr., F.E. Speizer, J.D. Spengler, D.W. Dockery, Y.M.M. Bishop, J.M. Wolfson, and C. Humble. Effects of sulfur oxides and respirable particles on human health: I. Methodology and demography of population in study. *Am. Rev. Respir. Dis.* 120, pp. 767–779, 1979.

Finklea, J.F., G.G. Akland, W.C. Nelson, R.I. Larsen, D.B. Turner, and W.E. Wilson. (1975). *Health Effects of Increasing Sulfur Oxide Emissions.* National Environmental Research Center, Office of Research and Development.

Lebowitz, M.D., M.K. O'Rourke, S. Gordon, D.J. Moschandreas, T. Buckley, and Nishioka. Population-based exposure measurements in Arizona: a Phase I filed study in support of the national human exposure assessment survey. *J. Exp. Anal. Environ. Epidemiol.* 5, pp. 297–326, 1995.

NRC. *Risk Assessment in the Federal Government: Managing the Process.* National Academy Press, 1983, p. 191.

NRC. *Human Exposure Assessment for Airborne Pollutants. Advance and Opportunities.* National Academy of Sciences, 1991, p. 321.

NRC. *Science and Judgment in Risk Assessment.* Committee on Risk Assessment of Hazardous Air Pollutants, Board of Environmental Studies and Toxicology. Commission on Life Sciences, National Research Council, Washington, DC, 1994.

Ott, W.R. Total human exposure. *Environ. Sci. & Technol.* 10, pp. 880–886, 1985.

Ozkaynak, H., P.B. Ryan, J.D. Spengler, and N.M. Laird. Bias due to misclassification of personal exposures in epidemiologic studies of indoor and outdoor air pollution. *Environ. Int.* 12, pp. 389–393, 1986.

Pellizzari, E., P. Lioy, J. Quackenboss, R. Whitmore, A. Clayton, N. Freeman, J. Waldman, K. Thomas, C. Rodes, and T. Wilcosky. Population-based exposure measurements in Region V: a Phase I filed study in support of the national human exposure assessment survey. *J. Exp. Anal. Environ. Epidemiol.* 5, pp. 327–358, 1995.

Ryan, P.B., M.L. Soczek, J.D. Spengler, and I.H. Billick. The Boston residential NO$_2$ characterization study: I. A preliminary evaluation of the survey methodology. *JAPCA* 38, pp. 22–27, 1988a.

Ryan, P.B., M.L. Soczek, R.D. Treitman, J.D. Spengler, and I.H. Billick. The Boston residential NO$_2$ characterization study: II. Survey methodology and population concentration estimates. *Atmos. Environ.* 22, pp. 2115–2125, 1988b.

Sexton, K. and P.B. Ryan. Assessment of human exposure to air pollution: Methods, measurement and models, in *Air Pollution, The Automobile, and Public Health*, Watson, A.Y., R.R. Bates, and D. Kennedy, Eds., National Academy Press, Washington DC, 1988.

Sexton, K., D.E. Kleffman, and M.A. Callahan. An introduction to the national human exposure assessment survey (NHEXAS) and related Phase I field investigations. *J. Exp. Anal. Environ. Epidemiol.* 5, pp. 229–232. 1995a.

Sexton, K., M.A. Callahan, E.F. Bryan, C.G. Saint, and W.P. Wood. Informed decisions about protecting and promoting public health: rationale for a national human exposure assessment survey. *J. Exp. Anal. Environ. Epidemiol.* 5, pp. 233–256, 1995b.

Shy, C.M., D.G. Kleinbaum, and H. Morgenstern. The effect of misclassification of exposure status in epidemiological studies of air pollution. *Bull. N.Y. Acad. Med.* 54, pp. 1155–1165, 1978.

Spengler, J.D., B.G. Ferris, Jr., D.W. Dockery, and F.E. Speizer. Sulfur dioxide and nitrogen dioxide levels inside and outside homes and implications on health effects research. *Environ. Sci. & Technol.* 13, pp. 1276–1280, 1979.

Spengler, J.D., D.W. Dockery, W.A. Turner, D.M. Wolfson, and B.G. Ferris, Jr. Long-term measurements of respirable sulfates and particles inside and outside homes. *Atmos. Environ.* 15, pp. 23–30, 1981.

Spengler, J.D., M. Schwab, P.B. Ryan, S.D. Colome, A.L. Wilson, I.H. Billick, and E. Becker. Personal exposure to nitrogen dioxide in the Los Angeles Basin: study design and results. *JAPCA* 44, pp. 39–47, 1994.

Stock, T.H., D.J. Kotchmar, C.F. Contant, P.A. Buffler, A.H. Holguin, B.M. Gehan, and L.M. Noel. The estimation of personal exposures to air pollutants for a community-based study of health effects in asthmatics—design and results of air monitoring. *JAPCA* 35, pp. 1266–1273, 1985.

Suter, G.W., II. *Ecological Risk Assessment.* Lewis Publishers, Boca Raton, FL, 1993, p. 538.

Wallace, L.A. *The Total Exposure Assessment Methodology (TEAM) Study: Summary and Analysis: Volume I.* EPA/600/6-87/002a. United States Environmental Protection Agency, Office of Acid Deposition, Environmental Monitoring and Quality Assurance, Washington, DC, 1987.

Ware, J.H., L.A. Thibodeau, F.A. Speizer, S.D. Colome, B.G. Ferris, Jr. Assessment of the health effects of atmospheric sulfur oxides and particulate matter: evidence from observational studies. *Environ. Health Perspect.* 41, pp. 255–276, 1981.

Wilson, A.L., S.D. Colome, P.E. Baker, and E.W. Becker. (1986). *Residential Indoor Air Quality Characterization Study of Nitrogen Dioxide.* Southern California Gas Company.

4 | Application of Simulation Models to the Risk Assessment Process

David A. Mauriello, Donald Rodier, and Francis Stay

INTRODUCTION

At the United States Environmental Protection Agency (EPA), the process of ecological risk assessment serves as a tool for carrying out the regulatory mandates of the various EPA Program Offices. This chapter will explore the development of a logical framework for Ecological Risk Assessment within EPA and the role of ecological models in facilitating the risk assessment process. The focus will be on the use of simulation models to estimate the risk to the environment from toxic chemicals and their potential role in chemical regulation by EPA.

A LOGICAL FRAMEWORK FOR ECOLOGICAL RISK ASSESSMENT

The NAS Red Book

The development of a logical framework for risk assessment in the federal government began, for all practical purposes, with the publication of the NRC report, *Risk Assessment in the Federal Government: Managing the Process* (NRC, 1983). This report, known as the "NAS Red Book," proposed a conceptual framework for managing *risks to human health* through a process incorporating research, risk assessment, and risk management. In this scheme, risk assessment was comprised of hazard identification, dose-response assessment, exposure assessment, and risk characterization.

Publication of the "Red Book" provided a strong impetus to risk assessment activities within federal regulatory programs, including those at EPA. However, the lack of similar assessment guidelines for *risks to the environment* created the perception that risks to human health were accorded a far higher priority than hazards to the environment.

The NAS Ecorisk Workshop and Report

In 1989, the National Research Council established a Committee on Risk Assessment Methodology within the Board on Environmental Studies and Toxicology of the Commission on Life Sciences. The task for the Committee was to identify important scientific and societal issues in risk assessment. Among the first issues considered by the Committee on Risk Assessment Methodology (CRAM) was the need to develop a conceptual framework for ecological risk assessment analogous to that for human health, as described in the 1983 "Red Book." Such a framework would identify the components of ecological risk assessment, define the relationship of ecological risk assessment to risk management, and facilitate the formulation of consistent technical guidelines.

The Committee on Risk Assessment Methodology conducted a workshop at the Airlie Foundation, Warrenton, Virginia, from February 26 to March 1, 1991. Workshop participants examined a set of six ecological risk case studies in an effort to identify common approaches and themes, as well as to assess the degree of consistency with the 1983 risk framework.

Based upon discussion and evaluation of the case studies presented, the workshop participants concluded that while the 1983 framework was generally applicable, there were a number of deficiencies with respect to ecological risk assessment. A consensus was reached concerning modifications needed to make the framework more suitable for ecological risk assessment problems (Figure 4.1). The participants also identified a number of scientific issues in ecological risk assessment for further research and analysis (NRC, 1993).

The EPA Risk Framework

A report of the EPA Science Advisory Board (EPA, 1990) emphasized the need for an analogous ecological risk assessment paradigm at EPA. An effort to develop ecological risk assessment guidelines began under the auspices of the EPA Risk Assessment Forum. The objective was to develop guidance that would incorporate a sound scientific approach to ecological risk assessment while meeting the regulatory needs of EPA. This effort culminated with the publication of *A Framework for Ecological Risk Assessment* by the Risk Assessment Forum (EPA, 1992). This document presented a paradigm for ecological risk assessment similar to that proposed by the NRC-CRAM workshop report (Figure 4.2).

Since publication of the Framework document, the Risk Assessment Forum has published several volumes of ecological risk assessment case studies (EPA, 1993) and a volume of issue papers (EPA, 1994). Additional ecological risk assessment guidelines are currently under development by the Risk Assessment Forum.

THE EPA RISK ASSESSMENT PARADIGM

The EPA *Framework for Ecological Risk Assessment* describes a three-part paradigm (Figure 4.2) which, while similar to that proposed by the NRC workshop, incorporates a unique set of objectives that address the diverse regulatory environment existing at EPA:

- Facilitate the formulation of regulatory objectives and identification of endpoints of ecological concern in order to improve the risk management process.
- Provide for an iterative approach to the assessment of complex ecological risk assessment issues at all levels of ecological organization.
- Provide a means of incorporating feedback into the decision-making process and for assimilating new risk assessment methodologies.

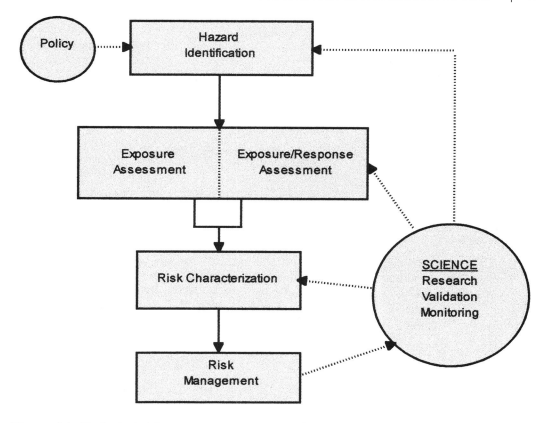

Figure 4.1. Ecological risk assessment paradigm. *Source:* NRC Committee on Risk Assessment Methodology.

Problem Formulation

The Problem Formulation phase (Figure 4.3) of the risk assessment process defines the objectives, scope, and scale of the ecological risk assessment. Objectives may be largely determined by regulatory, management, information, or societal constraints that lie outside the risk assessment process itself. Other considerations include the scale of the assessment or such factors as whether it will be generic or site-specific. In addition, the characteristics of the stressor and the ecosystem potentially at risk must be assessed, and potential system responses to the stressor must be identified. Negative consequences to the system from exposure to the stressor, or risk, are expressed in terms of impact on endpoints.

The selection of appropriate endpoints to characterize the potential risk is central to the process of risk assessment. Suter (1989) characterizes ecological risk assessment endpoints into the categories of *assessment* or *measurement:*

- Assessment endpoints are those attributes of the system in question which are of primary regulatory or ecological significance. They are generally integrated attributes of a system.
- Measurement endpoints are subsidiary characteristics of a system which may be directly observed or measured.

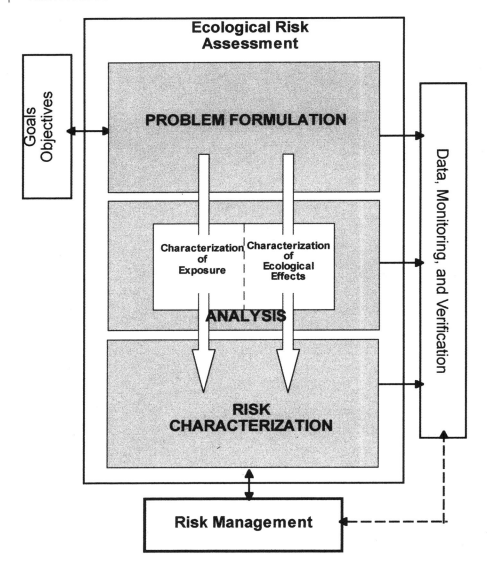

Figure 4.2. EPA Ecological Risk Assessment Paradigm.

Assessment endpoints should be chosen on the basis of biological relevance and values that are also of primary regulatory or societal significance. These are biological or ecological impacts of primary importance to the maintenance of populations or the structure and function of communities and ecosystems. Ideally, an assessment endpoint should incorporate the potential for both direct and indirect effects of the chemical stressor. Risk will ultimately be expressed in terms of the consequences to the assessment endpoints and demonstrated relevance to regulatory concerns.

A distinguishing feature of ecological risk assessment is that it is often necessary to utilize more than a single endpoint to characterize risk to populations or communities. Depending on the objectives of the risk assessment, endpoints may be selected from one or

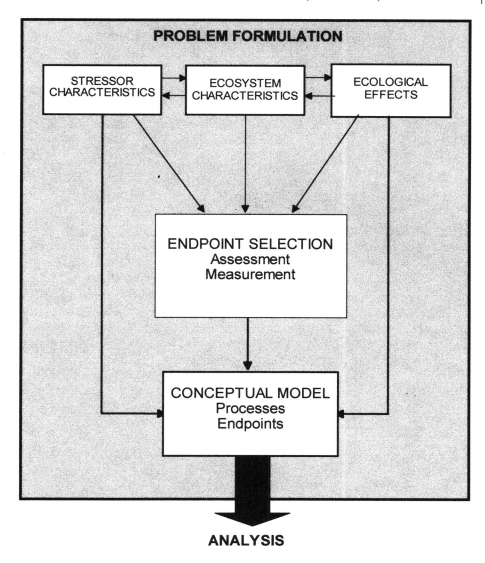

Figure 4.3. Problem Formulation under EPA Risk Assessment Framework.

more of the hierarchical levels of ecological organization. Figure 4.4 illustrates some examples of the assessment endpoints that may be used at each level of the ecological organization hierarchy. For example, growth, mortality, and behavior provide good indicators of the status of individual organisms, and indirectly, for populations. Changes in the abundance or distribution of species are useful endpoints at the community level.

The appropriate level for assessment endpoints depends on the objectives and the constraints imposed upon the risk assessment process. For ecological risk assessments of toxic chemicals, mortality data are usually collected from single species bioassays. While this is an appropriate measurement endpoint, assessment endpoints at the individual organism level are generally avoided because of the "so what" question. Many environmental regula-

ECOLOGICAL ENDPOINT HIERARCHY

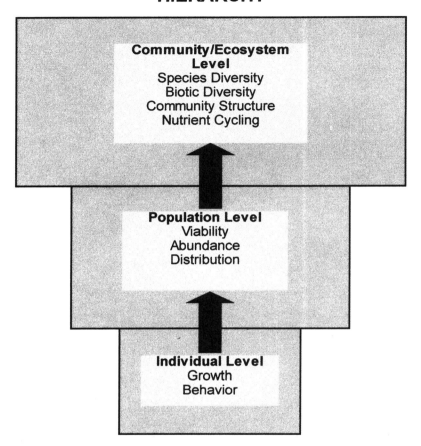

Figure 4.4. Ecological risk assessment endpoint hierarchy.

tions, such as the Endangered Species Act, operate explicitly at both the individual and population level, while others, such as the Toxic Substances Control Act (TSCA), do so implicitly.

Ferson et al. (1996) argue the necessity of conducting population level risk assessments and the feasibility of extrapolation from single-species toxicity bioassays to endpoints that predict risk at the population level. They suggest risk of decline in population level over specified time intervals as an appropriate population level assessment endpoint.

This practice provides a means of addressing the "so what" question frequently encountered in ecological risk assessment. The most sensitive indicator populations often have only marginal value as assessment endpoints, while suitable endpoints may not be sensitive to the chemical stressor; hence the "so what." However, the endpoint population may be dependent upon the indicator species as a food source and thus the indirect effects may influence the level of risk to the assessment endpoint to a far greater extent than would be predicted on the basis of direct toxic effects alone.

To fully realize the potential of populations as assessment endpoints requires the systematic application of a set of selection criteria to potential endpoint choices. While many different criteria could be identified for specific assessment circumstances, Kelly and Harwell (1989) describe some generally applicable guidelines:

- Sensitivity to stressor—Species with a high "signal-to-noise" ratio are preferable, as they provide a high level of response per unit of stress.
- Rapidity of response to stress—A rapid response is necessary to provide the early warning characteristic. Such responses generally occur at lower trophic levels among populations with short life cycles, such as phytoplankton.
- Specificity of response—Species that respond only to specific chemical stressors provide the mechanism for demonstrating causal stressor-response relationships.
- Ease and economy of measurement—While desirable, this criterion should not be elevated to primary importance in the selection process.
- Relevance to assessment goals—It underlies the necessity of answering the "so what" question fundamental to risk characterization.

Finally, the effective use of population level endpoints requires the availability of sufficient data to characterize the life histories of species of interest, define the dose-response relationships for all identified endpoints, and describe the fundamental population dynamics of the species selected as assessment endpoints. The procedure outlined above for identifying and selecting assessment endpoints is summarized in Figure 4.5. The selection process is driven from the top down, with the principal constraints being the objectives and the "regulatory endpoints." For population level assessments, candidate assessment endpoints are chosen and then filtered for specific indicator properties and for availability of sufficient data to support the assessment (Mauriello, 1993). It is unrealistic to assume that any single indicator would meet all the selection criteria or that all preferred assessment endpoints would possess exemplary indicator properties. Thus it is common to identify additional indicators along with those chosen as assessment endpoints.

The culmination of problem formulation is the development of a conceptual model which provides the logical framework binding the endpoints and the processes linking stressor and the ecosystem together. The conceptual model represents the intersection of the stressor(s) and the system potentially at risk. It describes the characteristics of the ecosystem, the stress regime, and the mechanisms by which the system may respond to exposure to the stressor. It links the endpoints in a logical fashion, providing a means of extrapolating between the measurement and assessment endpoints. It is a schematic for how the stressor may affect the system in question and describes the major and subsidiary risk hypotheses to be tested during the risk assessment process.

Analysis

The analysis phase of the ecological risk assessment process is concerned with the evaluation of the nature of exposure to the stressor and the potential effects resulting from that exposure (Figure 4.6).

Stressor characterization determines the distribution of the stressor over space and time. For chemical substances, this involves fate and transport analysis, and perhaps monitoring.

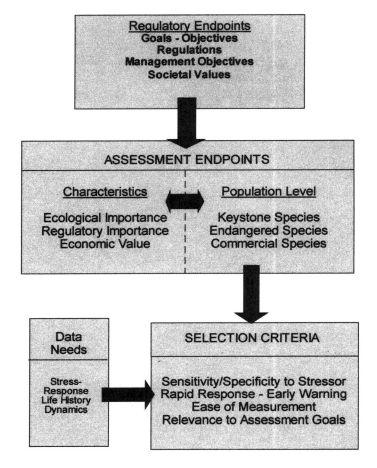

Figure 4.5. Selection process for assessment endpoints at the population level.

In addition, the biotic and abiotic attributes of the ecosystem under study that affect the distribution and nature of the stressor are evaluated. Ecological effects data are examined to determine what is known of the effects of stressors on ecological endpoints of concern. Such effects range from the sublethal to changes in the structure of whole communities. Cause-and-effect relationships between the stressor and measurement endpoints, based on exposure response scenarios developed in the conceptual model, are used to construct a stressor response profile, which can be linked to the assessment endpoints identified in the conceptual model.

Characterization of Ecological Risk

Risk characterization integrates the exposure and effects assessments and expresses the risk in terms of the magnitude and significance of the impacts upon the assessment endpoints (Figure 4.7). Risk estimation involves the comparison and integration of the exposure and stressor-response profiles into a risk summary, along with an estimate of the impacts of uncertainties upon the results. The risk description is an explanation of the magnitude

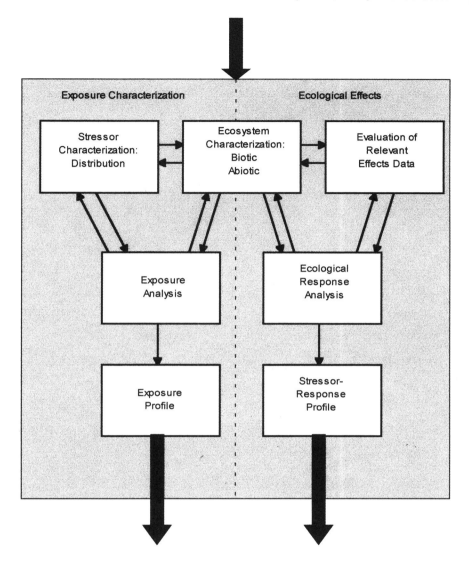

Figure 4.6. Analysis phase of ecological risk assessment process.

and the significance of the risks with respect to the assessment endpoints identified during problem formulation.

The modeling metaphor utilized in problem formulation is ideally suited to risk characterization, using simulation models—designed for use in ecological risk assessments (Bartell, 1990; Barnthouse, 1992; Bartell et al., 1992)—as direct analogs to the conceptual model. Models can be used to integrate the exposure and response profiles, summarize the risk, incorporate uncertainty and variability, and describe the risk in terms of the assessment endpoints.

One approach to coping with the problem of uncertainty is the use of ecological uncertainty analysis (O'Neill et al., 1982), which assigns probability distributions to population

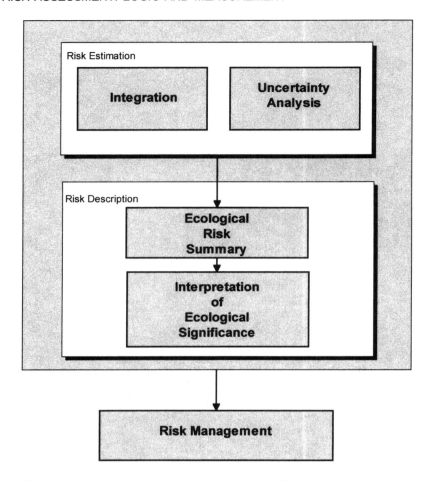

Figure 4.7. Risk characterization under EPA Ecological Risk Assessment Framework.

parameters and toxic responses. The simulation model is run over and over using different sets of parameter values selected from the probability distributions. This technique, known as Monte Carlo simulation, yields a distribution of potential impacts upon the system in question. An analysis of this distribution will provide a prediction of the probability that a given impact will occur for a particular level of exposure to the chemical.

In this context, ecological risk can be defined as the probability of an adverse impact upon an assessment endpoint. For example, an appropriate assessment endpoint might be the numbers or biomass of species of regulatory interest. Ginzburg et al. (1982) and Ginzburg et al. (1990) illustrate a method to characterize risks to populations as the probability that a population will fall below some minimum threshold over a specified time interval—the probability of quasi-extinction.

A convenient way of expressing probabilistic risk in graphical terms is exemplified by the risk-area curve (Mauriello, 1988; Rodier and Mauriello, 1993) shown in Figure 4.8. This graph defines zones of risk as the area set off by particular combinations of probability and the level of adverse impact, expressed as a percentage, upon the assessment endpoint of

Constant Risk Curves

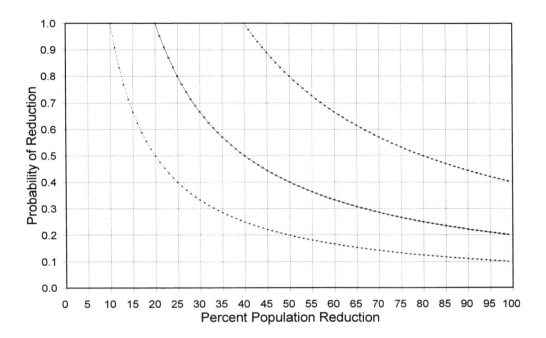

Figure 4.8. Risk-area graph. Probability of population or biomass reduction.

interest. Probability and endpoint axes are defined as percentages in order to normalize results. This makes it possible to directly compare the results using alternative models or completely different chemical exposure scenarios.

One important factor in determining the significance of an ecological risk is the potential for the population or ecosystem to recover after the chemical stressor is removed. For toxic substances, sublethal effects resulting from transformations, degradation products, or bioaccumulation may have far greater consequences to exposed populations over time than episodes of acute mortality.

Simulation models can be used to evaluate the potential for recovery after direct exposure to the toxic substance has ceased. A comparison of the perturbed and unperturbed trajectories of simulated populations can be used as a measure of the potential for recovery of a system from exposure. A simple "coefficient of convergence" (Mauriello, 1982) can be used as an index of recovery for model populations or other suitable endpoints. The coefficient (1) is the ratio of the absolute deviation between trajectories divided by the maximum deviation, and yields a number between 0 and 1. Recovery simulation scenarios thus integrate acute and chronic effects as well as fate of chemicals that are persistent in the environment to provide a dynamic picture of recovery potential over time. Bartell et al. (1992) discuss a number of approaches to assessing the potential for ecosystem recovery using ecological models.

$$R_c(t) = \frac{\left[y_p(t) - y_u(t)\right]_{abs}}{\left[y_p - y_u\right]_{max}}$$

y_u = Unperturbed Trajectory $\hspace{2cm}$ (1)

y_p = Perturbed Trajectory

RISK FROM 4-NONYLPHENOL IN AQUATIC ECOSYSTEMS

An example utilizing the chemical 4-nonylphenol will be used to illustrate the application of models to the ecological risk assessment process described above.

Nonylphenol was nominated by EPA as a candidate for rulemaking in 1987, based upon a rationale that included high toxicity to fish and aquatic invertebrates, high production volume, and high concentrations in sludge. Environmental fate and effects testing was recommended and EPA subsequently developed a Testing Consent Order, or Test Rule, for 4-nonylphenol. A two-tier testing scheme, including acute and chronic toxicity as well as chemical fate, was devised by EPA in 1990 (Rodier, 1996).

Annual production of 4-nonylphenol ranges from 80–93 million kilograms in the United States. Nonylphenol is used mainly in the synthesis of other compounds, primarily nonylphenol ethoxylates and alkyl phenol ethoxylates. Alkyl phenol ethoxylates are used extensively as emulsifiers, wetting agents, detergents, dispersants, and as inert constituents in pesticide formulations. While 4-nonylphenol is not discharged directly into aquatic environments, microbial activity can result in biodegradation of ethoxylate compounds into 4-nonylphenol. Studies indicate there is no significant biodegradation of 4-nonylphenol under either aerobic or anaerobic conditions.

Field Studies

The fate and effects of 4-nonylphenol were the subject of a series of field experiments conducted by the EPA Environmental Research Laboratory (Mid-Continent Ecology Division) in Duluth, Minnesota. The field studies utilized a testing protocol developed to examine the effects of toxic chemical compounds on enclosed littoral ecosystems under natural conditions. The field testing protocol involved a series of eighteen 5´10-m enclosures arrayed along the littoral zone of a small lake. In July, 1993, 4-nonylphenol was applied to the enclosures at nominal concentrations of 3, 30, 100, and 300 µg/L. Applications of the chemical were made every 48 hours for a period of twenty days. Each treatment level was replicated four times and there were two controls for each treatment level. Concentrations of 4-nonylphenol in the water and sediments were measured, as was bioconcentration in fish at the two lowest treatment levels (3 and 30 µg/L). Effects on algae, macrophytes, zooplankton, benthic organisms, and fish were recorded. Post-treatment monitoring continued through 1994, one year after the final application.

The results of the field studies are summarized below and described in detail elsewhere (Hanratty et al., submitted; Heinis et al., 1996; Liber et al., 1996; O'Halloran et al., 1996; Sheedy et al., 1996; Schnude et al., 1996).

During the application period, mean maximum concentrations in the water ranged from 75 to 80% of the nominal concentration for each treatment level, except the lowest (3 µg/L). The 4-nonylphenol dissipated rapidly from the water column, with a half-life of 0.74 days. The 4-nonylphenol partitioned rapidly to the enclosure wall, macrophytes, and the sediments within two days following an application. The ultimate sink for 4-nonylphenol appears to be the sediments, with a concentration of 1.97 mg/kg 440 days after initial treatment at a level of 300 µg/L. Thus, long-term persistence of 4-nonylphenol in aquatic environments seems likely.

All taxa of cladocera and copepods were severely reduced in abundance at the 300 µg/L treatment level (48–98%), while several taxa also showed declines at the 30 µg/L and 100 µg/L treatment levels. Populations recovered within four weeks after applications ceased. Rotifera were more resistant to the effects of exposure to 4-nonylphenol, while several ostracod taxa were reduced by up to 82%.

Benthic macroinvertebrates were also affected by exposure. Mollusks were the most severely impacted, with an 81% reduction in abundance at the 300 µg/L treatment level. More significantly, populations had not recovered to pre-treatment levels one year after the experimental treatments. While oligochaetes also experienced steep declines, their populations recovered within six weeks after treatments ceased. Chironomid populations only experienced high mortality at the 300 µg/L treatment level, but adult emergence was reduced by up to 92%, probably as the result of a surface film of surfactant in the enclosures.

Survival of bluegill sunfish was severely reduced at both 100 µg/L and 300 µg/L levels, with no fish being found at the highest treatment level. Overall growth of individual fish was not affected. At the 3 µg/L and 30 µg/L treatment levels, the bioconcentration factor (BCF) was 87.

Simulation Studies

Simulation studies of the littoral ecosystem experiments with 4-nonylphenol were conducted with two risk assessment models, LERAM and AQUATOX. The EPA review of 4-nonylphenol recommended model studies to further characterize the risk to aquatic ecosystems (Rodier, 1996). The models were used to analyze the dynamic behavior of the littoral enclosure ecosystems.

The Littoral Ecosystem Risk Assessment Model, LERAM (Hanratty and Stay, 1994; Hanratty and Liber, 1996), was developed by the EPA Mid-Continent Ecology Division in Duluth, Minnesota, to be used in conjunction with the littoral ecosystem field studies being conducted on pesticides and toxic chemicals. LERAM was modified from the Comprehensive Aquatic Simulation Model (CASM) (DeAngelis et al., 1989) to represent the dynamics of the littoral zone of a pond or lake.

The structure of LERAM is shown in Figure 4.9. The model links single-species toxicity data, derived generally from bioassays, to a bioenergetic ecosystem model. The Toxic Syndrome Calculator modifies growth parameters for the individual model populations in accordance with a toxic stress syndrome response to differing exposure concentrations. Risk projections are obtained from Monte Carlo simulation and are expressed as the probability of reductions in biomass. The aggregated populations included in LERAM are shown in Figure 4.10.

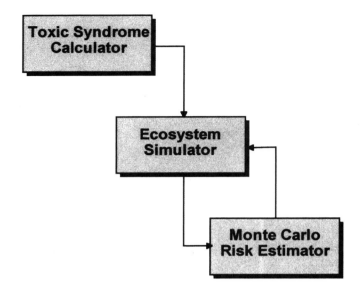

Figure 4.9. LERAM model structure and flow diagram.

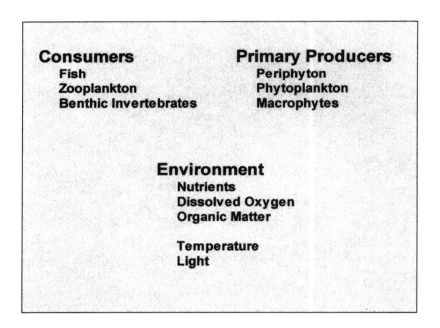

Figure 4.10. LERAM model structure and aggregated state variables.

In contrast to LERAM, the AQUATOX model was conceived as a general purpose risk assessment tool for aquatic ecosystems (Park et al., 1995; Park, 1996). The model can be used to simulate both the fate of chemical substances and mercury in streams, lakes, ponds, and reservoirs, and the effects on the dynamics of the constituent populations. The structure of the model, shown in Figure 4.11, is flexible and allows the user the latitude to modify

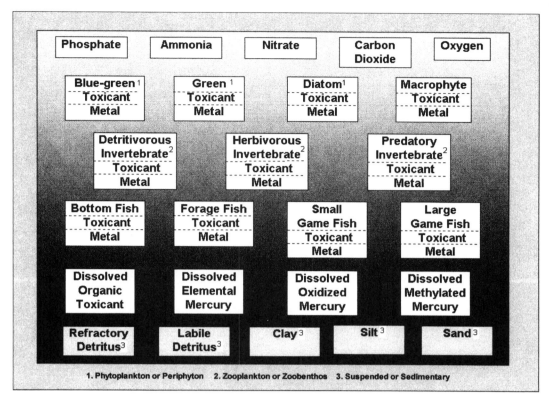

Figure 4.11. AQUATOX fate and effects aquatic ecosystem model structure and state variables.

characteristics of the site, the chemical, and the organisms to be included in the simulation experiments. Because AQUATOX combines both fate and effects, complex release and exposure scenarios can be simulated for both generic and site-specific risk assessments. For this study, the site characteristics and several of the biotic components of the model were modified to represent the littoral ecosystems.

The models were used in a complementary fashion to more completely characterize the risk from exposure to 4-nonylphenol and to assess the significance of those impacts. LERAM was used to generate probabilistic risk estimates for the major biotic components of the littoral ecosystems, while AQUATOX was employed to predict the dynamics of the major taxonomic groups for each of the 4-nonylphenol application scenarios. In addition, the recovery potential of the littoral enclosures, following the cessation of the 4-nonylphenol releases, was examined with the AQUATOX model.

Estimates of risk from LERAM simulations (Hanratty et al., submitted) of the control and the three highest 4-nonylphenol treatment levels are shown in Figure 4.12a–d. Risk is expressed as the probability of a specified percent decline in total population biomass over the course of the simulation, which lasted for the entire growing season. Also included is the probability of a 10% increase in biomass to account for natural variability and for those situations where a constituent population can be enhanced through either direct or indirect effects from exposure. The control simulation, Figure 4.12a, demonstrates a significant

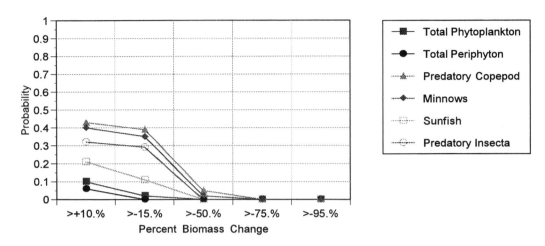

Figure 4.12a. Littoral enclosure risk simulation using LERAM. Control simulation. Probability of a change in biomass concentration.

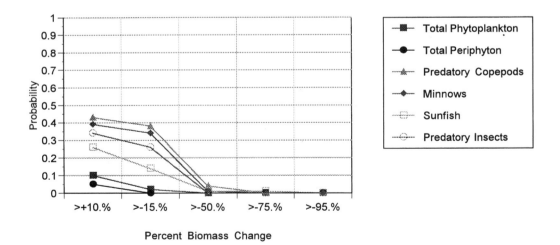

Figure 4.12b. Littoral enclosure risk simulation using LERAM. 4-Nonylphenol concentration 30 µg/L. Probability of a change in biomass concentration.

probability of both small increases and decreases in biomass for predatory copepods and insects in the littoral enclosures. The 30 µg/L treatment level, Figure 4.12b, yields similar

Aquatic Community Risk
Nonylphenol 100 µg/L

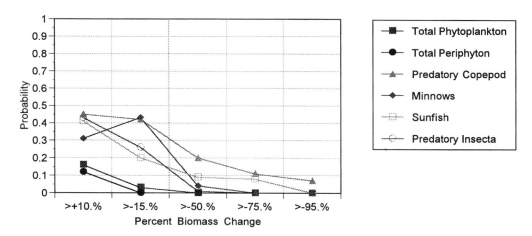

Figure 4.12c. Littoral enclosure risk simulation using LERAM. 4-Nonylphenol concentration 100 µg/L. Probability of a change in biomass concentration.

Aquatic Community Risk
Nonylphenol 300 µg/L

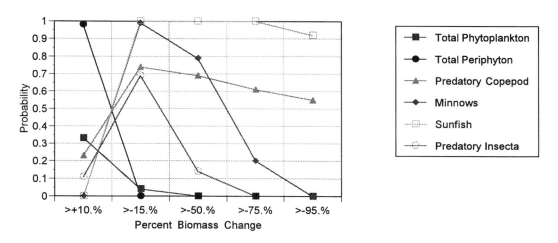

Figure 4.12d. Littoral enclosure risk simulation using LERAM. 4-Nonylphenol concentration 300 µg/L. Probability of a change in biomass concentration.

results. At the 100 µg/L treatment level (Figure 4.12c) the risk of a 50% reduction in the biomass of predatory copepods rises to 20%, while that for bluegill sunfish is also significantly greater than the control. The risk of a 95% reduction of bluegill sunfish biomass rises to 90% at the 300 µg/L treatment level (Figure 4.12d). Predatory copepods are also severely

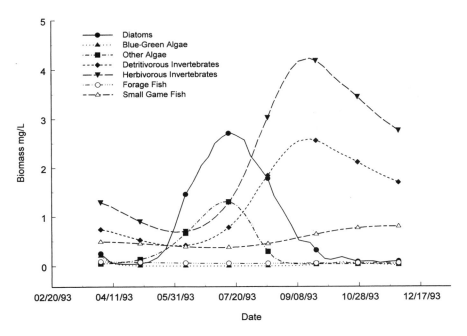

Figure 4.13a. Littoral enclosure ecosystem response simulation using AQUATOX. Control simulation.

impacted, as are minnows to a lesser extent. The latter outcome is probably the result of reduced predation by bluegill sunfish.

Deterministic simulations of the dynamics of the littoral enclosure ecosystems with AQUATOX are shown in Figure 4.13a–d. At the 30 µg/L nonylphenol treatment level, Figure 4.13b, the trajectories of major population groups are very similar to the control simulation. However, at the 100 µg/L (Figure 4.13c) and 300 µg/L (Figure 4.13d) treatment levels, the biomass of forage fish (minnows) and small game fish (bluegill sunfish) collapses. While algae and periphyton are relatively unaffected by exposure, diatom biomass declines and there is a bloom in blue-green algae. The herbivorous and detritivorous invertebrates maintain viable biomass levels largely as a result of reduced predation by fish.

Because 4-nonylphenol sorbs strongly to sediments, it has the potential to persist for long periods in aquatic ecosystems, raising the possibility of chronic exposures and impacts. To assess the potential for long-term system recovery following exposure to 4-nonylphenol, an additional AQUATOX simulation was run to assess system recovery for several years following treatment with 4-nonylphenol.

The littoral enclosure system was simulated for an additional two years, through 1995, to simulate the recovery of population biomass following exposure to a nominal 4-nonylphenol concentration of 50 µg/L. This treatment level was chosen in an effort to prevent the complete loss of populations from the model run. The recovery coefficient described earlier was plotted for three of the primary taxa in the littoral enclosure system; diatoms, herbivorous invertebrates, and forage fish (minnows). The results, shown in Figure 4.14, indicate that

Aquatox: Nonylphenol 30 µg/L

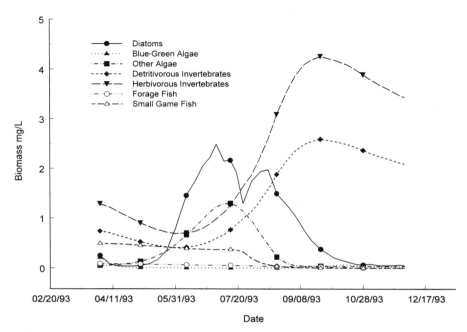

Figure 4.13b. Littoral enclosure ecosystem response simulation using AQUATOX. 4-Nonylphenol concentration 30 µg/L.

Aquatox: Nonylphenol 100 µg/L

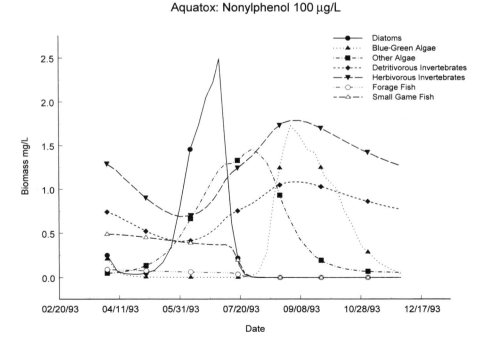

Figure 4.13c. Littoral enclosure ecosystem response simulation using AQUATOX. 4-Nonylphenol concentration 100 µg/L.

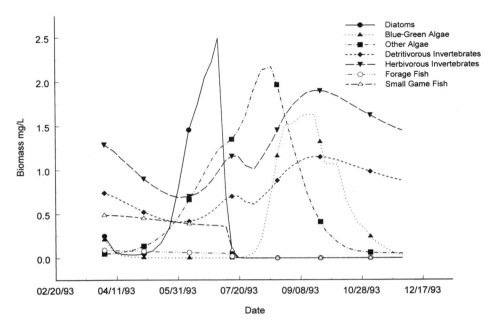

Figure 4.13d. Littoral enclosure ecosystem response simulation using AQUATOX. 4-Nonylphenol concentration 300 µg/L.

forage fish suffered 100% mortality, so that the recovery coefficient reaches an asymptote of 1. With the absence of fish from the system, the recovery coefficients for diatoms and herbivorous invertebrates oscillate about new asymptotes, indicative of a new stable state in the absence of fish.

APPLICATION TO ECOLOGICAL RISK ASSESSMENT

The close correspondence between the results of the littoral enclosure field studies and the modeling exercises with LERAM and AQUATOX demonstrates the capabilities of models designed specifically for estimating risks to aquatic ecosystems from toxic chemicals. Moreover, the models provide a framework for interpreting the results of the enclosure studies and for generating additional hypotheses for testing in future field experiments. In addition, the models provided quantitative estimates of the probability of biomass reductions and the potential for the pond ecosystems to recover from exposure to 4-nonylphenol. It is likely that the accuracy of model forecasts could be further improved by incorporating additional site-specific characteristics into the structure of the simulation models (Bartell, et al, 1992).

The role of simulation modeling in ecological risk assessment and regulatory decision-making for toxic chemicals has been very limited up to the present time. Regulatory actions are currently driven by a combination of human health concerns, ecological concerns derived from laboratory bioassay results, quantitative structure-activity predictions, and economic factors. The adoption of the Framework for Ecological Risk Assessment and the

Recovery Coefficient: Nonylphenol 50 µg/L

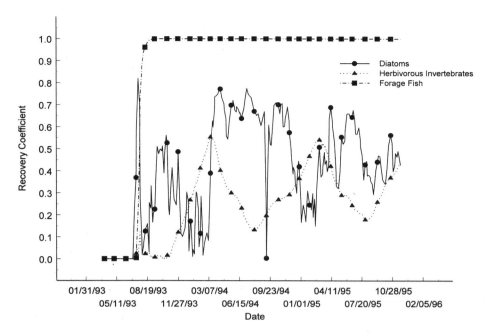

Figure 4.14. Littoral enclosure recovery simulation using AQUATOX. Recovery coefficient as a function of time since exposure for major ecosystem components.

publishing of additional Ecological Risk Assessment Guidelines are expected to provide an impetus for inclusion of more quantitative estimates of ecological risk into the regulatory decision making process at EPA. Thus, the role of simulation modeling in ecological risk assessment can be expected to increase as a result of the unique capability of models to address the critical elements of the assessment and decision-making processes:

  Models provide a logical mechanism for extrapolating from laboratory bioassay studies on individual organisms to population level assessment endpoints of primary concern to regulatory decision-makers.

  Models can forecast both direct and indirect effects while coupling multiple endpoint and exposure-response scenarios.

  Models can provide a temporal and spatial dimension to ecological risk assessment that is absent from current methodologies.

  Models can incorporate the variability in the responses of individual organisms to exposure to toxic chemicals as well as the natural fluctuations of ecosystems.

  We suggest that there is an important additional role for simulation models to play as the scope and complexity of ecological risk assessment and decision-making expands to encompass a number of outstanding issues not currently addressed.

  The majority of toxic chemical assessments are generic and incorporate no site-specific information. Models can be employed to provide site-specific and regional risk assessments.

Cumulative impacts of chemical releases are not currently addressed—impacts resulting from chemical mixtures or multiple chemical releases are not incorporated into the risk assessment or regulatory decision-making process.

There are no criteria for determining the magnitude of *"acceptable risk"* for ecosystems exposed to toxic chemical substances. The concept of *"acceptable risk"* is an integral part of risk assessments for human health. One possible approach is to use models to forecast the probability of system recovery to help determine the appropriate levels of risk that will result in regulatory action.

In conclusion, we look forward to an increasing degree of coevolution between ecological simulation models and the ecological risk assessment process for toxic chemicals, which should result in more consistent and logical regulatory decisions in the future.

REFERENCES

Bartell, S.M. Ecosystem context for estimating stress-induced reductions in fish populations. *Am. Fish. Soc. Symp.* 8, pp. 167–182, 1990.

Bartell, S.M., R.H. Gardner, and R.V. O'Neill. *Ecological Risk Estimation*. Lewis Publishers, Boca Raton, FL, 1992, p. 283.

Barnthouse, L.W. The role of models in ecological risk assessment: A 1990's perspective. *Environ. Toxicol. Chem.* 11, pp. 1751–1760, 1992.

DeAngelis, D.L., S.M. Bartell, and A.M. Brenkert. Effects of nutrient cycling and food chain length on resilience. *Am. Nat.* 134, pp. 778–805, 1989.

EPA. *Reducing Risk: Setting Priorities and Strategies for Environmental Protection*. EPA SAB-EC-90-021. Science Advisory Board. Washington, DC, 1990.

EPA. *A Framework for Ecological Risk Assessment*. EPA 630/R-92-001. Risk Assessment Forum, Washington, DC, 1992.

EPA. *A Review of Ecological Assessment Case Studies from a Risk Assessment Perspective*. EPA/630/R-92/005. Risk Assessment Forum, Washington, DC, 1993.

EPA. Ecological Risk Assessment Issue Papers. EPA/630/R-94/009. Risk Assessment Forum, Washington, DC, 1994.

Ferson, S., L.R. Ginzburg, and R.A. Goldstein. Inferring ecological risk from toxicity bioassays. *Water Air Soil Pollut.* 90, pp. 71–82, 1996.

Ginzburg, L.R., L.B. Slobodkin, K. Johnson, and A.G. Bindman. Quasiextinction probabilities as a measure of impact on population growth. *Risk Anal.* 21, pp. 171–181, 1982.

Ginzburg, L.R., S. Ferson, and H.R. Akcakaya. Reconstructibility of density dependence and the conservative assessment of extinction risks. *Conserv. Biol.* 4, pp. 63–70, 1990.

Hanratty, M.P. and F.S. Stay. Field evaluation of the Littoral Ecosystem Risk Assessment Model's predictions of the effects of chlorpyrifos. *J. Appl. Ecol.* 31, pp. 439–453, 1994.

Hanratty, M.P. and K. Liber. Evaluation of model predictions of the persistence and ecological effects of diflubenzuron in a littoral ecosystem. *Ecol. Modell.* 90, pp. 79–96, 1996.

Hanratty, M.P., K.A. McTavish, and K. Liber. Evaluation of model predictions of the ecological effects of 4-nonylphenol: Before and after model refinement. *Ecol. Modell.* 1996.

Heinis, L.J., M.L. Knuth, K. Liber, B.R. Sheedy, R. Purnell, and G.T. Ankley. Persistence and distribution of 4-nonylphenol following repeated applications to littoral enclosures. *Environ. Toxicol. Chem.* 1996.

Kelly, J.R., and M.A. Harwell. Indicators of ecosystem response and recovery, in *Ecotoxicology: Problems and Approaches,* Levin, S.A., M.A. Harwell, J.R. Kelly, and D.K. Kimball, Eds., Springer-Verlag, New York, 1989.

Liber, K., J.A. Gangl, T.D. Corry, L.J. Heinis, and F.S. Stay. Effects of 4-nonylphenol on bluegill sunfish in littoral enclosures. *Environ. Toxicol. Chem.* (accepted for publication).

Mauriello, D.A. Global stability concepts for complex ecological models, in *State-of-the-Art in Ecological Modelling*, Lauenroth, Skogerboe, and Flug, Eds., Elsevier Scientific Publishing Company, New York, 1982.

Mauriello, D.A. (1988). Uncertainty and alternative models in ecological risk assessment, presented at the *Use of Models in a Regulatory Environment Symposium*, Annual Meeting, International Society for Ecological Modelling, Davis, CA.

Mauriello, D.A. Stress to indicator species as assessment endpoints. *J. Hazard. Mater.* 35, pp. 273–282, 1993.

NRC (National Research Council). *Risk Assessment in the Federal Government: Managing the Process.* National Academy Press, Washington, DC, 1983, p. 191.

NRC. A paradigm for ecological risk assessment, in *Issues in Risk Assessment*, Committee on Risk Assessment Methodology, National Academy Press, Washington, DC, pp. 243–356. 1993.

O'Halloran, S., K. Liber, J. Gangl, and M.L. Knuth. Effects of repeated exposure to 4-nonylphenol on the zooplankton community in littoral enclosures. *Environ. Toxicol. Chem.* (submitted for publication).

O'Neill, R.V., R.H. Gardner, L.H. Barnthouse, and G.W. Suter. Ecosystem risk analysis: A new methodology. *Environ. Toxicol. Chem.* 1, pp. 167–177, 1982.

Park, R.A., B. Firlie, R. Camacho, K. Sappington, M. Coombs, and D. Mauriello. Aquatox, a general fate and effects model for aquatic ecosystems, in *Toxic Substances in Water Environments Proceedings*, Water Environments Federation, Alexandria, VA, pp. 3-7 to 3-17, 1995.

Park, R.A. *Aquatox for Windows: A Modular Toxic Effects Model for Aquatic Ecosystems.* Contract: 68-C5-0051,1-13. U.S. Environmental Protection Agency, Washington, DC, 1996, p. 139.

Rodier, D. (1996). *Draft Rm-1 Document for Para-nonylphenol. U.S. Environmental Protection Agency.* Office of Pollution Prevention and Toxics. Chemical Screening and Risk Assessment Division.

Rodier, D.J. and D.A. Mauriello. The quotient method of ecological risk assessment and modeling under TSCA: A review, in *Environmental Toxicology and Risk Assessment*, ASTM STP 1179, Landis, W.G., J.S. Hughes, and M.A. Lewis, Eds., American Society for Testing and Materials, Philadelphia, PA, pp. 80–91, 1993.

Sheedy, B.R., L.J. Heinis, M.L. Knuth, and G.T. Ankley. Determination of 4-nonylphenol concentrations in sediment from a littoral enclosure study. *Environ. Toxicol. Chem.* (submitted for publication).

Schnude, K.L., K. Liber, T.D. Corry, and F.S. Stay. The effects of 4-nonylphenol on benthic macroinvertebrates and insect emergence in littoral enclosures. *Environ. Toxicol. Chem.* (submitted for publication).

Suter, G.W. Ecological endpoints, in *Ecological Assessments of Hazardous Waste Sites: A Field and Laboratory Reference Document,* EPA 600/3-89/013, Warren-Hicks, W., B.R. Parkhurst, and S.S. Baker, Jr., Eds., Washington, DC, 1989.

5 ||| Ecological Models and Ecological Risk Assessment

Kym Rouse Campbell and Steven M. Bartell

INTRODUCTION

This chapter reviews the basic ecological modeling process and summarizes current capabilities in ecological modeling. Assessment of current capabilities includes both a discussion of the range of ecological systems that have been the subject of model development and implementation, and a discussion of the accuracy and precision of the models developed to address individual organisms, populations, communities, ecosystems, and larger-scale systems (e.g., watersheds, habitats, landscapes, and biomes). This chapter also outlines current modeling needs and suggests future modeling approaches in support of ecological risk assessment, a discipline which dates at least to the early 1980s (e.g., Goldstein and Ricci, 1981; O'Neill et al., 1982; Bartell et al., 1983) and continues to evolve. The need is to know not only where ecological risk assessment is now, but also where it is headed in relation to improvements in the accuracy and precision of models used to characterize ecological risks.

A detailed history of the role of mathematical models in ecology lies clearly beyond the scope of this chapter, but some general observations are possible. Differential equations have been used to describe and understand the dynamics of ecological systems since the 1920s (Lotka, 1925). Riley et al. (1949) developed a mathematical model to describe plankton dynamics, and Olson (1963) used a mathematical model to simulate forest production. In the 1950s and 1960s, further development of population dynamics models took place (Jorgensen et al., 1996). Perhaps the greatest expansion in ecological modeling activities resulted from the International Biological Programme of the 1960s and 1970s (McIntosh, 1985; Golley, 1993). Ecological models developed during the 1970s are often characterized as being too complex, maybe due to the revolution in computer technology; it was easy to write programs for a computer that could handle rather complex models (Jorgensen et al., 1996). During the mid-1970s, it became increasingly clear that the limitations in modeling were not due to computers or mathematics but to the lack of data and knowledge about ecosystems and ecological processes, and modelers became more critical in their acceptance and use of models (Jorgensen et al., 1996). Parallel to the development of ecological models, ecologists became more quantitative in their approach, which was critically important to improving the quality of ecological models (Jorgensen et al., 1996). Ecological modeling continues to play an important role in the advancement of theoretical and applied ecology (e.g., chaos theory, hierarchy theory, community assembly, biological invasions, propagation of disturbance, infectious disease transmission, food webs, nutrient cycling) and in the development of quantitative ecology (Allen and Starr, 1982; Pimm, 1982; 1991; Post and Pimm, 1983; Swartzman and Kaluzny, 1987; Gleick, 1988; Drake, 1988; 1989; 1990; 1991; Drake et al., 1989; 1993; Anderson and May, 1991; DeAngelis, 1992; Allen and Hoekstra, 1992; Luh and Pimm, 1993; Russel et al., 1995; Yodzis; 1995).

Mathematical models have become effective tools for estimating and managing ecological risks. In risk estimation, ecological models can be applied to project or forecast future potential risks (e.g., climate change). Ecological models may be the only recourse for estimating risks under circumstances where the desired field experiment cannot be performed (e.g., release of a novel toxic chemical into the environment, release of a genetically engineered organism, introduction of an exotic species). Simulation models can be used to perform a very large number of numerical "experiments" using different assumptions when empirical testing of hypotheses is either impractical or destructive (Fleming et al., 1994). In addition, ecological models can be used to reconstruct past ecological risks for situations where evidence of exposure or effects cannot be currently demonstrated (e.g., habitat loss or degradation). Models can also provide a means to assess the implication of gaps in information (Dale and Rauscher, 1994). Finally, risk management can benefit from the use of ecological models. For example, models can be used to evaluate ecological risks in a "what if" context in scenario-consequence analysis. Ecological models can also be implemented to assess and evaluate the efficacy of different treatment technologies for reducing risk in a general approach to risk-based remediation.

It is important to note that models are used in a fundamentally different way in applied ecology (e.g., risk assessment) than in basic research and development. A fair evaluation of the contribution of ecological models to the risk assessment process is based on an understanding of the basics of ecological modeling. The real issue in determining whether models can contribute to ecological risk assessment is not validity, but credibility; establishing credibility for models used in risk assessment can be done by experimental testing, successful use in other applications, and peer-reviewed publication (Barnthouse, 1992).

Background

Risk is the conditional probability of a specified event occurring, combined with some evaluation of its consequences (Kaplan and Garrick, 1981). Kaplan and Garrick (1981) introduced a concise model for the quantitative characterization of risk:

$$\text{Risk} = [x_i, p_i, s_i], \qquad i = 1, n$$

where x_i delineates one of n specified ecological effects that is the subject of the assessment,

p_i is the conditional probability of x_i occurring and represents an integration of exposure and effects assessment in estimating risk, and

s_i evaluates the consequences of x_i.

This ordered triplet provides a simple template for risk assessment, although the characterization of each risk component will likely involve difficult measurements and complex models that push assessment capabilities to their scientific limits. An *ecological risk* is the conditional probability of a specified ecological event occurring, coupled with some statement of its consequences. Examples of ecological events include climate change, reduced biodiversity, loss of commercially–valuable resources, species extinction, biological invasions, habitat alteration, thermal pollution, and contamination by inorganic and organic chemicals, radionuclides, and pesticides.

Understanding the various ecological risk assessment approaches or frameworks that have been developed to standardize the risk assessment process is essential to effectively select and apply ecological models most appropriate for specific assessments. The descriptions of the components in these frameworks are deliberately general so that the risk assessment process can be applied to a wide variety of ecological stressors (Figure 5.1). Ecological models can be used in any of the components of the risk assessment frameworks. Organized frameworks for risk assessment should provide for *coherent* risk assessment and risk management (Apostolakis, 1989). However, these frameworks must also permit the scientific flexibility necessary to address the ecologically unique aspects of individual assessments. Several methodologies for assessing ecological risks are represented in Figure 5.1; they include the U.S. Environmental Protection Agency (EPA) Framework, the National Research Council (NRC) Framework, the Canada Council of Ministers of the Environment (CCME) Framework, and the Water Environment Research Foundation (WERF) Methodology (EPA, 1992; 1996; NRC, 1992; CCME, 1994; Parkhurst et al., 1995).

THE BASIC APPROACH TO ECOLOGICAL MODEL DEVELOPMENT

Ecological models are used in fundamentally different ways when applied to basic research and development than when used to assess and manage ecological risk. This distinction is important in order to effectively incorporate models into the risk assessment process and must be understood before reviewing the basic approach to developing ecological models.

Ecological Models in Basic Research and Development

Models have become valuable conceptual and operational aids in understanding the natural world, and in recent years, they have become a core component of ecology (Hastings, 1997). The modeling process forces the scientist to identify and describe essential ecological structures and functions that can either be quantified or can be predicted even if incompletely understood. Modeling provides the opportunity to specify the simplifying assumptions that are appropriate to the ecological phenomena of interest. Knowing what significantly affects the ecological process allows the scientist to focus on understanding the key components to develop methods and measurements, to design critical experiments, and to increase quantitative understanding of the ecological process.

In basic ecological research and development, models are constructed with the express purpose of invalidating them. Critical new experiments or field measures can be identified by forcing a model to failure (i.e., model predictions disagree with observations). Failure

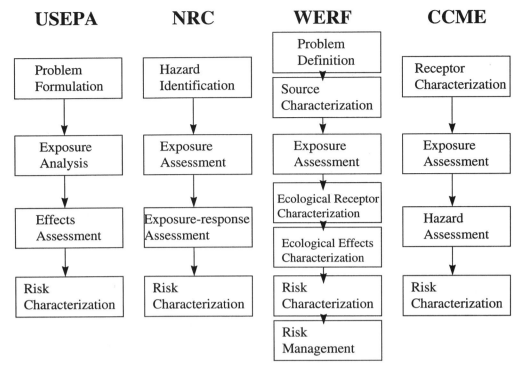

Figure 5.1. A representation of several methodologies [the U.S. Environmental Protection Agency (USEPA) Framework, the National Research Council (NRC) Framework, the Water Environment Research Foundation (WERF) Methodology, and the Canada Council of Ministers of the Environment (CCME) Framework] that have been developed for assessing ecological risks (EPA, 1992; 1996; NRC, 1992; CCME, 1994; Parkhurst et al., 1995).

often is achieved by systematically probing model sensitivity to changes in the underlying assumptions, parameter values, model structure, and process formulations. Basic scientific advancement can proceed as a result of repeated iterations of model failure, the generation of new information, and model revisions. A model's domain of applicability is expanded through the model evaluation process (Mankin et al., 1975).

Ecological Models in Ecological Risk Assessment and Management

In contrast, risk assessors and risk managers desire ecological models that can be reliably and confidently used to predict potential or impending risks, as well as to reconstruct past ecological effects. Simply stated, assessors and managers demand "validated" models that are accurate and precise. Model failure is something to be avoided—not pursued—by the risk assessment and management community.

Before investing time and resources in implementing a particular model for assessing ecological risk, the assessor must evaluate previous performances of the model for circumstances similar to those of the present assessment. Performance is measured primarily as the degree of similarity between the model predictions and the corresponding actual measurements of the ecological system of interest. Many alternatives for quantifying the degree of similarity have been suggested (Mankin et al., 1975; Schaeffer, 1981; Bartell et al., 1986b).

Model performance must be evaluated in the context of the model's potential contribution to the overall risk assessment and decision-making process. The most effective way to use models in ecological risk assessment is to start with an explicit description of the decision process and an answer to the question of "how good" the model has to be to allow for an appropriate alternative decision. Without an answer, the risk assessor may inadvertently disqualify a perfectly useful model or select an inadequate model. Once this question is answered, candidate models can be intelligently evaluated. Working backward from the decision-making process to model evaluation and selection is more effective than searching for the "validated" model independent of its intended use. In a strict sense, ecological models cannot be validated any more than statistical hypotheses can be proven, and any ecological model can be pushed to failure. As a result, the fundamental context for evaluating model performance includes its usefulness in the decision-making process in combination with its scientific defensibility.

The difference between the role of ecological models in basic research and development and their role in risk assessment and risk management suggests a potential impasse: How can reliable risk assessment models be produced by ecological researchers devoted to invalidating their models? How can risk assessors use models that have been shown to be invalid? Quite simply, as the model evaluation ("validation") process operates to expand its domain of applicability, a model can still be reliably (i.e., with known confidence) used to assess risks for circumstances that lie well within this domain (Mankin et al., 1975). Therefore, basic research and ecological risk assessment can proceed simultaneously.

Building an Ecological Model

To effectively use ecological models in risk assessment, it is important that models be neither under- nor oversold in relation to their capabilities in forecasting or reconstructing ecological risks. Ecological models are simplifications of the natural systems around us. They are, by definition, imperfect (Mankin et al., 1975); however, the key issue here is their potential utility in assessing ecological risks. Briefly reviewing the basic approach to developing an ecological model can prove useful when evaluating ecological models for use in risk assessment.

Detailed descriptions of the basic approach to developing an ecological model have been provided elsewhere (e.g., Odum, 1971; Patten, 1971; Swartzman and Kaluzny, 1987). The purpose here is to emphasize that what might be important is not necessarily just the model, but the capability of ecological risk assessors to construct models specific to the assessment issue at hand. The steps in the ecological modeling process include the following.

Identification of the System

The focus of the modeling effort—the ecological phenomena of interest—should be defined as precisely as possible (e.g., population dynamics, energy flow, material cycling, system stability). This step identifies the necessary and sufficient physical, chemical, and biological structures and processes necessary to characterize the phenomena of interest.

The complexity of the model is determined largely in this initial step; the critical simplifying assumptions are made, and aggregation errors are addressed (Gardner et al., 1982). In addition, the appropriate spatial and temporal scales are defined for the model. A routine part of this step is the production of a compartmental schematic or flowchart that illustrates the model state variables (objects of interest) and functional interrelations among the state variables (i.e., a "box and arrow" diagram).

Formulation of the Model Equations

After definition of the variables that characterize the state of the ecological system of interest, the next step is to formulate the equations that determine the changes in state. There should be one such governing equation per state variable. The mathematics of these equations will establish the fundamental dynamics of the system and can include, for example, linear and nonlinear differential equations, difference equations, and/or simple algebraic equations. Traditional approaches emphasize conservation of mass/energy as a basic prescription in developing the governing equations. The fundamental currency of the model (e.g., biomass, kilojoules, grams phosphorus) is also reflected in the governing equations.

Definition of Any External Inputs or Forcing Functions

Ecological models nearly always require inputs of data or information considered external to the system. This data or information is critical for describing the modeled system's dynamics (e.g., for models of aquatic system production, examples include surface light intensity, nutrient loadings, and water temperature). Chemical loading values or exposure concentrations are model inputs of interest in developing models for ecological risk assessments of toxic chemicals. Such external forcings often take the form of data files that are accessed by the computer software used to develop the model.

Estimation of Model Initial Conditions and Parameters

Values that quantify the beginning state of the model—usually one value per state variable—are needed to begin the computations. These initial conditions are ideally based on site-specific measurements. In their absence, published data from similar systems are usually used. Values must also be estimated for all of the model parameters, which are the mathematical terms in the governing equations. The sources of these values might include site-specific measurements, laboratory and field experiments, theoretical derivations, and published estimates for similar ecological systems. In any case, the source and justification for all initial conditions and parameter values should be properly documented as part of the modeling process.

Solution of the Equations

Once the system has been described, defined, formulated, and quantified, the equations must be solved in order to explore the implications of all the previous steps in model development. In some cases, the model might have a simple analytical solution that can be evaluated by hand. But in most situations, model complexity requires either commercial software for simulation modeling or coding the model in some higher level programming language (e.g., BASIC, Fortran, C/C++). Verification that the method of solving the equations is accurate and that results are numerically correct is an important part of this step.

Evaluation of Model Performance

The results of the verified model should be evaluated by comparing the calculated values with values measured for the system of interest or for a similar system. The criteria for model performance should be specified during system identification. It is not uncommon for model calculations to be adjusted (i.e., calibrated) by modifying initial conditions, parameter values, forcing functions, or all three to obtain acceptable agreement between simulated and observed system dynamics. Methods of sensitivity and uncertainty analysis are particularly useful in understanding model operation and determining if model performance is consistent with the original conceptualization of the system. Evaluation of model performance may lead to model acceptance, rejection, or more often, model revision. Each of the preceding steps in the ecological modeling process will likely be repeated in a continued effort to revise and refine model performance.

Selecting a Model for Ecological Risk Assessment

When selecting or developing a model for estimating ecological risks, the risk analyst should consider the following questions.

1. What are the ecological risks? What are the specific ecological effects of interest in the overall analysis? [Most risk assessors use the word "endpoint." Endpoint is a curious term in this context; Webster defines an endpoint as "a point marking the completion of a process..." (Bartell, 1996). In terms of the risk assessment process, selecting an endpoint is the beginning of the process, not the end (Bartell, 1996). For this reason, we prefer the use of the word "effect" or the words "ecological effect," instead of endpoint.] Ecological effects measured at different levels of organization could be included. What are the relevant space-time dimensions for the selected ecological effects?
2. What information is available to use in the analysis? This includes ecological data as well as toxicological data. Sufficient information must be available to set up and execute the model. What are the sources and magnitudes of uncertainties associated with these data?
3. What are the (clear and precisely defined) decision criteria for assessing unacceptable risk? Acceptable degrees of model uncertainty that still permit a decision regarding ecological risk should be specified at the outset.
4. How is exposure translated into the ecological effect of interest? The model must have an operational way to do this. Examples of exposure-response functions include direct mortality; alteration of transition probabilities in a matrix (demographic) model; and changes in rates of processes that determine growth, reproduction, energy capture, nutrient cycling, and decomposition.

Selecting a model will depend on how closely the model outputs address the ecological effects of interest and on whether there is sufficient information to use the model in the particular application. Selecting a basic research model for use in ecological risk estimation will often depend on how easily the model can be modified.

MODELS FOR ECOLOGICAL RISK ASSESSMENTS

Current capabilities to model ecological risks are largely unknown and undocumented, and a comprehensive review and evaluation of ecological models that could be used to assess ecological risks remains to be performed. Limited and directed reviews by Barnthouse et al. (1986) and Emlen (1989) emphasized population and ecosystem models available in the mid-1980s; however, the principal emphasis in these reviews was on the applicability of these models and not on their usefulness in risk assessment. A review of environmental and ecological modeling was recently completed by Jorgensen et al. (1996) but again, not in the context of risk assessment. The role of models in ecological risk assessment was reviewed by Barnthouse (1992); he focused on individual-based models (IBMs) and regional models. Models that address the risk of climate change on forests were reviewed at four levels of biological organization: global, regional or landscape, community, and tree (Dale and Rauscher, 1994). Since ecological models have been developed at an increasing rate during the last thirty or so years, a comprehensive analysis of modeling capabilities is not a trivial exercise; however, it is important to the future of ecological risk assessment. Such a comprehensive review and evaluation is beyond the scope of this chapter; however, examples of the most recent advances in ecological modeling that could be modified for assessing ecological risks are presented. In addition, examples of models that have been developed or adapted for assessing ecological risks in recent years also will be discussed.

An organization of ecological structure as presented in textbooks provides a convenient framework for discussing different kinds of ecological models that could be (and in some instances, have already been) used to assess ecological risks. The following sections describe and discuss approaches to modeling of (1) individual organisms, (2) populations, (3) communities, (4) ecosystems, (5) landscapes, and (6) biological invasions in relation to ecological risk assessment. However, it should be emphasized that each of these different categories results from different views of the same natural world. Each viewpoint has its ecological phenomena of interest, specified or implied corresponding hypotheses, derived measures or metrics, and models. No one viewpoint can be logically argued as the correct one. The important point is to choose or construct a model that is appropriate in scale, simplifying assumptions, and ecological complexity to address a particular risk assessment problem.

Models of Individual Organisms

Most individual organisms are conveniently bounded on space and time scales relevant to human observers. Individuals of threatened or endangered species can be justifiably recognized as ecological resources worthy of protection. The ecological importance of individuals can also be assessed in terms of their contribution to the persistence and adaptability of their species; some individuals (e.g., reproductive females) are ecologically and genetically more important than others. A proper focal level in ecotoxicology must consider the individual organism's response; chemical effects occur at the level of the individual (Hallam et al., 1990). The exposed, affected individual is the appropriate reference point for extrapolation to the population level or level of a biological system (Hallam et al., 1990). Therefore, models of individual organisms can contribute substantially to understanding the ecological response to toxic chemicals and other disturbances (Hallam et al., 1990; DeAngelis and Gross, 1992). Model algorithms for individuals can effectively combine bioenergetics, behavior, and life-history characteristics.

Individual-Based Models (IBMs)

An IBM describes population dynamics by following the fate of each individual; all or a representative fraction of the members of a population are simulated simultaneously (Fleming et al., 1994). Several examples of IBMs that could easily be modified for use in ecological risk assessment or have been used to assess ecological risks are discussed below. Acceptance criteria for using IBMs to make management decisions have been developed by Bart (1995); these criteria could be used as a guideline for evaluating IBMs proposed for use in ecological risk assessment.

IBMs have been developed to describe PCB accumulation and variability in lake trout (*Salvelinus namaycush*), rainbow trout (*Oncorhynchus mykiss*), coho salmon (*Oncorhynchus kisutch*), and chinook salmon (*Oncorhynchus tshawytscha*) and dieldrin contamination in lake trout in Lake Michigan (Madenjian et al., 1993a; 1993b; 1994; Stow and Carpenter, 1994). These models accounted for the individual variability in contaminant concentrations; the results indicated that the IBM has the potential to be applicable to a variety of organochlorine contaminants and should prove useful for risk assessments of contaminants in fish.

An IBM was calibrated for use to compare random harvest and size limit management strategies for a lotic population of brook trout (*Salvelinus fontinalis*) (Power and Power, 1996). This model could be used to assess the risks of a collapse in a fishery resulting from overharvesting.

Dunham and Overall (1994) developed individual-based, physiologically-structured models to predict population responses of the lizard, *Sceloporus merriami*, to two different climate change scenarios. This particular population of *S. merriami* has been intensively studied over 20 years, and together with the appropriate physiological studies, provided the basis for the construction of individual-based, physiologically-structured models of population response to environmental disturbance.

Collecting birds and analyzing their tissue yields unclear information on local contaminants because birds are highly mobile, and toxicity assays with captive birds ignore changes in food intake with varying activity and weather experienced by wild birds (Lovvorn and Gillingham, 1996). To provide for a more realistic ecological risk assessment, an IBM was developed to simulate the intake of contaminated food by diving ducks for varying conditions such as weather, water depth, food dispersion, and the size and digestibility of food (Lovvorn and Gillingham, 1996).

Top-level carnivores, which include many threatened and endangered species, are vulnerable to changes that affect the natural landscape (Pimm and Gilpin, 1989). The landscape is a mosaic of habitat patches covering a region occupied by a population or a community of interest; the size of the landscape depends on the questions being asked (Fleming et al., 1994). Spatial extent and spatial heterogeneity are landscape characteristics that must be taken into account in describing a population (Fleming et al., 1994). Fleming et al. (1994) developed an IBM for the endangered wood stork (*Mycteria americana*) to assess the importance of landscape heterogeneity in the Florida Everglades. The model simulated the behavior and energetics of each individual wood stork in a breeding colony and included a landscape submodel, a prey-based submodel, behavior and energetics submodels of potential nesting adults, and energetics and growth submodels of nestlings. IBMs, such as the one developed by Fleming et al. (1994), are a promising method of assessing the risks of extinction to wildlife species, especially those that are threatened or endangered due to habitat or landscape alteration.

In the process of evaluating ecological and economic consequences of human activities, ecologists and economists usually do not work together, and policy-makers are often in a dilemma because of the contradictory recommendations from these two sides (Liu, 1993). A well-known example is the heated argument about the impacts of timber harvesting on the northern spotted owl (*Strix occidentalis*) and the economy in the northwestern United States (Simberloff, 1987). To address this problem, Liu (1993) has developed an ecological-economic model (ECOLECON) that simulates animal population dynamics and economic cash flows in response to landscape structure and timber harvest in managed forests. ECOLECON is a spatially-explicit, individual-based, and object-oriented program that contains both ecological and economic algorithms and information (Liu, 1993). The model has been parameterized using growth and yield functions for loblolly pines (*Pinus taeda*) and using ecological information of Bachman's sparrow (*Aimophila aestivalis*), a species of management interest occurring in pine forests in the southeastern United States (Liu, 1993). This model can be useful in assessing the ecological risks of timber harvest to individual species, while providing a link to forest economics through forest structure and management.

An IBM was used to simulate the threatened grizzly bear (*Ursus arctos horribilis*) response to region-level management practices represented by ownership patterns and watershed-level changes in habitat availability due to proposed harvests and road building in southwestern Montana (Boone and Hunter, 1996). Risks to large mammals because of changes in land use can be demonstrated by using IBMs similar to that developed by Boone and Hunter (1996).

Turner et al. (1994) developed a spatially-explicit IBM to explore the effects of fire scale and pattern on the winter foraging dynamics and survival of free-ranging elk (*Cervus elaphus*) and bison (*Bison bison*) in northern Yellowstone National Park. The effects of fire on ungulate (hooved mammal) survival became important when winter conditions were average to severe, and effects were apparent in both the initial and later postwinter fires (Turner et al., 1994). The interaction between fire scale and spatial pattern suggests that knowledge of fire size alone is not always sufficient to predict ungulate survival (Turner et al., 1994).

The relationship between alpine plant distributions and environmental conditions was evaluated to predict the spatial distribution of major alpine plant communities found along the Front Range of the Rocky Mountains in the western United States by modifying the individual plant-based STEPPE model (Humphries et al., 1996). Similar models could be used to assess the risks of environmental change to plants or predict the effects of changes in environmental conditions on plants.

Dale and Rauscher (1994) included models of individual tree responses to climate change in a model review and a conceptual framework of spatially-explicit models that address the impacts of climate change on forests. The response of individual trees to climate change occurs through phenological, reproductive, and physiological processes on time scales ranging from minutes to decades and spatial scales ranging from cells to that of a large tree; and individual tree models provide information on carbon and nitrogen fluxes, leaf area, and growth, death, and reproduction rates (Dale and Rauscher, 1994).

An IBM, such as those developed by Hill et al. (1995) and Higgins et al. (1996), could be used to assess the risk of invasion of unwanted, exotic, or introduced tree species. A spatially-explicit IBM was used to predict rates and patterns of the spread of exotic pine trees (*Pinus* spp.) in South Africa (Higgins et al., 1996). Critical life-history stages and ecological processes responsible for the variation in the rates of tree invasion in utility right-of-ways were calculated using an IBM (Hill et al., 1995).

Physiologically Based Pharmacokinetics (PBPK) Models

PBPK models can simulate the distribution of contaminants throughout an organism and estimate the rate of accumulation in target organs (Gerlowski and Jain, 1983). PBPK models, which were developed in the medical sciences, define an organism in terms of its anatomy, physiology, and biochemistry (Newman, 1995). These models have been used in ecotoxicology in bioaccumulation studies, primarily for organic contaminants; bioaccumulation can be linked to such processes as gill exchange, bivalve filtration rates, or temperature-dependent metabolic rate (Newman, 1995). In recent years, PBPK models have been used in human health risk assessments for carcinogens (Welsch et al., 1995; Yang et al., 1995; Casanova et al., 1996). PBPK models also could be used in ecological risk assessments. For example, modeled toxic effects could be related to growth or survival of contaminated individuals, and more realistic integration of individual variability in exposure and response might result; such integration could provide more realistic estimates of risk.

Population Models

Populations have received the most attention as ecological responses of interest in the development of ecological risk assessment (e.g., Emlen, 1989; Bartell, 1990; Cardwell et al., 1993; Suter, 1993). Mathematical models of population dynamics have been used to characterize ecological risks posed by toxic chemicals (e.g., Barnthouse et al., 1990; Bartell, 1990; Hallam et al., 1993). Population models recently have been used to determine the risk of extinction of metapopulations (all the subpopulations that make up the larger population) of the endangered Leadbeater's possum (*Gymnobelideus leadbeateri*) that inhabits ensembles of habitat patches in timber production areas in southeastern Australia (Lindenmayer and Lacy, 1995; Lindenmayer and Possingham, 1996).

Traditionally, mortality of organisms has been used as the ecological effect of toxicant stress in risk assessments. However, other population-level effects are increasingly being used. For example, Hallam et al. (1993) coupled a physiological, energetics-based model of an individual daphnid with an exposure model to investigate the sublethal effects of a nonpolar narcotic on individual growth and the resulting manifestation at the population level.

Population has come to mean a localized collection of individuals of the same species. Indeed, toxicity assays routinely imply this characterization of a population, that is, a population in a statistical sense. This view of a population has led to the selection of various fractional decreases in standing crop as responses in risk assessment (O'Neill et al., 1982; Barnthouse et al., 1990; Bartell, 1990; Bartell et al., 1992). These typically annual decreases in population size must be related to future production, maintenance, or survival of the assessed populations.

Populations can also be described in terms of the life history of the species in question. This perspective introduces, of course, considerations of age (size) classes, demographics, life tables, net reproduction, and reproductive value, which are all of potential utility in defining ecologically meaningful responses for risk assessment. In 1990, Barnthouse et al. focused on a 20% decrease in the recruitment of largemouth bass (*Micropterus salmoides*) in relation to disturbance, yet this choice of response was not developed in a quantitative context of the production dynamics or sustainability of the bass population. Diminished net reproductive rates or reproductive values that carry an unacceptable risk of local extinction appear to be the only ecologically justifiable responses for ecological risk assessment using this definition of population; however, societal concerns can certainly justify selection of other population-level effects for assessments.

In addition, populations can also be described genetically (Roughgarden, 1979). The ability of some populations to survive in a polluted environment has focused attention on contaminants as selective agents and factors contributing to differential tolerance (Woodward et al., 1996). Changes in the genetics of populations have been demonstrated in relation to toxic chemical exposures (Landis et al., 1996); for example, genetic and demographic changes occurred in mosquitofish (*Gambusia holbrooki*) populations chronically exposed to mercury (Mulvey et al., 1995). If populations are composed of individuals that vary in their tolerance to a toxicant, and this tolerance has a genetic basis, tolerant genotypes should increase in frequency as stress continues; tolerance of a toxicant can be manifested as differential viability, growth, or reproduction (Mulvey et al., 1995). Physiological differences due to allozyme polymorphisms for single genes can be reflected in differences in individual growth, fecundity, timing of reproduction, behavior, or other critical aspects of life history (Mulvey et al., 1995). It has been suggested that sensitive individuals in a population might be genetically similar; the survivors might, as a result, compose a population with decreased capacity for adaptation. Diminished genetic diversity might also portend an even greater response to the next disturbance. Changes in the population genetics of species that suggest an unacceptable risk from decreased adaptability may serve as the most important response in ecological risk assessment.

Although population-level responses to a disturbance are basic to the understanding of the long-term effects of sublethal contamination in many habitats, few methods are currently available (Newman, 1995). However, genetic responses to disturbance have begun to be seriously addressed in ecological risk assessment in recent years. A note of caution has been expressed by Woodward et al. (1996) regarding the use of allozymes to determine population effects of contaminants: studies must define population structure, naturally occurring variation within and among populations, and relevant ecological factors in conjunction with the presence of a contaminant.

Demographic models and bioenergetics models are two approaches to modeling population dynamics that have been used to assess ecological risks; they are discussed in the following paragraphs.

Demographic Models

Demographic models that describe populations as a set of age (size) classes have been well-developed in ecology (Caswell, 1989) and wildlife management (Starfield and Bleloch, 1991). These models are commonly used to simulate the time-varying number of individuals in each class for a given population. The numbers in each class change through time as a function of age (size)-specific survivorship and reproductive contributions to the population. Demographic models have been used to quantify the effects of nutrient enrichment, blue-green algae, and hydrocarbons on survival, reproduction, and age at maturity of estuarine polychaetes (*Capitella* sp. and *Streblospio benedicti*), and hence on population growth rate (Levin et al., 1996). The risk of competitive fishing on the population dynamics of largemouth bass and walleye (*Stizostedion vitreum*) was predicted using a demographic model (Hayes et al., 1995).

Implementing demographic models for ecological risk assessment requires basic knowledge of the dynamics of the population (i.e., life table data) and quantitative relationships between chemical exposure and changes in mortality and fecundity (Barnthouse et al., 1990). Establishing these exposure-response relations for toxic chemicals has been challenging when demographic models are applied to ecological risk assessments.

Bioenergetics Models

An alternative approach considers population dynamics as a function of physiological processes that determine growth, reproduction, and survival (Kitchell et al., 1974). The equations are commonly developed in terms of bioenergetics. Implementing the models requires quantitative estimates of rates of feeding (or photosynthesis), respiration, ingestion, excretion, reproduction, non-predatory mortality, and other processes, depending on the particular model. This bioenergetics approach is also amenable to modeling entire food webs and ecosystems (e.g., O'Neill et al., 1982; Bartell et al., 1988; 1992).

Applying bioenergetics models in risk assessment requires establishing relationships between exposure and changes in rates of the physiological processes that determine population dynamics (e.g., Bartell, 1990). These models typically lack the detailed age or size class structure of the demographic models, but they are based on parameters that can be determined experimentally. The sensitivity of these models to parameter uncertainty has been well-characterized (e.g., Kitchell et al., 1974; Bartell et al., 1986a). Bioenergetics models have been used to develop exposure-response relationships required to use the demographic models in risk assessment.

Communities

Biotic communities have been described as "...any assemblage of populations living in a prescribed area or physical habitat...with a compositional unity in that certain species will live together" (Odum, 1971). Significant research efforts have been directed toward describing, measuring, and understanding the relationships (e.g., food web interactions, competition, mutualism, interference, resilience, variability, resistance, persistence, and assembly) that determine observed correlations (+/–) among species composition in space and time (Pimm, 1982; 1991; Post and Pimm, 1983; Hairston, 1990; Luh and Pimm, 1993; Russel et al., 1995; Yodzis, 1995). Ecological communities are portrayed as highly dynamic systems; therefore, the dynamics of an entire community should be described, not just the dynamics of particular populations or small slices of the community (Pimm, 1991). Different sequences of assembly, prior toxic effects, as well as other types of stressors, can alter the subsequent dynamics of a community (Landis et al., 1996). Community ecology focuses on the response of species distributions to complex environmental gradients and on the ecological interactions among species that forge community structure. Such emphasis has produced complex methods to describe community structure and organization (Whittaker, 1978a; 1978b).

Methods to Assess Community Structure and Organization

Indices that characterize community structure and diversity have been developed to describe and quantify community structural responses to disturbances. The potential strength and usefulness of these indices for estimating ecological risk lies in summarizing complex community structure in a single number, which is also their major weakness (Hurlbert, 1971). Issues similar to those confronting community ecologists in determining the most appropriate measure of diversity must be considered in the development and application of community indices used to advance current capabilities in ecological risk estimation.

Several approaches exist for addressing the effects of stressors or disturbances on community structure. Methods for assessing changes in community structure have traditionally included diversity indices, other indices designed to evaluate community status [e.g., the Index of Biotic Integrity (IBI) (Karr et al., 1986)], and multivariate statistical techniques (e.g., cluster analysis, factor analysis, discriminant analysis). These alternative methods differ primarily in the structural attributes of the community that they describe or summarize, in the way these data are manipulated, and in the expression and presentation of the results.

Diversity indices quantify the number of species (richness or alpha diversity) and the relative apportionment of individual organisms among those species (equitability or beta diversity) in assessing community structure. Several diversity measures have been developed and applied to describe community structure (Krebs, 1989). Using diversity indices requires criteria for selecting among available indices and the data necessary for calculating diversity. The implicit hypothesis is that stress by a disturbance decreases species diversity and that systems of greater diversity are more ecologically desirable. Using diversity as an assessment tool requires a reference or benchmark diversity in order to meaningfully evaluate differences in diversity measured in communities affected by a disturbance.

Indices of community integrity are based on comparisons of species composition in reference communities to the composition of communities in disturbed areas. For example, the community structure of benthic invertebrates sampled from relatively pristine aquatic ecosystems can be contrasted with the species that dominate communities affected by toxic chemicals (Karr et al., 1986). Dissimilarities in community structure are summarized by a single number. The necessary data are similar to those used in calculating diversity (i.e., abundance data by species), although organisms can be classified among different functional groups instead of taxonomic divisions. In addition to data demands, the challenge of using these indices in ecological risk assessments is selecting reference communities and calibrating the indices to these references for different geographic locations and community types (e.g., plants, birds, fish, decomposers, etc.).

Wiens et al. (1996) used community-level measures, such as species richness, species diversity, and species occurrence, to obtain a broader perspective of the effects of the Exxon Valdez oil spill on marine bird communities than could be obtained from studies directed toward individual species of concern. The results suggested that at the community level recovery was well under way, and regional processes should be considered in evaluating recovery following environmental accidents (Wiens et al., 1996).

Multivariate statistical methods differ from the diversity and community integrity indices in that they do not reduce the data to a single number. The methods have been designed to summarize variance in the data in ecologically meaningful ways. These statistical methods reduce the dimensionality of the data, yet retain the essential community characteristics. Multivariate classification and ordination methods have been used since the early 1950s by ecologists to describe and summarize community structure (Whittaker, 1978a; 1978b). These methods have made their way into environmental toxicology and ecological risk assessment (e.g., Johnson and Bartell, 1988). Using multivariate methods as alternatives for priority substance assessment requires competence in multivariate statistics, a working knowledge of the different methods (including their strengths and weaknesses in describing community structure), and experience in presenting and interpreting the results of these methods. The required data are similar to those used in the calculation of diversity and other community indices. Using these methods to assess changes in community structure also requires some reference community data to include in the analysis. The effects of a disturbance may be assessed as differences (i.e., distances) between reference and exposed communities, summarized in an n-dimensional species space, where n refers to the number of species used in the overall community comparison. Multivariate methods can be used to assess ecological risk; however, their use is data-intensive and requires considerable competence and experience on the part of the assessor. If data are available, these methods should be explored for ecological risk assessment.

Landis et al. (1996) proposed an alternative framework—the community conditioning hypothesis—for estimating the risks from stressors or disturbances on ecological systems. The community conditioning hypothesis states that ecological communities tend to retain information about all events in their history and that the information about past events can be stored in a variety of layers of biological interaction, from the molecular to the landscape (Landis et al., 1996). The historical aspects of the formation of a biological community, which are critical in the forecast of environmental effects, can be stored in the genetic structure of populations, in the producer-consumer cycles, as well as in the chemical contamination of tissue; the ties between effects at the molecular level and at other organismal and ecological levels are implicit (Landis et al., 1996). The complexity and nonlinear dynamics of a biological community may create long latency periods between observable cause and effects, and these effects may be a categorical change in the structure of a community (Landis et al., 1996). The observation of the changes of a community due to additional disturbances may be necessary to differentiate two similar systems (Landis et al., 1996).

Community Models

Selected examples of community-level models that have been used or could possibly be used to assess ecological risks are presented in the paragraphs below.

Communities develop through a process—community assembly—in which species invade, persist, or become extinct, and community assembly is a sequence of different community states (Luh and Pimm, 1993). An ecological event could affect community assembly and alter the transition between community states. Many community assembly models have been developed; and modification for use in ecological risk assessment is possible (e.g., Post and Pimm, 1983; Drake, 1988; 1989; 1990; 1991; Drake et al., 1993; Luh and Pimm, 1993).

Manipulative experiments establish causal linkages between different components of natural communities (Schmitz, 1997). The issue of using field experiments to quantify the dynamic couplings—and the form and strengths of both direct and indirect species interactions—is a current focal point of food web ecology (Yodzis, 1995). Press perturbation experiments, in which one or more species are experimentally altered and held at higher or lower levels, is a common field approach used to understand community dynamics (Schmitz, 1997). Results of these experiments could be used to construct community-level models for use in ecological risk assessment.

Bioenergetics models have been used to model entire food webs and to simulate changes in food webs due to a stressor or disturbance (e.g., Bartell et al., 1988; 1992). The Comprehensive Aquatic Systems Model (CASM) is a bioenergetics food web model that links data on single species to a bioenergetic model of the trophic structure of a system to simulate community-level effects of stressors; Monte Carlo iterations of these simulations are used to calculate probabilistic ecological risk assessments of stressors (DeAngelis et al., 1989). CASM can be easily modified for use in ecological risk assessments of the particular aquatic system of interest.

A spatially-explicit forest community model was used to perform an ecological risk assessment of the long-term adaptation potential of central European mountain forests to climate change (Kienast et al., 1996). The model generated estimates of the potential natural vegetation for the entire potential forest area of Switzerland under today's as well as under altered climate regimes (Kienast et al., 1996). Dale and Rauscher (1994) included models of community responses to climate change in a model review and a conceptual framework of spatially-explicit models that address the effects of climate change on forests.

Bayesian models provide a structure for studying collections of parameters such as are considered in the investigation of communities, ecosystems, and landscapes (Link and Hahn, 1996). They also provide an alternative modeling methodology for use in ecological risk assessments of communities. Bayesian procedures recently have been used to assess the risks of rain forest logging on birds and small mammals (Crome et al., 1996) and to estimate the risks of brown-headed cowbird (*Molothrus ater*) parasitism on bird species in a forest community (Link and Hahn, 1996). This model structure allows for improved estimation of individual parameters by considering them in the context of a group of related parameters; individual estimates are differentially adjusted toward an overall mean, with the magnitude of their adjustment based on their precision (Link and Hahn, 1996). Consequently, Bayesian estimation allows for a more reliable ranking of parameters and a more credible identification of extreme values from a collection of estimates (Link and Hahn, 1996).

Ecosystem Models

Methods in ecological risk analysis can successfully address ecosystem-level responses to disturbance. Ecosystems are fundamental units, not only of basic ecology (Odum, 1969; Smith, 1975) but also of ecological risk analysis (O'Neill et al., 1982; 1983; Bartell, 1990; Bartell et al., 1992). The understanding of ecosystem interactions continues to evolve (Forbes, 1887; Tansley, 1935; Lindemann, 1942; Hutchinson, 1948; Odum, 1969; Allen and Starr, 1982; O'Neill et al., 1986; Allen and Hoekstra, 1992). Ecosystems are more than habitat; the interactions and feedbacks among the physical, chemical, and biological components are the essential ingredients. One logical consequence of this emphasis is that ecosystem-level responses to disturbance should, by definition, address alterations to these interactions and feedback mechanisms. Ecosystems may or may not be out there in the real world; what is important is that they appear to be helpful conceptions that lend predictive power (Allen and Hoekstra, 1992). Examples of models that have been used in ecological risk assessments of various ecosystem types are presented below.

The Littoral Ecosystem Risk Assessment Model (LERAM) is a bioenergetic ecosystems effects model that links single species toxicity data to a bioenergetic model of the trophic structure of an ecosystem to simulate community- and ecosystem-level effects of chemical stressors; it uses Monte Carlo iterations of these simulations to calculate probabilistic ecological risk assessments of chemical stressors (Hanratty and Stay, 1994). Hanratty and Stay (1994) evaluated LERAM using field data from a littoral enclosure study of the effects of the insecticide chlorpyrifos. Data from enclosure studies (i.e., mesocosm studies) are commonly used to calibrate bioenergetics ecosystem-level models. Other similar bioenergetics ecosystem models have been developed (e.g., O'Neill et al., 1982; Bartell et al., 1988; 1992; DeAngelis et al., 1989).

The relationship between nonpoint-source phosphorus loading and the trophic status of an aquatic ecosystem was identified using the Spatially Integrated Model for Phosphorus Loading and Erosion (SIMPLE); a risk-based paradigm was used, along with SIMPLE, to characterize the probability of exceeding a critical phosphorus loading to a lotic ecosystem (Matlock et al., 1994). SIMPLE was used to identify probable phosphorus sources in a watershed, simulate the phosphorus loading to streams, and analyze the relationships between input variables and their ecological impact (Matlock et al., 1994).

Coral reefs, the most complex and diverse of all aquatic ecosystems, are currently facing widespread devastation; massive species extinctions and loss of biodiversity are expected if the process is not soon reversed (Stone et al., 1996). For this reason, Stone et al. (1996) used a modification of a model developed by Tilman et al. (1994) to examine the species extinction debt incurred as the habitat destruction increased, and he attempted to predict the characteristics of those species at greatest risk. The spatial model was originally designed for coral reef communities, where species have been understood to interact in a hierarchical web of competitive interactions (Stone et al., 1996). This model could be modified for use in other ecosystems to predict the risk of species extinction due to habitat alteration or loss.

Lowrance and Vellidis (1995) developed a model for quantifying the ecological risk and associated uncertainties to the water quality function of bottomland hardwood riparian ecosystems due to a variety of physical, chemical, and biological stressors. The stressors are related to nonpoint-source pollution from a variety of land uses, especially agriculture, the conversion of bottomland hardwood riparian ecosystems, and the encroachment of domestic animals into bottomland hardwood riparian ecosystems (Lowrance and Vellidis, 1995). Society chooses to value the functions of the bottomland hardwood riparian ecosystem, as evidenced by laws, regulations, or common usage (Lowrance and Vellidis, 1995).

To investigate the risk of frost damage to Scots pine (*Pinus sylvestris*) in northern regions under climatic warming, a submodel for such damage to trees was included in a forest ecosystem model of gap type (Kellomaeki et al., 1995). The risk of frost implies uncertainty in yield expectations from boreal forest ecosystems in the event of climatic warming (Kellomaeki et al., 1995).

Habitat, Regional, and Landscape Models

Habitat Models

Together with the concept that habitat is composed of proximal and/or superimposed landscape elements, the fact that wildlife move within and among habitat areas makes spatial analysis a vital part of evaluating habitat (Rickers et al., 1995). Habitat Suitability Index (HSI) models have been developed for a large number of terrestrial and aquatic species (USFWS, 1981); but, due to the lack of objective data concerning the habitat requirements of many species, HSI models are based on expert opinion (Rickers et al., 1995). However, HSI models can be used in ecological risk assessment. With regard to habitat-related risks, effects might reasonably be measured as changes in habitat quality or quantity, as reflected in differences in the value of an HSI for the species of concern. In the context of HSI models, risk can be evaluated by assigning the model parameters to statistical distributions (or fuzzy sets) and repeatedly solving the HSI model for different degrees of stress or disturbance. The results of this characterization are a distribution of probable HSI values for each degree of disturbance; for each disturbance, the probability of an HSI value being less than an acceptable value can be calculated from the distribution. HSI models can also be modified for use in risk assessment.

Rickers et al. (1995) combined a Geographic Information System (GIS) and an HSI model for ruffed grouse (*Bonasa umbellus*) to assess the risks of habitat suitability over time due to timber harvesting. Models, such as those developed by Rickers et al. (1995), and Boone and Hunter (1996), could be used in ecological risk assessment to assess the risks of habitat availability, change, or loss due to a disturbance or changes in land use for a particular wildlife species. Changes in the size, number, and spatial configurations of patches of potentially suitable habitat can be used to quantify the risks of extinctions for wildlife species (Lindenmayer and Lacy, 1995; Lindenmayer and Possingham, 1996).

Regional and Landscape Models

Landscape ecology is now commonly used to describe ecological studies large enough to include the heterogeneity in ecosystems. From landscape ecology comes a new understanding of the spatial arrangement of ecosystems, the flow of energy, materials, and organisms among ecosystems, and the changes in structure and function of landscapes resulting from both anthropogenic and natural events (Cairns and Niederlehner, 1996). Many of our most important environmental problems involve disturbances that are distributed over large areas, even globally; examples include the spread of toxicants, acid rain, ozone, and climate change. Many models have been developed in recent years to quantify the ecological risks of large-scale disturbance; numerous examples are summarized below.

A regional-scale risk assessment was demonstrated by Graham et al. (1991). They linked spatial and stochastic modeling methods, and terrestrial and aquatic systems, to quantify the risks associated with elevated ozone in Adirondack forests. Spatially explicit simulation models, such as BACHMAP, provide a powerful method for modeling landscape and population changes at large spatial scales and can suggest modifications to current management plans that would increase the probability of population persistence for species of special concern in managed landscapes (Pulliam et al., 1992). BACHMAP was developed by Pulliam et al. (1992) for the Bachman's sparrow, a species of management concern that inhabits the Savannah River Site, South Carolina, an area managed for timber production.

ECOLECON, a model mentioned earlier, developed by Liu (1993), predicts population dynamics, spatial distribution, extinction probability of the species under consideration, future landscape structure, and economic income from timber harvest based on current tax and timber market situations. Defoliation by the gypsy moth (*Lymantria dispar*) was related to components of the landscape in order to elucidate processes operating at the regional scale (Liebhold et al., 1994). Dale and Rauscher (1994) included regional or landscape models in a model review and a conceptual framework of spatially-explicit models that address the impacts of climate change on forests.

A landscape approach, using remote sensing and GIS technologies, was developed to discriminate between villages at high and low risk for malaria transmission; the most important landscape elements in terms of explaining vector abundance were transitional swamp and unmanaged pasture (Beck et al., 1994). Beck et al. (1994) stated that their landscape approach could be applied to vector-borne diseases in areas where (1) the landscape elements critical to vector survival are known and (2) these elements can be detected at remote sensing scales. Brusven and Walker (1995) integrated models of agricultural nonpoint-source pollution with a GIS for investigating the ecological risks of various farming practices. Maps developed with a GIS of dieldrin soil contamination at a Superfund site were integrated with tissue concentrations and food web information to estimate the exposure across the foraging range of the burrowing owl (*Athene cunicularia*) (Clifford et al., 1995). A spatially-explicit, quantitative index was developed by Poiani et al. (1996) to assess the risks of nonpoint-source subsurface nitrogen pollution to wetlands using landscape characteristics.

A GIS-based spatial model (DISPATCH) was used to simulate the effects of the initial density of patches on the rate of response to alteration of a disturbance regime, the effects of global warming and cooling, and the effects of fragmentation and restoration on the structure of a generalized temperate-zone forested disturbance landscape (Baker, 1995). Starfield and Chapin (1996) have developed the first model of ecosystem change in response to transient changes in climate, disturbance regime, and recruitment over the next 50 to 500 years. Scenarios of vegetation change from arctic tundra to boreal forest were developed in response to global changes in climate; seed availability, tree growth rate, and probability of fire were the model parameters that most strongly influenced the balance between tundra and boreal forest in transitional climates (Starfield and Chapin, 1996).

The impact of pesticides on populations in spatially-complex landscapes was modeled by Maurer and Holt (1996), since wildlife populations often exist in landscapes created by fragmentation of natural habitats. The addition of pesticides to a complex landscape may create a system of populations with refuge habitats (where individuals are not exposed to pesticides and populations tend to persist) interspersed with sink habitats (where individuals are exposed to pesticides and populations are faced with extinction) (Maurer and Holt, 1996).

Cairns and Niederlehner (1996) promote the field of landscape ecotoxicology and state that the strongest evidence of effects of toxic substances on landscapes integrates four kinds of information: studies of damaged natural systems, toxicity tests, simulation models, and biomonitoring. Since recent developments in landscape-level ecological modeling rest upon poorly understood behavioral phenomena, Lima and Zollner (1996) are encouraging the development of a behavioral ecology of ecological landscapes. Such a "landscape-conscious" behavior undertaking would revolve around understanding, at the scale of ecological landscapes, the sort of information available to an animal as it moves through its environment and how this information is used in selecting a patch or habitat (Lima and Zollner, 1996).

Models of Biological Invasions

The movement of organisms beyond their natural range can have consequences that are ecologically or even economically devastating (Kareiva, 1996). Most exotic invaders never take hold when entering a new territory, or if they do manage to survive, are barely noticeable (Kareiva, 1996). However, some biological invasions have gotten our attention. For example, the introduced kudzu (*Pueraria lobata*) chokes native forests in the southeastern United States; zebra mussels (*Dreissina polymorpha*), which most likely invaded through ship ballast water, have caused significant ecological and economic damage over an increasing area of the Great Lakes and Mississippi River drainages; and the island of Guam has lost almost all of its forest birds to the introduced brown tree snake (*Boiga irregularis*).

Ecologists are increasingly being called upon to predict the consequences of species additions and/or deletions (e.g., extinction of a top-level carnivore), since these are occurring at an increasing rate and often have had unanticipated consequences (Abrams, 1996). The last decade has seen an explosion of quantitative and theoretical studies aimed at making invasion ecology a predictive science (Drake et al., 1989; Kareiva, 1996). Abrams (1996) argues that current theory is insufficiently developed to provide guidance in predicting species or population-level consequences of evolutionary changes initiated by species addition or deletion (although there is little doubt that evolution in one species in response to a species addition or deletion can cause population densities of other species to change). Hastings (1996) and Kot et al. (1996) describe models that have been developed to describe the spread of invading organisms. Kareiva et al. (1996) discuss the environmental risks of genetically engineering organisms and methods that could be used to quantify invasiveness of a genetically engineered organism. Recently developed models focusing on the spread of invading organisms, such as the models discussed above, could be used to assess the ecological risks of biological invasions and the release of genetically engineered organisms.

Models already described in the IBM section of this chapter (Hill et al., 1995; Higgins et al., 1996) could be used to assess the risk of invasion of unwanted, exotic, or introduced tree species. There is no doubt that advances in invasion ecology will continue, providing modeling tools that can be used in ecological risk analysis.

KEY ISSUES REGARDING MODELS AND ECOLOGICAL RISK ASSESSMENT

Many of the same considerations that apply to model development apply to model selection. The principal advantage of model development is that the user may design the model specifically for the problem at hand. The key is to precisely specify the ecological risk(s) and to determine the subsequent information requirements. It makes little sense to design a model for risk assessment that cannot be used for lack of information. Incorporating ecological complexity for its own sake in model construction may inhibit model performance to the same degree as oversimplification. A complex model with many parameters may be able to match any set of data with sufficient calibration efforts, and agreement between observations and the model might be produced by more than one set of parameter values (i.e., nonuniqueness).

Models that are too simple might produce the general pattern of the data but fail to agree with any specific observations. A prudent procedure is to explore different model structures and add complexity only as it moves the model results closer to the observations for reasons consistent with process-level understanding of the ecological and toxicological phenomena represented in the model. The important issue is not simple versus complex models, but models of necessary and sufficient detail to meet the risk estimation objectives.

The use of models in ecological risk assessment will likely benefit from the development of a set of differently scaled models (e.g., Kercher et al., 1981), that is, models scaled appropriately to effects at different levels of ecological measurement. No single model will likely provide accurate estimates across a wide range of relevant spatial and temporal scales associated with different ecological effects (e.g., individual organisms, populations, communities, ecosystems, landscapes) for risk estimation and risk management. Spatially explicit models that include the spatial complexity in model projections, such as the many examples mentioned in this chapter, can be used to assess ecological risks at various temporal and spatial scales. For example, Dale and Rauscher (1994) reviewed models that assessed the effects of climate change at various spatial and temporal scales of forests at four levels of biological organization: global, regional or landscape, community, and species. A hierarchical approach involves the nesting of models with information (or of models and data) with information at finer resolution being aggregated to a broader scale (King, 1991). For example, Cherrill et al. (1995) developed a model to predict the distribution of plant species at the regional scale; the model linked four levels (landscape, land cover type, community, and species) in an ecological hierarchy using a series of matrices. Combining both approaches, Lavore et al. (1995) developed a spatially-explicit model of the dispersal of annual plants in landscapes that were hierarchically structured (i.e., the spatial pattern of suitable sites was nested and scale-dependent).

Since mathematical models are likely to play an increasingly important role in ecological risk assessment, risk assessors and risk managers must be able to evaluate the reliability of the models. For those skeptical of the entire ecological and environmental modeling enterprise, reliability becomes the key, and the reality is that any model can be pushed to the point of failure. However, a systematic rigorous evaluation of the performance of these models has never been undertaken, nor have criteria for model rejection been developed for risk managers. A fairer evaluation is that model developers and model users do not really know the extent of the general applicability and accuracy of these tools. One critical issue in improving current capabilities in risk assessment concerns an intensive and organized effort in model evaluation, such as that being undertaken for global climate change. Corresponding efforts in human health risk assessment have been under way for several years and continue with an international level of participation (e.g., BIOMOVS I and II) (BIOMOVS, 1993; IAEA, 1995; 1996; BIOMOVS II, 1996).

Integrating Model Results into the Overall Assessment Process

Finally, the results of the risk characterization phase (Figure 5.1), including the discussion of uncertainties, enter back into the risk management process. The risk manager(s) determine if the results are useful and consistent with the overall assessment, particularly in terms of choosing among management alternatives. If decisions are possible, the risk assessment may stop at this point. If not, the process can begin again at the problem formulation phase and iterate through the entire procedure using refined assumptions, parameters, etc. (Figure 5.1). These iterations may continue until management alternatives can be defensibly selected.

Coincidental with all phases of the assessment is the acquisition of additional data, verification of the analyses, and monitoring. One important and distinguishing aspect of the EPA framework is the separation of risk assessment from risk management, except during the initial problem formulation and final risk communication (EPA, 1992; 1996).

The ability to evaluate the consequences of an occurrence (i.e., s_i in the model from Kaplan and Garrick, 1981) should play an important part in defining an ecological effect for risk assessment. The answer to "so what?" should be apparent and accepted by risk managers, decisionmakers, and other stakeholders. Surprisingly enough, justifying the selection of ecological effects on purely ecological grounds has proven difficult. This difficulty arises in part from an unspecified reference environment and different underlying models that might influence the ecological characteristics of such a reference (e.g., Holling, 1986). Human perceptions of the natural world will influence the definition of reference environments, the selection of ecological risks, and consequently, the effectiveness of any assessment. In the context of environmental regulation, the implicit model seems to be one of a constant environment and some corresponding ecological *status quo*. However, the ecological significance of measured disruption caused by chemical poisoning cannot be easily or convincingly assessed in relation to such a static model of nature.

The formulation of each ecological risk assessment should include consideration, and hopefully specification, of the required accuracy and precision for the risk estimates in order to be a tool in decisionmaking. Simply put, how good do the risk estimates have to be for them to contribute meaningfully to decisions concerning, for example, remediation and restoration? Importantly, defensible decisions might be made with different degrees of confidence in the assessment results for different ecological effects.

Risk-based decisionmaking is not necessarily the same as statistical hypothesis testing, although statistical rigor certainly plays a role in any assessment. Therefore, different degrees of accuracy and precision might be specified for different ecological risks. These differing degrees might perhaps be determined by the anticipated magnitudes of ecological and management consequences of errors in risk estimation. It is important to recognize that one principal way to increase the economy and effectiveness of ecological risk assessment is to specify in advance the accuracy and precision required of the risk estimates to provide meaningful input to the risk management process.

Current capabilities in ecological risk assessment have been limited largely to single stressors. However, real-world applications are seldom limited to single stressors. Sources of ecological stress, whether point or nonpoint sources, exist on a complex landscape with diverse land use patterns. Larger watersheds (e.g., the Chesapeake Bay watershed, Florida Bay) contain agricultural, urban, and industrial lands and facilities that jointly contribute to habitat loss, and complex effluents consisting of agrochemicals, organic contaminants, and domestic and industrial wastes. In some instances, complex mixtures of effluents have been assessed (Mount, 1989; DiToro et al., 1988). In these few instances, the toxicity of the mixture has been either dominated by one of two constituents, the toxicity has been essentially additive, or special toxicity assays have been performed using the mixture as the test substance. Increased ability to assess the overall risks posed by a composite stressor remains a critical need for ecological risk assessment.

Such increased ability will undoubtedly require the implementation of GISs that contain the locations and extent of point and nonpoint sources of contamination, characterization of waste streams, comprehensive toxicity data, location of valuable habitat and distribution of ecological resources at risk, environmental characteristics (e.g., soil type, elevation, vegetation, weather data, surface water, and groundwater resources), and human demographic data. These GISs will likely be integrated with dynamic models of fate and transport of contaminants, ecological models, and risk-based decision tools, all customized for the specific region.

However, the success of this integration relies on developing the basic quantitative understanding of how different stressors combine in their potential effects on ecological resources. Currently, risk assessors lack the ability to forecast the combined effects of, for example, increased fertilization, increased sedimentation, temperature change, and additions of organics and metals on the production dynamics of aquatic ecological systems. Substantial fundamental research and development are necessary to acquire the necessary understanding to make such forecasts. But this is certainly one direction toward which ecological risk analysis must advance.

Efficient Data Acquisition

Ecological risk analysis is data intensive. Quantitative estimates of site-specific risks quickly exhaust available ecological and toxicological data. In applying models for risk estimation, derivation of critical model parameters from values reported in the primary and secondary literature is time-consuming, costly, and fraught with uncertainty. Repeated derivations of the same parameter in independent assessments compounds costs and frustration. With the exception of notable examples (e.g., Jorgensen, 1979), compendia of ecological and toxicological data are largely absent. Clearly, one avenue for advancing capabilities in risk estimation lies in increased effort being directed at constructing and maintaining electronic databases. An important beginning has taken the form of a compendium of environmental data routinely collected by different government agencies and available for electronic access (EPA, 1992).

Data describing the toxicity of chemical compounds to terrestrial organisms are particularly lacking. The toxic effects of several pesticides and lead have been examined and quantified for selected avian species (Eisler, 1988; Chambers, 1994); however, very little is known about the effects of organic contaminants and metals on small or large mammals, reptiles, or amphibians. In performing assessments involving such animals, toxicity data measured for domestic animals are routinely used (e.g., SENES, 1994—see McArthur River Project mining exploration case study). Owing to the difficulty in obtaining permission to perform the needed experiments, it is not likely that these critical data will be forthcoming. Increased efforts in developing rules for extrapolating from the meager domestic animal database appear as the only recourse for the present.

Long-term environmental monitoring can generate data sets for use in ecological risk assessment. The EPA Environmental Monitoring and Assessment Program (EMAP) was established to monitor the status and document trends of ecological resources. The results of monitoring will be summarized annually. While the EMAP was not designed for purposes of ecological risk assessment, the EPA recognizes the potential contribution of EMAP data to the risk assessment process. Organizers of the EMAP also embrace the framework for ecological risk analysis as one approach for evaluating the results of long-term monitoring. A critical issue in the continued development and implementation of the EMAP will be the ease of access to monitoring data. Online access to EMAP data may be extremely useful in developing regional site characterizations, deriving model parameter values, or estimating initial conditions for ecological models used in risk estimation. Landscape monitoring, using the landscape metrics recommended by Riitters et al. (1995), can monitor environmental change at the broad spatial scale of landscapes.

SUMMARY

Nature challenges us with ecological systems that are inherently complex. The strength of each ecological perspective (i.e., level) is in the nature of its set of simplifying assumptions that stimulate certain kinds of measurements and interpretation. But using these ecological concepts to their full advantage in ecological risk assessment, particularly in identifying what is important to measure (i.e., ecological effects), requires greater ecological sophistication than that offered by a simple model. Simplifying assumptions or simplifying models does not simplify nature. Ecological modeling represents a powerful and more realistic alternative to simple quotient calculations and pathway models of dose currently being used to assess ecological risks.

ACKNOWLEDGMENTS

We thank E.W. Reed and an anonymous reviewer for their comments on an earlier draft of this manuscript. This work was supported in part by the NOAA under contract II6-0460-67SORI and the USEPA under contract DE-FG01-95EW55085 (subcontract 55085-SENES-1 to The Cadmus Group, Inc.).

REFERENCES

Abrams, P.A. Evolution and the consequences of species introductions and deletions. *Ecology* 77, pp. 1321–1328, 1996.

Allen, T.F.H. and T.W. Hoekstra. *Toward a Unified Ecology*. Columbia University Press, New York, NY, 1992, p. 384.

Allen, T.F.H. and T.B. Starr. *Hierarchy: Perspectives for Ecological Complexity*. University of Chicago Press, Chicago, IL, 1982, 310 p.

Anderson, R.M. and R.M. May. *Infectious Diseases of Humans: Dynamics and Control*. Oxford University Press, Oxford, England, 1991, p. 757.

Apostolakis, G. Expert opinion in probabilistic safety analysis, in *Risk-Based Decision Making in Water Resources*, Haimes, Y.Y. and E.Z. Stakhiv, Eds., ASCE, New York, NY, 1989.

Baker, W.L. Longterm response of disturbance landscapes to human intervention and global change. *Landscape Ecol.* 10, pp. 143–159, 1995.

Barnthouse, L.W. The role of models in ecological risk assessment: A 1990's perspective. *Environ. Toxicol. Chem.* 11, pp. 1751–1760, 1992.

Barnthouse, L.W., R.V. O'Neill, S.M. Bartell, and G.W. Suter II. Population and ecosystem theory in ecological risk assessment, in *Aquatic Toxicology and Environmental Fate: Ninth Volume*, Poston, T.M. and R. Purdy, Eds. STP 921, American Society for Testing and Materials, Philadelphia, PA, 1986.

Barnthouse, L.W., G.W. Suter II, and A.E. Rosen. Risks of toxic contaminants to exploited fish populations: influence of life history, data uncertainty, and exploitation intensity. *Environ. Toxicol. Chem.* 9, pp. 297–311, 1990.

Bart, J. Acceptance criteria for using individual-based models to make management decisions. *Ecol. Appl.* 5, pp. 411–420, 1995.

Bartell, S.M. Ecosystem context for estimating stress-induced reductions in fish populations. *Am. Fish. Soc. Symp.* 8, pp. 167–182, 1990.

Bartell, S.M. Ecological risk assessment and sustainable environmental management, in *Interconnections Between Human and Ecosystem Health*, DiGiulio, R.T. and E. Monosson, Eds., Chapman & Hall, London, 1996.

Bartell, S.M., R.V. O'Neill, and R.H. Gardner. Aquatic ecosystem models for risk assessment, in *Analysis of Ecological Systems: State-of-the-Art in Ecological Modeling*, Lauenroth, W.K., G.V. Skogerboe, and W. Flug, Eds., Elsevier, Amsterdam, Holland, 1983.

Bartell, S.M., A.L. Brenkert, J.E. Breck, and R.H. Gardner. Comparison of numerical sensitivity and uncertainty analyses of bioenergetic models of fish growth. *Can. J. Fish. Aquat. Sci.* 43, pp. 160–168, 1986a.

Bartell, S.M., R.H. Gardner, and R.V. O'Neill. The influence of bias and variance in predicting the effects of phenolic compounds in ponds, in *Supplementary Proceedings for the 1986 Eastern Simulation Conference,* Simulation Councils, Inc., San Diego, CA, pp. 173–176, 1986b.

Bartell, S.M., R.H. Gardner, and R.V. O'Neill. An integrated fates and effects model for estimation of risk in aquatic systems, in *Aquatic Toxicology and Hazard Assessment: 10th Volume,* ASTM STP 971, American Society for Testing and Materials, Philadelphia, PA, 1988.

Bartell, S.M., R.H. Gardner, and R.V. O'Neill. *Ecological Risk Estimation*. Lewis Publishers, Boca Raton, FL, 1992, p. 233.

Beck, L.R., M.H. Rodriquez, S.W. Dister, A.D. Rodriquez, E. Rejmankova, A. Ulloa, R.A. Meza, D.R. Roberts, J.F. Paris, M.A. Spanner, R.K. Washino, C. Hacker, and L.J.

Legters. Remote sensing as a landscape epidemiologic tool to identify villages at high risk for malaria transmission. *Am. J. Trop. Med. Hyg.* 51, pp. 271–280, 1994.

Biospheric Model Validation Study (BIOMOVS). *BIOMOVS Technical Report 15, Final Report,* 1993.

BIOMOVS II. *An Overview of the BIOMOVS II Study and its Findings.* Swedish Radiation Protection Institute, BIOMOVS II Technical Report No. 17, Stockholm, Sweden, 1996.

Boone, R.B. and M.L. Hunter, Jr. Using diffusion models to simulate the effects of land use on grizzly bear dispersal in the Rocky Mountains. *Landscape Ecol.* 11, pp. 51–64, 1996.

Brusven, M.A. and D.J. Walker. Integrated ecologic-economic assessment of agricultural nonpoint source pollution on aquatic resources in a watershed and landscape perspective in Idaho. *Bull. N. Am. Benth. Soc.* 12, pp. 88, 1995.

Cairns, J., Jr. and B.R. Niederlehner. Developing a field of ecotoxicology. *Ecol. Appl.* 6, pp. 790–796, 1996.

Canada Council of Ministers of the Environment (CCME). *An Introduction to the Ecological Risk Assessment Framework for Contaminated Sites in Canada, Final Report.* The National Contaminated Sites Remediation Program, 1994.

Cardwell, R.D., B. Parkhurst, W. Warren-Hicks, and J. Volosin. Aquatic ecological risk assessment and clean-up goals for metals arising from mining operations, in *Applications of Ecological Risk Assessment to Hazardous Waste Site Remediation*, Bender, E.S. and F.A. Jones, Eds., Water Environment Federation, Alexandria, VA, pp. 61–72, 1993.

Casanova, M., R.B. Conolly, and H.D.A. Heck. DNA-protein cross links (DPX) and cell proliferation in $B6C3F_1$ mice but not Syrian golden hamsters exposed to dichloromethane: Pharmacokinetics and risk assessment with DPX as dosimeter. *Fundam. Appl. Toxicol.* 31, pp. 103–116, 1996.

Caswell, H. *Matrix Population Models.* Sinauer Associates, Sunderland, MA, 1989, p. 328.

Chambers, J.E. Toxicity of pesticides, in *Basic Environmental Toxicology*, Cockerham, L.G. and B.S. Shane, Eds., CRC Press, Boca Raton, FL, 1994.

Cherrill, A.J., C. McClean, P. Watson, K. Tucker, S.P. Rushton, and R. Sanderson. Predicting the distribution of plant species at the regional scale: A hierarchical matrix model. *Landscape Ecol.* 10, pp. 197–207, 1995.

Clifford, P.A., D.E. Barchers, D.F. Ludwig, R.L. Sielken, J.S. Klingensmith, R.V. Graham, and M.I. Banton. An approach to quantifying spatial components of exposure for ecological risk assessment. *Environ. Toxicol. Chem.* 14, pp. 895–906, 1995.

Crome, F.H.J., M.R. Thomas, and L.A. Moore. A novel Bayesian approach to assessing the impacts of rain forest logging. *Ecol. Appl.* 6, pp. 1104–1123, 1996.

Dale, V.H. and H.M. Rauscher. Assessing impacts of climate change on forests: The state of biological modeling. *Clim. Change* 28, pp. 65–90, 1994.

DeAngelis, D.L. *Dynamics of Nutrient Cycling and Food Webs.* Chapman & Hall, London, 1992, p. 270.

DeAngelis, D.L. and L.J. Gross. *Individual-Based Models and Approaches in Ecology.* Chapman & Hall, New York, NY, 1992, p. 525.

DeAngelis, D.L., S.M. Bartell, and A.L. Brenkert. Effects of nutrient recycling and food-chain length on resilience. *Amer. Nat.* 134, pp. 778–805, 1989.

DiToro, D.M., J.A. Halden, and J.L. Plafkin. Modeling *Ceriodaphnia* toxicity in the Naugatuck River using additivity and independent action, in *Toxic Contaminants and Ecosystem Health: A Great Lakes Focus*, Evans, M.S., Ed., John Wiley & Sons, New York, NY, 1988.

Drake, J.A. Models of community assembly and the structure of ecological landscapes, in *Mathematical Ecology*, Hallam, T., L. Gross, and S. Levin, Eds., World Press, Singapore, 1988.

Drake, J.A. Communities as assembled structures: do rules govern patterns? *TREE* 5, pp. 159–163, 1989.

Drake, J.A. The mechanics of community assembly and succession. *J. Theo. Biol.* 147, pp. 213–234, 1990.

Drake, J.A. Community assembly mechanics and the structure of an experimental species ensemble. *Amer. Nat.* 137, pp. 1–26, 1991.

Drake, J.A., H.A. Mooney, F. diCastri, R.H. Groves, F.J. Kruger, M. Rejmanek, and M. Williamson, Eds. *Biological Invasions: A Global Perspective*. Scope 37, Wiley, Chichester, 1989, p. 525.

Drake, J.A., T.E. Flum, G.J. Witteman, T. Voskuil, A.M. Hoylman, C. Creson, A.A. Kenny, G.R. Huxel, C.S. Larue, and J.R. Duncan. The construction and assembly of an ecological landscape. *J. Anim. Ecol.* 62, pp. 117–130, 1993.

Dunham, A.E. and K.L. Overall. Population responses to environmental change: life history variation, individual-based models, and the population dynamics of short-lived organisms. *Amer. Zool.* 34, pp. 382–396, 1994.

Eisler, R. (1988). *Lead Hazards to Fish, Wildlife, and Invertebrates: A Synoptic Review*. U.S. Fish and Wildlife Service (USFWS), Biological Report 85(1.14), U.S. Department of the Interior, Washington, DC.

Emlen, J.M. Terrestrial population models for ecological risk assessment: a state-of-the-art review. *Environ. Toxicol. Chem.* 8, pp. 831–842, 1989.

EPA. *Framework for Ecological Risk Assessment*, EPA/630/R-92/001. Risk Assessment Forum, Washington, DC, 1992.

EPA. *Proposed Guidelines for Ecological Risk Assessment*, EPA/630/R-95/002B, August 1996.

Fleming, D.M., W.F. Wolff, and D.L. DeAngelis. Importance of landscape heterogeneity to wood storks in the Florida Everglades. *Environ. Manage.* 18, pp. 743–757, 1994.

Forbes, S.A. The lake as a microcosm. *Bull. Sc. A. Peoria.* 1887. (Reprinted in *Ill. Nat. Hist. Surv. Bull.* 15, pp. 537–550, 1925).

Gardner, R.H., W.G. Cale, and R.V. O'Neill. Robust analysis of aggregation error. *Ecology* 63, pp. 1771–1779, 1982.

Gerlowski, L.E. and R.K. Jain. Physiologically based pharmacokinetic modeling: principles and applications. *J. Pharm. Sci.* 72, pp. 1103–1126, 1983.

Gleick, J. *Chaos: The Making of a New Science*. Penguin, New York, NY, 1988, p. 352.

Goldstein, R.A. and P.F. Ricci. Ecological risk uncertainty analysis, in *Energy and Ecological Modeling*, Mitsch, W.J., R.W. Bosserman, and J.M. Klopatek, Eds., Elsevier, Amsterdam, 1981.

Golley, F.B. *A History of the Ecosystem Concept in Ecology: More Than the Sum of the Parts*. Yale University Press, New Haven, CT, 1993, p. 254.

Graham, R.L., C.T. Hunsaker, R.V. O'Neill, and B.L. Jackson. Ecological risk assessment at the regional scale. *Ecol. Appl.* 1, pp. 196–206, 1991.

Hairston, N.G. *Ecological Experiments: Purpose, Design and Execution*. Cambridge University Press, Cambridge, England, 1990. p. 370.

Hallam, T.G., R.R. Lassiter, J. Li, and W. McKinney. Toxicant-induced mortality in models of *Daphnia* populations. *Environ. Toxicol. Chem.* 9, pp. 597–621, 1990.

Hallam, T.G., G.A. Canziani, and R.R. Lassiter. Sublethal narcosis and population persistence: a modeling study on growth effects. *Environ. Toxicol. Chem.* 12, pp. 947–954, 1993.

Hanratty, M.P. and F.S. Stay. Field evaluation of the Littoral Ecosystem Risk Assessment Model's predictions of the effects of chlorpyrifos. *J. Appl. Ecol.* 31, pp. 439–453, 1994.

Hastings, A. Models of spatial spread: is the theory complete? *Ecology* 77, pp. 1675–1679, 1996.

Hastings, A. *Population Biology Concepts and Models*. Springer-Verlag, New York, NY, 1997, p. 220.

Hayes, D.B., W.W. Taylor, and H.L. Schramm, Jr. Predicting the biological impact of competitive fishing. *N. Amer. J. Fish. Manage.* 15, pp. 457–472, 1995.

Higgins, S.I., D.M. Richardson, and R.M. Cowling. Modeling invasive plant spread: the role of plant-environmental interactions and model structure. *Ecology* 77, pp. 2043–2054, 1996.

Hill, J.D., C.D. Canham, and D.M. Wood. Patterns and causes of resistance to tree invasion in rights-of-way. *Ecol. Appl.* 5, pp. 459–470, 1995.

Holling, C.S. The resilience of terrestrial ecosystems: local surprise and global change, in *Sustainable Development of the Biosphere*, Clark, W.C. and R.E. Munn, Eds., International Institute for Applied Systems Analysis, Laxenburg, Austria, 1986.

Humphries, H.C., D.P. Coffin, and W.K. Lauenroth. An individual-based model of alpine plant communities. *Ecol. Model.* 84, pp. 99–126, 1996.

Hurlbert, S.H. The nonconcept of species diversity: a critique and alternative parameters. *Ecology* 52, pp. 577–586, 1971.

Hutchinson, G.E. On living in the biosphere. *Scient. Monthly.* 67, pp. 393–398, 1948.

IAEA (International Atomic Energy Agency). *Validation of models using Chernobyl fallout data from the Central Bohemia region of the Czech Republic—Scenario CB*, First Report of the VAMP Multiple Pathways Assessment Working Group, IAEA-TECDOC-795, 1995.

IAEA. *Validation of models using Chernobyl fallout data from southern Finland—Scenario S.* Second Report of the VAMP Multiple Pathways Assessment Working Group, IAEA-TECDOC-904, 1996.

Johnson, A.R. and S.M. Bartell. *Dynamics of Aquatic Ecosystems and Models Under Toxicant Stress: State Space Analysis, Covariance Structure, and Ecological Risk.* ORNL/TM-10723, Oak Ridge National Laboratory, Oak Ridge, TN, 1988.

Jorgensen, S.E., Ed. *Handbook of Environmental Data and Ecological Parameters*. International Society for Ecological Modeling, Copenhagen, Denmark, 1979.

Jorgensen, S.E., B. Halling-Sorensen, and S.N. Nielsen, Eds. *Handbook of Environmental and Ecological Modeling*. CRC Lewis Publishers, Boca Raton, FL, 1996, p. 672.

Kaplan, S. and B.J. Garrick. On the quantitative definition of risk. *Risk Anal.* 1, pp. 11–27, 1981.

Kareiva, P., Special Features Ed. Developing a predictive ecology for non-indigenous species and ecological invasions. *Ecology* 77, pp. 1652–1697, 1996.

Kareiva, P., I.M. Parker, and M. Pascual. Can we use experiments and models in predicting the invasiveness of genetically engineered organisms? *Ecology* 77, pp. 1670–1675, 1996.

Karr, J.R., K.D. Fausch, P.L. Angermeier, P.R. Yant, and I.J. Schlosser. *Assessing Biological Integrity in Running Waters: A Method and Its Rationale*. Illinois Natural History Survey Special Publication No. 5., Champaign, IL, 1986.

Kellomaeki, S., H. Haenninen, and M. Kolstroem. Computations on frost damage to Scots pine under climatic warming in boreal conditions. *Ecol. Appl.* 5, pp. 42–52, 1995.

Kercher, J.R., M.C. Axelrod, and G.E. Bingham. Hierarchical set of models for estimating the effects of air pollution on vegetation, in *Energy and Ecological Modelling*, Mitsch, W.J., R.W. Bosserman, and J.M. Klopatek, Eds., Elsevier, New York, NY, 1981.

Kienast, F., B. Brzeziecki, and O. Wildi. Long-term adaptation potential of Central European mountain forest to climate change: a GIS-assisted sensitivity assessment. *For. Ecol. Manage.* 80, pp. 133–153, 1996.

King, A.W. Translating models across scales in the landscape, in *Quantitative Methods in Landscape Ecology*, Turner, M.G. and R.H. Gardner, Eds., Springer-Verlag, New York, NY, 1991.

Kitchell, J.F., J.F. Koonce, R.V. O'Neill, H.H. Shugart, J.J. Magnuson, and R.S. Booth. Model of fish biomass dynamics. *Trans. Amer. Fish. Soc.* 103, pp. 786–798, 1974.

Kot, M., M.A. Lewis, and P. van den Driessche. Dispersal data and the spread of invading organisms. *Ecology* 77, pp. 2027–2042, 1996.

Krebs, C.J. *Ecological Methodology*. Harper Collins Publishers, New York, NY, 1989, p. 654.

Landis, W.G., R.A. Matthews, and G.B. Matthews. The layered and historical nature of ecological systems and the risk assessment of pesticides. *Environ. Toxicol. Chem.* 15, pp. 432–440, 1996.

Lavore, S., R.H. Gardner, and R.V. O'Neill. Dispersal of annual plants in hierarchically structured landscapes. *Landscape Ecol.* 10, pp. 277–289, 1995.

Levin, L., H. Caswell, T. Bridges, C. DiBacco, D. Cabrera, and G. Plaia. Demographic responses of estuarine polychaetes to pollutant: life table response experiments. *Ecol. Appl.* 6, pp. 1295–1313, 1996.

Liebhold, A.M., G.A. Elmes, J.A. Halverson, and J. Quimby. Landscape characterization of forest susceptibility to gypsy moth defoliation. *For. Sci.* 40, pp. 18–29, 1994.

Lima, S.L. and P.A. Zollner. Toward a behavioral ecology of ecological landscapes. *TREE* 11, pp. 131–135, 1996.

Lindemann, R.L. The trophic-dynamic aspect of ecology. *Ecology* 23, pp. 399–418, 1942.

Lindenmayer, D.B. and R.C. Lacy. Metapopulation viability of the Leadbeater's possums, *Gymnobelideus leadbeateri*, in fragmented old-growth forests. *Ecol. Appl.* 5, pp. 164–182, 1995.

Lindenmayer, D.B. and H.P. Possingham. Modelling the inter-relationships between habitat patchiness, dispersal capability and metapopulation persistence of the endangered species, Leadbeater's possum, in south-eastern Australia. *Landscape Ecol.* 11, pp. 79–105, 1996.

Link, W.A. and D.C. Hahn. Empirical Bayes estimation of proportions with application to cowbird parasitism rates. *Ecology* 77, pp. 2528–2537, 1996.

Liu, J. ECOLECON: An ECOLogical-ECONomic model for species conservation in complex forest landscapes. *Ecol. Model.* 70, pp. 63–87, 1993.

Lotka, A.J. *Elements of Physical Biology*. Williams and Wilkins, Baltimore, MD, 1925 (Reprinted in 1956 by Dover Publications, New York, NY as *Elements of Mathematical Biology*), p. 465.

Lovvorn, J.R. and M.P. Gillingham. A spatial energetics model of cadmium accumulation by diving ducks. *Arch. Environ. Contam. Toxicol.* 30, pp. 241–251, 1996.

Lowrance, R. and G. Vellidis. A conceptual model for assessing the ecological risk to water quality function of bottomland hardwood forests. *Environ. Manage.* 19, pp. 239–258, 1995.

Luh, H. and S.L. Pimm. The assembly of ecological communities: a minimalist approach. *J. Anim. Ecol.* 62, pp. 749–765, 1993.

Madenjian, C.P., S.R. Carpenter, G.W. Elk, and M.A. Miller. Accumulation of PCBs by lake trout (*Salvelinus namaycush*): an individual-based model approach. *Can. J. Fish. Aquat. Sci.* 50, pp. 97–109, 1993a.

Madenjian, C.P., S.R. Carpenter, and G.E. Noguchi. Individual-based model for dieldrin contamination in lake trout. *Arch. Environ. Contam. Toxicol.* 24, pp. 78–82, 1993b.

Madenjian, C.P., S.R. Carpenter, and P.S. Rand. Why are the PCB concentrations of salmonine individuals from the same lake so highly variable? *Can. J. Fish. Aquat. Sci.* 51, pp. 800–807, 1994.

Mankin, J.B., R.V. O'Neill, H.H. Shugart, and B.W. Rust. The importance of validation in ecosystem analysis, in *Simulation Council Proceedings Series*, La Jolla, CA, pp. 63–71, 1975.

Matlock, M.D., D.E. Storm, J.G. Sabbagh, C.T. Haan, M.D. Smolen, and S.L. Burks. An ecological risk assessment paradigm using the Spatially Integrated Model for Phosphorus Loading and Erosion (SIMPLE). *J. Aquat. Ecosyst. Health.* 3, pp. 287–294, 1994.

Maurer, B.A. and R.D. Holt. Effects of chronic pesticide stress on wildlife populations in complex landscapes: processes at multiple scales. *Environ. Toxicol. Chem.* 15, pp. 420–426, 1996.

McIntosh, R.P. *The Background of Ecology, Concept and Theory*. Cambridge University Press, New York, NY, 1985, p. 383.

Mount, D.I. *Methods for Aquatic Toxicity Identification Evaluations: Phase III Toxicity Confirmation Procedures*. EPA-600/3-88-036, U.S. Environmental Protection Agency (USEPA), Duluth, Minnesota, 1989.

Mulvey, M., M.C. Newman, A. Chazal, M.M. Keklak, M.G. Heagler, and L.S. Hayes, Jr. Genetic and demographic responses of mosquitofish (*Gambusia holbrooki* Girard 1859) populations stressed by mercury. *Environ. Toxicol. Chem.* 14, pp. 1411–1418, 1995.

National Research Council (NRC). *Issues in Risk Assessment, Vol. 3—Ecological Risk Assessment*. National Academy Press, Washington, DC, 1992, p. 374.

Newman, M.C. *Quantitative Methods in Aquatic Ecotoxicology*. Lewis Publishers, Boca Raton, FL, 1995, p. 426.

Odum, E.P. The strategy of ecosystem development. *Science* 164, pp. 262–270, 1969.

Odum, E.P. *Fundamentals of Ecology*. W.B. Saunders, Philadelphia, PA, 1971, p. 574.

Olson, J.S. Energy storage and the balance of producers and decomposers in ecological systems. *Ecology* 44, pp. 322–311, 1963.

O'Neill, R.V., R.H. Gardner, L.W. Barnthouse, G.W. Suter II, S.G. Hildebrand, and C.W. Gehrs. Ecosystem risk analysis: a new methodology. *Environ. Toxicol. and Chem.* 1, pp. 167–177, 1982.

O'Neill, R.V., S.M. Bartell, and R.H. Gardner. Patterns of toxicological effects in ecosystems: a modeling study. *Environ. Toxicol. Chem.* 2, pp. 451–461, 1983.

O'Neill, R.V., D.L. DeAngelis, J.B. Waide, and T.F.H. Allen. *A Hierarchical Concept of Ecosystems*. Princeton University Press, Princeton, NJ, 1986, p. 186.

Parkhurst, B.R., W. Warren-Hicks, T. Etchison, J.B. Butcher, R.D. Cardwell, and J. Voloson. *Methodology for Aquatic Ecological Risk Assessment*. RP91-AER-1, Water Environment Research Foundation, Alexandria, VA, 1995.

Patten, B.C. A Primer for Ecological Modeling and Simulation with Analog and Digital Computers, in *Systems Analysis and Simulation in Ecology, Volume 1*, Patten, B.C., Ed., Academic Press, New York, NY, 1971, p. 607.

Pimm, S.L. *Food Webs*. Chapman & Hall, London, 1982, p. 219.

Pimm, S.L. *The Balance of Nature? Ecological Issues in the Conservation of Species and Communities.* University of Chicago Press, Chicago, IL, 1991, p. 434.

Pimm, S.L. and M.E. Gilpin. Theoretical issues in conservation biology, in *Perspectives in Ecological Theory*, Roughgarden, J., R.M. May, and S.A. Levin, Eds., Princeton University Press, Princeton, NJ, 1989.

Poiani, K.A., B.L. Bedford, and M.D. Merrill. A GIS-based index for relating landscape characteristics to potential nitrogen leaching to wetlands. *Landscape Ecol.* 11, pp. 237–255, 1996.

Post, W.M. and S.L. Pimm. Community assembly and food web stability. *Math. Biosci.* 64, pp. 169–192, 1983.

Power, M. and G. Power. Comparing minimum-size and slot limits for brook trout management. *N. Amer. J. Fish. Manage.* 16, pp. 49–62, 1996.

Pulliam, H.R., J.B. Dunning, Jr., and J. Liu. Population dynamics in complex landscapes: a case study. *Ecol. Appl.* 2, pp. 165–177, 1992.

Rickers, J.R., L.P. Queen, and G.J. Arthaud. A proximity-based approach to assessing habitat. *Landscape Ecol.* 10, pp. 309–321, 1995.

Riley, G.A., H. Stronnel, and D.F. Bumpus. Quantitative ecology of the plankton of the western Atlantic. *Bull. Bingham Oceanogr. Coll.* 12, pp. 1–169, 1949.

Riitters, K.H., R.V. O'Neill, C.T. Hunsacker, J.D. Wickham, D.H. Yankee, S.P. Timmons, K.B. Jones, and B.L. Jackson. A factor analysis of landscape pattern and structure patterns. *Landscape Ecol.* 10, pp. 23–39, 1995.

Roughgarden, J. *Theory of Population Genetics and Evolutionary Ecology.* MacMillan Publishing Co., Inc., New York, NY, 1979, p. 634.

Russel, G.J., J.M. Diamond, S.L. Pimm, and T.M. Reed. A century of turnover: community dynamics at three timescales. *J. Anim. Ecol.* 64, pp. 628–641, 1995.

Schaeffer, D.L. A model evaluation methodology for cost-risk analysis in the statistical validation of simulation models. *Commun. Assoc. Comput. Machinery.* pp. 190–197, April 24, 1981.

Schmitz, O.J. Press perturbations and the predictability of ecological interactions in a food web. *Ecology* 78, pp. 55–69, 1997.

SENES Consultants Limited (SENES). *Screening Level Ecological Risk Assessment—McArthur River Project.* Report prepared for Cameco Corporation, 1994.

Simberloff, D. The spotted owl fracas: mixing academic, applied, and political ecology. *Ecology* 68, pp. 766–772, 1987.

Smith, F.E. Ecosystem and evolution. *Bull. Ecol. Soc. Amer.* 56, pp. 2–6, 1975.

Starfield, A.M. and A.L. Bleloch. *Building Models for Conservation and Wildlife Management.* Burgess International Group, Edina, MN, 1991, p. 253.

Starfield, A.M. and F.S. Chapin III. Model of transient changes in arctic and boreal vegetation in response to climate and land use change. *Ecol. Appl.* 6, pp. 842–864, 1996.

Stone, L., E. Eilam, A. Abelson, and M. Ilan. Modelling coral reef biodiversity and habitat destruction. *Mar. Ecol. Prog. Ser.* 134, pp. 299–302, 1996.

Stow, C.A. and S.R. Carpenter. PCB accumulation in Lake Michigan coho and chinook salmon: individual-based models using allometric relationships. *Environ. Sci. & Technol.* 28, pp. 1543–1549, 1994.

Suter, G.W., II. *Ecological Risk Assessment.* Lewis Publishers, Boca Raton, FL, 1993, p. 538.

Swartzman, G.L. and S.P. Kaluzny. *Ecological Simulation Primer.* MacMillan Publishing, New York, NY, 1987, p. 370.

Tansley, A.G. The use and abuse of vegetational concepts and terms. *Ecology* 16, pp. 284–307, 1935.

Tilman, D., R.M. May, C.L. Lehman, and M.A. Nowak. Habitat destruction and the extinction debt. *Nature* 371, pp. 65–66, 1994.

Turner, M.G., Y. Wu, L.L. Wallace, W.H. Romme, and A. Brenkert. Simulating winter interactions among ungulates, vegetation, and fire in northern Yellowstone Park. *Ecol. Appl.* 4, pp. 472–496, 1994.

USFWS. *Standards for the Development of Habitat Suitability Index Models.* Ecological Services Manual 103, U.S. Department of the Interior, Fish and Wildlife Service, Division of Ecology Services, Government Printing Office, Washington, DC, 1981.

Weins, J.A., T.O. Crist, R.H. Day, S.M. Murphy, and G.D. Hayward. Effects of the *Exxon Valdez* oil spill on marine bird communities in Prince William Sound, Alaska. *Ecol. Appl.* 6, pp. 828–841, 1996.

Welsch, F., G.M. Blumental, and R.B. Conolly. Physiologically based pharmacokinetic models applicable to organogenesis: extrapolation between species and potential use in prenatal toxicity risk assessments, in *Proceedings of the International Congress of Toxicology VII*, Reed, D.J., Ed., Seattle, WA, pp. 539–547, 1995.

Whittaker, R.H. *Classification of Plant Communities.* Dr. W. Junk Publishers, The Hague, Netherlands, 1978a, p. 408.

Whittaker, R.H. *Ordination of Plant Communities.* Dr. W. Junk Publishers, The Hague, Netherlands, 1978b, p. 388.

Woodward, L.A., M. Mulvey, and M.C. Newman. Mercury contamination and population-responses in chironomids: can allozyme polymorphism indicate exposure? *Environ. Toxicol. Chem.* 15, pp. 1309–1316, 1996.

Yang, R.S.H., H.A. El-Masri, R.S. Thomas, and A.A. Constan. The use of physiologically-based pharmacokinetic/pharmacodynamic dosimetry models for chemical mixtures, in *Proceedings of the International Congress of Toxicology VII*, Reed, D.J., Ed., Seattle, WA, 82–83, pp. 497–504, 1995.

Yodzis, P. Food webs and perturbation experiments: theory and practice, in *Food Webs: Integration of Pattern and Dynamics*, Polis, G.A. and K.O. Winemiller, Eds., Chapman & Hall, New York, NY, pp. 192–200, 1995.

6 Canonical Modeling as a Unifying Framework for Ecological and Human Risk Assessment

Eberhard O. Voit and Mary K. Schubauer-Berigan

INTRODUCTION

The realization that our well-being depends on the quality of our surroundings is not new. Hippocrates, still considered the father of medicine after more than 2000 years, was acutely aware of the role the environment plays in human health. The Romans knew that lead in water pipes is detrimental to human health, and throughout the middle ages and modern times, more and more natural and man-made compounds were identified as hazardous to our health.

Given the anthropocentric attitude of ancient and modern societies, it is not surprising that the interest in our own health has superseded concerns for other creatures, and for our and their physical environment. Until rather recently, the endangerment of animal and plant species was seen as a natural part of global evolution, and the oceans and skies seemed infinite and their tolerance unlimited. Human survival and well-being received priority, when slowly the realization set in that the human race critically depends on other species, that it is part of a finite and sometimes fragile environment, and that humans have developed so much power that their actions can actually cause permanent damage at local and even global scales.

With the realization of the finiteness of the earth's resources and the potential irreversibility of human impact, interest in the environment grew rapidly within the scientific community and society at large. The first actions resulting from this new awareness concentrated on observing and monitoring environmental trends, but it became clear that these alone were insufficient. Quantitative assessments were required that would demonstrate the immediate and indirect effects of human actions and that would allow predictions of environmental conditions under current policies and under best case and worst case scenarios.

Pursuing these goals, researchers from ecology, toxicology, and other branches of science began to develop a methodological framework for quantitatively assessing the current status of the environment and of its trends. This framework is still in its infancy, the methods are quite fragmented and, interestingly, they are often rather dissimilar to procedures in human health risk assessment, even though some of the underlying questions are the same.

On one hand, this is understandable, given the different histories of human and ecological risk assessment and the oftentimes differing goals. On the other hand, it appears that approaches to human and to ecological risk assessment could benefit from each other and that an integrative approach to assessing environmental hazards with respect to human and environmental health would be better balanced and more cost-effective than two separate analyses.

Considering the enormous complexity of environmental assessments, integrated approaches may appear to be overwhelmingly difficult. Numerous branches of science and mathematics are involved, and beyond scientific questions, issues of policies, politics, management, and ethics are just as daunting. Nonetheless, if one appreciates the value of each component of human health assessment on one hand, and of ecological assessments on the other, it appears that a close comparison, integration, and standardization of the two are highly desirable.

This chapter takes a small step in this direction. It is not intended to be comprehensive, and while it will touch on several issues of relevance, it will primarily focus on the role of modeling in the integration of human and ecological assessments. This focus overemphasizes the relative importance of modeling in comparison to biochemical, ecological, and toxicological research and to pressing societal issues, but was chosen as most pertinent in the context of this book.

The chapter begins with a discussion of similarities and differences between human and ecological assessments. This section is followed by a brief review of the components of human health risk assessment which demonstrates the diverse roles that modeling can assume in integrated assessments. The second part of the chapter elaborates in more detail on the modeling needs and requirements, and the third part proposes a comprehensive modeling framework that—in our opinion—has great potential as a tool in the development of integrated approaches to quantitative risk assessment. It will be demonstrated that this *canonical* modeling approach unifies many *ad hoc* models and provides a mathematical framework within which the various traditional modeling approaches of human health risk assessment, ecological assessment, or both become related.

ECOLOGICAL AND HUMAN RISK ASSESSMENT: SIMILARITIES AND DIFFERENCES

On one hand, ecological risk assessment and human risk assessment have the same overall goal in mind—namely, to analyze our surroundings and to determine whether there is cause for concern. On the other hand, the specific goals of the two types of assessment are rather different, and so are the methods of analysis. The differences can be roughly categorized into two classes: *science* and *importance*.

The category of science includes the intrinsic differences between humans and other organisms, as they are manifest in time and size scales and in our biological interactions with the environment. There are—and always will be—differences in exposure to contaminants between humans and fishes, for example, simply because the exposure pathways and durations are different. The effects of some agents will always be different between mice and men, simply because these species can be expected to have different sensitivities and may even have different metabolic pathways. The characterization of scientific differences is a matter of science and fact finding. It may be very difficult, but it is not likely to be overly contentious.

The issue of importance is of a different nature, because society assigns more value to a human being than to any member of another species. This value system has far-reaching consequences that cannot be treated with scientific methods alone. Indeed, these issues of importance have been a driving force for scientific research in general. The first significant difference is the selection of endpoints for risk assessments. In human risk assessment, the individual receives attention, and potentially fatal hazards are considered along with comparably less significant consequences, such as rashes or headaches. The profusion of scientific discoveries of the effects of environmental contaminants on human physiology is informed by the degree of importance attached to preventing gross or subtle harm to human individuals.

In contrast, ecological risk assessment is characterized by the sheer number and diversity of populations, species, and ecosystem-level interactions for which some level of protection is desired. This distinction is the crucial difference between human and ecological risk assessment, and necessitates somewhat different approaches to these two types of risk assessment. Obviously, the approach to risk assessment for an endangered large mammal such as the Florida panther will be more similar to that for humans than will a risk assessment for an assemblage of nonendangered cladocerans.

Ecological assessments often target the survival of entire populations or the sustainability of an ecosystem, and individual deaths or illnesses are rather readily tolerated. Even the local extinction of populations is often accepted, unless these populations belong to globally endangered species. Changes in ecosystem composition are monitored, but are not necessarily cause for concern in themselves. The basis for evaluating the significance of such changes is often poorly understood and still more poorly incorporated into environmental legislation.

These issues of importance require societal and political attention. Societies and their representatives must decide on the value of an organism, a population, a species, and even an ecosystem in relation to human interests. These decisions must include the establishment of endpoints that address individual organisms, species, and complete ecosystems. They must determine what damages are tolerable or unacceptable. Clearly, the existence of the human race in its present form puts a burden on the global ecosystem, and it amounts to an ethical question for the global society how much we are entitled to alter this ecosystem in order to maintain the *status quo* and its future developments.

Even though these difficult questions of ethics and value are nonscientific, science and mathematics can and must contribute to the decision process by providing facts and making predictions on possible scenarios. Science holds the diagnostic tools and, in concert with mathematical modeling, allows us to assess the potential future consequences of today's actions. Of course, our current diagnostics or predictions are not always correct; however, without scientific and mathematical methods, objective compilations of relevant information in determining risk of environmental threats are not possible. Even a critical assessment of science in general will come to the conclusion that progress has been made toward understanding nature.

With that we turn to the question: What can ecological and human health risk assessors learn from each other? What can we do to improve our diagnostic and predictive tools? Human health risk assessment, with all its pitfalls, has been reasonably successful because at least some of the specific goals are clearly defined: Protect as many individuals as possible, yet accept excess risks up to a certain point, such as a *de minimis* risk. Based on these tenets, "acceptable numbers," such as RfDs and water and air quality indices, can be—and have been—determined, and these are easily checked by industries and regulators against a given environmental situation. The development of relevant endpoints and characteristic numbers for ecological assessments appears to us as crucial (similar comments are found throughout Suter (1993)).

The focus of ecological risk assessment has been the protection of individual species, usually from mortality, morbidity, or decreased fecundity. Although some research has been conducted on contaminant effects on species assemblages (e.g., Cairns, 1990; Weber et al., 1992), there still exists a great need for risk assessment practices for the protection of ecologically important processes, interactions, and even whole ecosystems (Burger, 1994). Attempts at quantifying and simplifying such complex properties into a single set of numeric values has met with some resistance. An example is the *Index of Biotic Integrity* (Karr, 1991) for ascertaining the health of fish species assemblages in streams. Although the reduction of such complex interactions to a manageable scale of acceptability is attractive to the risk assessor, the methods have been criticized for oversimplification. However simplistic current techniques may be, the development of defensible quantitative methods for such emergent properties of ecosystems as stability or integrity could be a very valuable tool for the ecological risk modeler, assessor, and manager.

Ecological risk assessment has made great strides in characterizing the effects of contaminants on integrated environmental systems. By contrast, human health risk assessment often pursues the reductionist paradigm in that it tries to associate a disease with a single cause, while system thinking is usually absent. It appears that a more holistic approach would benefit human risk assessment, in philosophy as well as methodology. Such an approach would consider human health as special, yet still a part of a healthy ecosystem. Human and ecological assessments would be executed simultaneously and by the same group of assessors, using the same data and methodologies, as far as feasible, and result in overall risk numbers.

Modifications of the 1983 risk assessment paradigm of the National Research Council (NRC, 1983) have been proposed that incorporate a holistic approach to both ecological and human risk assessment (Barnthouse, 1994; Harvey et al., 1995). Proponents of holistic risk assessment argue that simultaneous ecological and human risk assessment affords maximum consistency when differentiating among competing risks, allows comparable exposure assessment and dose-response methods, and may even permit the identification of previously unknown routes of exposure and effects on humans (e.g., Colborn, 1994).

Although the use of holistic risk assessment is new to prospective risk analysis, it is quite commonly used in retrospective risk assessment (EPA, 1993a), in which the effects of past or continuing pollution are observed and quantified across a spectrum of human and ecological endpoints. Examples of the successful application of holistic procedures to contaminant-specific risk assessment are the development of criteria for dioxin (EPA, 1993b) and the determination of the effects of methyl-mercury (Harvey et al., 1995). The use of a consistent framework in problem formulation for ecological and human risk assessment by risk managers can only help the modeler reduce the vast assortment of potential modeling approaches to the best possible tools for the problem at hand. Since models may use similar structures and pathways in both ecological and human risk assessment, maximum efficiency is provided by simultaneous consideration of a variety of endpoints. An example is the analysis of the bioaccumulation of organochlorines and organometalloids in food chains.

COMPONENTS OF RISK ASSESSMENT

Health risk assessment, whether it focuses on humans or wildlife, is as multidisciplinary and interdisciplinary as one can imagine. Obviously, biology, biochemistry, chemistry, geology, physics, and many of their subdisciplines are at the heart of environmental assessments, but even beyond the natural sciences, it is difficult to find a branch of human knowledge that would not be able to contribute to the multifaceted enterprise of understanding and controlling our environment. There are questions of quantification and prediction that call for mathematical and statistical tools; there are unsolved ethical aspects; and there are uncounted problems dealing with social, political, and psychological issues.

Even within the "hard sciences," the overlap among contributors to environmental assessments is quite pronounced, and it may be useful to highlight the roles that various disciplines play in human and ecological risk assessment. The representation in Table 6.1 attempts to illustrate some interrelationships. It is given in the form of an "adjacency matrix," but in fact, the boundaries are fluid. Furthermore, the right side and the left side should be adjacent, as should be top and bottom, mathematically speaking, thus requiring representation on the surface of a torus—a risk assessment doughnut.

The rows of the adjacency matrix were chosen according to the standard human risk assessment paradigm, which was first publicized by the National Research Council (NRC, 1983), and which is still in use, albeit with occasional slight modifications. The columns of the matrix show scientific disciplines that share some of their expertise with the risk assessor. This classification may not fairly depict the relative importance of disciplines, but was chosen to emphasize topics most relevant to the present, quantitative context.

Many of the topics listed in Table 6.1 and their role in risk assessment are self-explanatory. Others are so deeply embedded in their mother disciplines that any further discussion would greatly exceed the scope of this chapter. For instance, every entry in the column marked *Science* in itself would command a volume of documentation in a comprehensive discussion of risk assessment. Similarly, statistics has contributed in manifold ways that are not justly represented here. However, the relevant statistical topics are readily found in the literature and need not be discussed here. Epidemiology is currently a cornerstone of human health risk assessment and its concepts are readily found in the literature. By contrast, epidemiology is still considered "unconventional" in ecological risk assessments (Suter, 1993).

Modeling represents the primary focus of this chapter. Modeling in itself is interdisciplinary, and it is arguably the least developed of the contributing branches of risk assessment. In contrast to statistical analysis, which by and large follows generally accepted procedures, modeling is not (yet) guided by a canon of good practice. It appears that "anything goes" in modeling, as long as it is mathematically defensible.

Given the diffuse and fractionated character of mathematical modeling, it is necessary to describe where models fit into the overall picture of risk assessment. We will elucidate the role of modeling with a diagrammatic description of human health risk assessment that is quite different from the standard NRC paradigm and develop from these diagrams the niche

Table 6.1. Contributions of Science, Epidemiology, Statistics, and Modeling to the Four Components of the NRC Risk Assessment Paradigm (see text for explanations).

Discipline Paradigm Component	Science (Toxicology, Physiology, Biochemistry,....)	Epidemiology	Statistics	Modeling
Hazard Assessment	Animal Experiments Cell Lines Bioindicators (Accidents, Acute Exposure)	Disease Clusters Deformity Clusters Measures of Occurrence Measures of Association Risk Perception	Correlation Regression Fault/Event Trees Spatial Statistics Meta-Analysis	Biochemical Systems Carcinogenesis Sensitivity Analysis QSAR Models
Dose-Response Analysis	Animal Experiments Cell Lines (Accidents, Acute Exposure)	Measures of Association Trend Analysis	Correlation Regression "Mega-Mouse" Study Significance ANOVA	Dose-Response Curves Extrapolation Multi-Hit Models Multi-Stage Models PBPK Models Carcinogenesis
Exposure Assessment	Biomarkers: Blood, Urine, Hair, Breast Milk, Enzyme Profiles (CYP 450), Genetic Markers	Occupational Studies Consumption Surveys Frequency of Outcomes Study Designs	Sampling Strategies Data Quality Spatial Statistics Geographic Information Systems	Pathway Analysis Transport Models Fugacity Models Food Chain Models Catastrophic Events (Worst Case Scenarios)
Risk Characterization	Potency Factors Risk Factors	Odds Ratio Relative Risk Causation Generalization	Survival Analysis Significance Uncertainties	Integrated Models Risk Models Causation Generalization Uncertainties

for mathematical models. In fact, we will demonstrate that almost every arrow in the following diagrams corresponds to some type of mathematical model.

Steps of Human Health Risk Assessment

The ultimate goal of human health risk assessment is to predict reliably, for an individual in any group or population, the probability of disease, given that environmental, occupational, and intrinsic susceptibility factors are known. The initial attempt at this deduction is represented schematically in Figure 6.1.

In this representation, G and G_i correspond to a population group and Dz and Dz_i to an associated disease outcome. For example, the disease outcomes could be smoking-adjusted lung cancer prevalence, and the population groups could represent different geographic locations in the United States. This type of health risk assessment corresponds to a correlational or "ecological" study and is generally the first step in the epidemiological elucidation of disease etiology. In this first step, disease clusters are observed, and the assessor attempts to identify general characteristics of the populations that may be used to differentiate the disease occurrence in the subgroups. The groups could be characterized by geographic location, occupation, race, or sex; the grouping characteristics are presumed to account for both exposure to disease agents and intrinsic susceptibility. Measures of occurrence (prevalence) or association (relative risk, if incidence rates are known) may then be used to test for significant differences among the various populations. The arrows in Figure 6.1 signify an epidemiological or statistical model, typically in the form of a correlation, regression, or associational (e.g., Chi-square) model.

Beyond this simple association, one is interested in why particular groups or individuals are susceptible to the disease and others are not. The next step is thus to characterize the group of interest with respect to special, relevant features, which may be physiological (internal) or environmental (external). Such features are often called *risk factors*. They are included in the next diagram (Figure 6.2), along with a box H which signifies that human metabolism in members of the exposed group "converts" these risk factors into disease, with a certain probability.

In the jargon of modeling and systems science, the arrows into and out of the box H are inputs and outputs, and H is characterized by some transfer function, which one would expect to be rather complicated. The arrows again correspond to mathematical models. In this case, such models include biochemical, metabolic, and physiological models that elucidate the mechanisms of disease and may also include dose-response relationships.

The next step of health risk assessment is the quantification of the risk factors R_i. This can be accomplished directly, for instance by means of diaries, questionnaires, or exposure monitors, or indirectly, through measuring biomarkers, such as blood, hair, breast milk, DNA adducts, or enzyme profiles, to name but a few.

It is *a priori* unknown which intrinsic features of the exposed individuals or groups are actually related or maybe even causal to disease development. The next step is therefore an attempt to deduce associations between disease and a minimal number of individual risk factors, or between disease and particular combinations of risk factors which address multiple exposures, confounders, and protectors. The underlying methods of analysis are simple or complicated correlation or regression models, such as Cox's (1972) proportional hazard model.

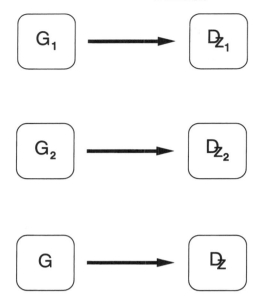

Figure 6.1. Disease frequencies, Dz, are associated with human populations or groups, G.

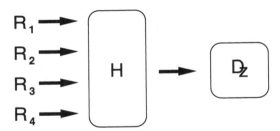

Figure 6.2. Disease frequencies Dz are associated with many potential risk factors R. These risk factors are thought possibly to affect human metabolism H in members of the exposed group or subpopulation. In this simple scenario, risk factors are shown as independent. In reality, they may act independently, synergistically, antagonistically, or in sequence.

If these steps are successful and yield reliable results, the discovered associations are used to predict risk of disease development in other population groups with similar characteristics. These extrapolations are rather safe if they pertain to closely related groups. For instance, based on results of smoking-related illnesses in white males, one may predict similar disease developments in white females or in nonwhite individuals. However, if extrapolations are to be made between rather different groups, such as adults and fetuses, it may be better to resort to mathematically based scaling, for instance, with a physiologically based pharmacokinetic (PBPK) model. These models are discussed in later sections.

In the reality of environmental health assessment, many of the associations between risk factors and human disease cannot be determined with epidemiological methods alone. For instance, there is rarely a direct ability to determine causation of disease by internal or

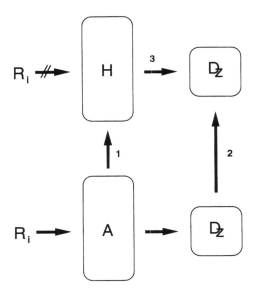

Figure 6.3. The association between disease frequencies Dz and specific risk factors is inconclusive in humans and therefore subject to animal testing. Arrow 1: Allometry, biochemical and metabolic systems analysis, PBPK models; Arrow 2: Animal-human extrapolation, safety factors; Arrow 3: Prediction of human Dz development based on Dz in animals.

external risk factors in humans—an exception being the Japanese atomic bomb survivors, for whom exposure characteristics were fairly well established. Instead, it is necessary to employ substitutes, which are typically animals, cell cultures, or even bacteria. This type of substitution is diagramed in Figure 6.3.

In such surrogate tests, an animal is exposed to a comparable risk, disease outcomes and frequencies are observed in the animal, and from the results, deductions are made on human disease. These deductions may follow the path of arrows 1 and 3 or the path of arrow 2. Again, these arrows signify the involvement of mathematical models. Arrow 1 relates human metabolism to the metabolism of the animal used in the corresponding exposure experiment. It requires metabolic and physiological models that can be scaled and adjusted to account for interspecies size differences and other variations. The simplest models of this type are allometric relationships, while more complicated models are biochemical systems models and PBPK models, as will be discussed later.

Arrow 2 relates disease occurrence in humans directly to disease occurrence in animals. If the disease mechanisms were fully understood, this extrapolation could be performed with some reliability. As it stands, not enough is known, and it is routine to use *safety factors* that artificially inflate human risks in order to err on the safe side. These safety factors—and their employment—are based on empirical observations of the relationship between disease in humans and the modeled animal. Since the degree of similarity between the disease processes in humans and model animals is generally unknown, this method is subject to great uncertainties.

Arrow 3 predicts human disease occurrence based on mechanistic ideas about human metabolism that are deduced from the corresponding mechanisms in animals. As a conceptual example, one could attempt to measure in an animal species all parameter values necessary for identifying a two-stage tumor model, such as cell proliferation rates, death rates, and transition rates between stages. Using the same model structure, but with human parameter values, one could then make predictions about carcinogenesis in humans.

In contrast to pharmacology, doses of chemical agents are usually very small in an environmental setting. In fact, they are typically so small that the expected disease occurrence is somewhere between 0 and 1 in several thousand. This causes considerable difficulties in the implementation of dose-response experiments, since the *de minimis* target for human excess risk may range from 1 in 10,000 to 1 in 1 million.

To differentiate such a low level of outcome from background risks, literally millions of animals would have to be tested to obtain sufficient statistical power. While this was done once in the *mega-mouse study* (e.g., Staffa and Mehlman, 1979; SOT Task Force, 1981; Brown and Hoel, 1983), such effort is not feasible for all of the thousands of natural and anthropogenic chemicals around. Instead, laboratory animals are exposed to much higher doses than expected in the environment, and the researcher extrapolates dose-response curves from measured high-dose data toward the desired but unavailable low-dose range (cf., Figure 6.4). This extrapolation can be based on various algebraic models that may be as complicated as a log-logistic or Weibull function or as simple as a linear function with zero intercept. The former are chosen because of their overall shape, while the latter is justified as being the Taylor approximation of any of the multistage cancer models for very low doses. High-dose to low-dose extrapolations are depicted as arrow 4 in Figure 6.4. It is rather obvious that this type of extrapolation not only is associated with a high degree of uncertainty but that its outcome has a very high impact on the ultimate characterization of risk. In the future, mathematical models capturing the mechanisms of disease development may allow the risk assessor to circumvent this step, but at this point no viable alternatives are available.

Finally, it is not always possible to administer doses high enough to cause significant disease outcomes via the most relevant exposure route. This situation is most often encountered in inhalation studies in which the ambient air is saturated below the desired dose of the chemical agent. In such cases, the desired elevated dose is administered via gavage, i.e., forced ingestion (cf., Figure 6.5). Again, this change in conditions amounts to an extrapolation (arrow 5) that is affected by some uncertainty. PBPK models have the potential of assisting in this extrapolation and thus in reducing uncertainty (cf., Chen and Blancato, 1987).

All diagrams so far have addressed associations between risk factors and disease frequencies. In addition to these analyses, it is necessary to associate quantitatively the risk factors with risk sources or hazards. Different sources can be, or can affect, risk factors, and it is a matter of exposure and pathway analysis to quantify their relationships. Fate-and-transport models, probabilistic exposure and scenario models, and geographic information systems are common tools for assessing exposure.

Combining the associations between sources (hazards) and risk factors with associations between risk factors and disease frequencies, we obtain an overall diagram for human health risk assessment (Figure 6.6). In this diagram, several diseases or disease components Dz_i may combine to form a disease complex Dz (arrow 6). Almost independent of the quantity and quality of scientific evidence, man-made environmental sources are often considered by lay people with suspicion, and this may result in an incorrect or skewed perception of imminent disease risk. The diagram therefore includes a regulatory feedback that, based on real or perceived disease risks, affects the mitigation of risk sources.

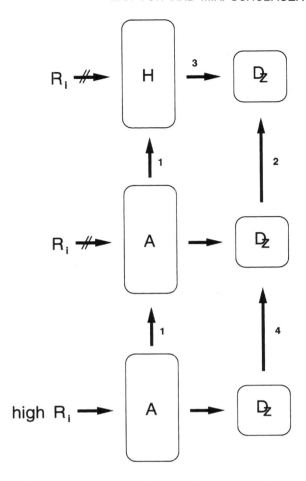

Figure 6.4. The association between disease frequencies and specific risk factors at (relevant) low levels is inconclusive in animals. Response tests are executed with high doses instead. Arrow 4: High-dose low-dose extrapolation. Other arrows: See Figure 6.3.

The models discussed above can also be applied to ecological risk assessments, where it may not be possible directly to assay the target populations, particularly if they belong to endangered or nonculturable species. It must be realized in such situations that problems with interspecies modeling and scaling may be exacerbated due to poor understanding of target species metabolism and toxicology.

MODELING NEEDS

The previous sections suggest that models play an important role in human health risk assessment and that their involvement is likely to be similar in ecological risk assessments. The question then becomes: What types of models are best suited to deal with the issues at hand? To answer this question, it is useful to list the specific modeling needs in human and environmental risk assessment. We concentrate on biomathematical models and will not

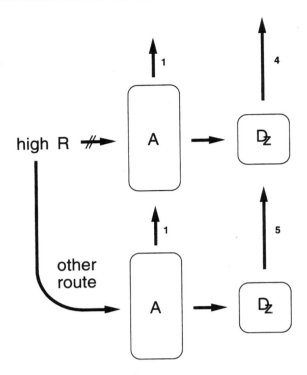

Figure 6.5. Sufficiently high doses cannot be administered via the most relevant route and are administered in a different manner. Arrow 5: Route-to-route extrapolation. Other arrows: See previous figures.

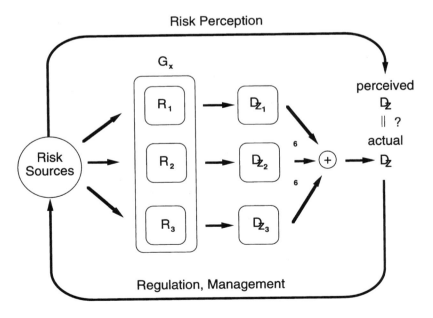

Figure 6.6. Overall representation of human health risk assessment. Arrow 6: Computation of total risk of disease development. (See previous figures and text for details.)

discuss purely statistical and epidemiological methods, since these are readily available in standard textbooks.

One may classify the current modeling needs and tools into four groups:

- High-level risk assessment models
- Models that account for the use of surrogates
- Models of underlying mechanisms and dynamics
- Meta-models for evaluating other models

These categories are neither mutually exclusive nor exhaustive. Nonetheless, they provide a useful classification scheme for analyzing modeling needs in risk assessment. Following a discussion of these modeling approaches, we propose a *canonical* modeling strategy that subsumes many of the necessary models into one unifying mathematical framework.

High-Level Risk Assessment Models

These models use measured parameters to compute numbers that are useful in risk characterization. Three prototype models are

- Risk models
- Exposure models
- Transport models

Risk Models

The simplest risk model quantifies risk in terms of dose and potency:

$$\text{Risk} = \text{Dose} \times \text{Potency} \tag{1}$$

While trivial in appearance, this product is loaded with very challenging problems of estimating the contributing factors. The dose component is the result of often complicated and uncertain transport and exposure models, while the potency component summarizes in one number the entire toxicology of an agent and includes all uncertainties associated with the various extrapolations involved. The greatest appeal of this simple model is that potency numbers are often published by agencies such as the U.S. EPA, and that researchers are permitted to use these numbers without further justification. Consequently, risk estimation in such cases essentially reduces to estimating dose.

More complicated risk models include the logistic model (e.g., Kelsey et al., 1986) and Cox's (1972) proportional hazard model. Both are algebraic models based on logarithmic, exponential, and logistic functions.

The logistic model relates the probability or risk of disease development, P, during some specified time interval, to risk factors Y_i. The relationship follows a multivariate logistic function of a weighted sum of the risk factors:

$$P = \frac{1}{1 + e^{-(a_0 + a_1 Y_1 + a_2 Y_2 + \ldots + a_n Y_n)}} \tag{2}$$

The coefficients a_1, ..., a_n quantify the impact of the respective risk factors on disease development, while the coefficient a_0 is not associated with a particular risk factor but reflects the background risk. In epidemiological interpretation, it corresponds to the logarithm of the odds for disease among the unexposed.

If a significant number of individuals is lost to followup or to unrelated death, the use of the logistic function, which models cumulative incidence (probability), becomes problematic. Cox (1972) suggested instead to base risk estimates on person-years of survival and to compute an *incidence density* or *hazard rate*. Cox's proportional hazards model in the typical case reads

$$\ln(hazard\ rate) = a(t) + a_1Y_1 + a_2Y_2 + ... + a_nY_n \tag{3}$$

The first term, $a(t)$, describes a time-dependent "baseline" effect, which can be complicated but is unaffected by the risk factors Y_1, ..., Y_n. The risk factors may be dichotomous, categorical, or continuous, and they may be constant or time-dependent during the observation period. The coefficients a_1, ..., a_n account for the contribution of each risk factor to the overall disease incidence rate. Cox's formulation implies proportionality of hazards, since these are given as

$$h(t) = h_0(t)e^{a_1Y_1+a_2Y_2+...+a_nY_n} \tag{4}$$

and if we compare the hazards of a population exposed to Y_i and one not exposed to Y_i, their hazard rates are related as

$$\frac{hazard\ rate\ (exposed\ to\ Y_i)}{hazard\ rate\ (not\ exposed\ to\ Y_i)} = e^{a_iY_i} \tag{5}$$

For a dichotomous variable Y_i with values 0 and 1, the logarithms of the hazards rates thus simply differ by a_i.

Recently, Jiang and Succop (1996) analyzed the relationship between blood lead and environmental lead, and suggested as the best suited mathematical representation of this relationship a "log-log regression model" of the form

$$\ln(PbB_i) = \beta_0 + \beta_1\ln(PbH_i) + \beta_2\ln(PbD_i) + \beta_3\ln(PbSS_i) + \beta_4\ln(XRF_i) + \varepsilon_i \tag{6}$$

where PbB_i is the blood lead level of the i-th individual and the right-hand side contains various environmental lead concentrations and an X-ray fluorescence measurement of lead in interior paint. The parameters β_i are regression coefficients and the ε_i denotes the random error associated with individual i.

Exposure Models

Exposure models may be sophisticated, but in the majority of cases they are ultimately sums of products. The products quantify parameters like dose, duration, body weights, and conversion factors accounting for such aspects as bioavailability or changes in physical state, while the sum of these products accounts for different exposure routes and different exposure scenarios. More complicated models of exposure assessment account for spatial and temporal aspects of source and contact. Exposure models feed their results into risk models.

A typical example, in rather general terms, is an exposure equation that accounts for differences in concentrations and exposure durations among different microenvironments (cf., Ryan, 1991),

$$E_{total} = \sum_i f_i (P_i C_{amb} + S_i) \tag{7}$$

which expresses the total exposure as the sum of partial exposures in microenvironments; each partial exposure is determined by the *fraction* f_i *of time* spent in the i-th microenvironment, the *ambient concentration* C_{amb}, the *effective penetration factor* P_i for the ambient pollution into the microenvironment, and the *effective source strength* S_i of the pollutant.

As in the case of risk models, the simplicity of exposure computations as sums of products is deceiving in that it is often extremely difficult to obtain valid measurements that enter into the terms of an exposure function. For instance, the parameters P_i and S_i above may be related to the numerous physical features of the microenvironment (Ryan, 1991).

Recently, van Veen (1996) suggested the use of differential equation models for uptake and exposure, arguing that these processes are functions of space and time. Under slightly simplifying assumptions, the spatial component was eliminated, yielding a system of linear differential equations in the form of a compartment model. One of the state variables described exposure, and from this variable, uptake was computed as an algebraic function.

Transport Models

Transport models can be considered tools of exposure assessment in that they describe how agents move from hazard sources to susceptible organisms or to other target sites of interest. Transport models are often used to characterize critical sampling points, e.g., at the boundary of a polluted site. Because these models, by necessity, address changes in time and location, they are usually formulated as partial differential equations. These cannot often be evaluated analytically and require considerable computational support. Transport models feed into exposure models by predicting doses or concentrations at desired locations and points in time, and are especially useful if these cannot be measured directly. Depending on their purpose, transport models focus on different aspects, such as overall mass balance, intercompartmental movements, spills, or remediation, and subsequently result in mathematical structures that may be as simple as linear algebraic equations or as complicated as full systems of partial differential equations (e.g., Burns and Baugham, 1985; Clark et al., 1988; Mackay and Paterson, 1993; EPA, 1994). Transport models may also incorporate fate, i.e., chemical or biological transformation and fugacity, as well as contaminant bioaccumulation and biomagnification through food webs.

Even if transport models use partial differential equations, the underlying processes are often described as systems of ordinary differential equations which, in a secondary step, are combined with spatial phenomena. Consider, for instance, EPA's standard model for two-dimensional contaminant transport that is affected by microbial activities, BIOPLUME II (e.g., Rifai et al., 1989). The model simulates the dynamics of contaminant concentration due to convection, advection, dispersion, sorption, mixing, biodegradation, and reaeration. Its theoretical basis is a model by Borden and Bedient (1986) which describes the dynamics of microbial growth and the removal of hydrocarbons and oxygen. These processes are formulated as ordinary differential equations whose right-hand sides are composed of sums of linear functions and products of Monod terms. The latter are saturated algebraic functions that are mathematically equivalent with Michaelis-Menten rate laws. This fundamental component of BIOPLUME II is mathematically indistinguishable from a simple biochemical or physiological systems model. The ordinary differential equations become spatially dependent by combining them with an advection-dispersion equation (Bear, 1979), which captures the spatial characteristics of the contaminant and oxygen gradients, and with a spatial model of microbial movements (Freeze and Cherry, 1979). It appears that BIOPLUME II is among the more sophisticated transport models currently used by EPA (cf., EPA, 1994). At the same time, the underlying mathematical structures are not complicated, and they are comparable to simple, unregulated biochemical systems models.

If transport models are simplified to ordinary differential equations, the result is often that of a linear mass balance model, and if this model operates at steady state, the result is simply a set of linear equations (e.g., Mackay and Paterson, 1993). Mass balance models may be applied on either a large or small scale for a system, depending on the degree of resolution required for the concentration distributions and the patchiness of contaminant sources and sinks. Some models assume steady-state conditions for the relevant control volume, obviating the need for differential equations. Other models use probability-based models to account for environmental distribution. For example, Gaussian plume models are frequently used to model point source discharge into the atmosphere (Stern et al. 1994).

Transient mass balance models are also used and often take the general form

$$d(\text{Stored Mass}) / dt =$$
$$\text{Transport In} + \text{Source Production} - \text{Transport Out} - \text{Sink Elimination} \qquad (8)$$

(Hemond and Fechner, 1994). As a variation, they may address a smaller scale, using an infinitely small control volume which represents the change in chemical mass with respect to time at any small unit of the two- or three-dimensional space. In this case, advection and/or Fickian diffusion/dispersion are added to the above model.

Typically, human and ecological exposure assessments incorporate several of the transport models described above (e.g., EPA, 1993a), with the primary constraint being the source of exposure of the population or species to the contaminant of concern. In general, transport modeling can occur in an identical manner for ecological and human health risk assessment; indeed, these are often conducted concomitantly for retrospective risk assessments where the same exposure pathway is operative in both humans and wildlife. An example is food chain bioaccumulation of organochlorines and heavy metals (e.g., EPA, 1993a).

Models that Account for the Use of Surrogates

The previous sections showed that modeling plays a major role when direct measurements are not possible, unfeasible, or unethical to obtain. The models relevant in this context can be categorized as

- Scaling models
- Extrapolation models

Scaling Models

Scaling models make the explicit or implicit assumption that the same mechanisms are at work in different species, but that their magnitude or speed is related to other physiological parameters that characterize the species. The most frequently used parameter is body weight. It is known that smaller organisms have higher metabolic and physiological rates, and in very many cases, body weight and these rates are rather accurately related by power-law functions of the type

$$Y = \alpha X^g \tag{9}$$

where X is the body weight and Y some other physiological variable. Such relationships are also called *allometric*, indicating the different (*allo-*) scales (*metric*) at which these processes occur. α and g are parameters that are process-specific. Allometric relationships are mathematically very convenient. On one hand, they are nonlinear and tend to describe biological phenomena with greater accuracy than linear functions. On the other hand, these relationships are linear when represented in logarithmic coordinates. Specifically, Equation 9 is equivalent to the linear function

$$\ln(Y) = \ln(\alpha) + g \cdot \ln(X) \tag{10}$$

There are hundreds of examples of this relationship, including blood clearance, respiratory activity, and cardiac cycles, which are all allometrically related to body mass for numerous species (e.g., Needham, 1942; Adolph, 1949). A directly relevant example in this context is the relationship between LD_{50} and the time to metabolize one body mass equivalent of O_2, which follows a power-law function that many species obey (Lindstedt, 1987).

Even "higher-order" characteristics often are related allometrically. For instance, the relationship between the sizes and numbers of trees in a plot very often follows the "3/2 rule," which asserts a power-law function with power 3/2 (e.g., White, 1981; Voit, 1988). Similarly, the number of wildlife species appears to be allometrically related to the inhabited area, even though the scatter in this case is often considerable (e.g., Kooijman, 1987).

An example at a different biological scale is provided by quantitative structure-activity relationships (QSARs). QSARs have the mathematical form of power-laws and are used to predict rate constants of unknown chemical reactions from corresponding known reactions or known reactions in structurally similar compounds (cf., Mill, 1993). QSARs are also of vital importance in modeling the toxicity of mixtures of contaminants, which may act synergistically or antagonistically (e.g., Arnold et al., 1996).

An allometric relationship is not necessarily a function of only one independent variable. For instance, in the context of risk assessment and interspecies scaling, Hayton (1989) summarized findings showing that the systemic clearance of chemicals is a power-law function of both body weight and brain weight. Suter (1993, p. 185) discusses dose-response models that are power-law functions in both dose and duration.

Expanding the terminology of Equations 9 and 10, a two-variable scaling function reads

$$Y = \alpha X_1^{g1} X_2^{g2} \tag{11}$$

or, equivalently,

$$\ln(Y) = \ln(\alpha) + g_1 \cdot \ln(X_1) + g_2 \cdot \ln(X_2) \tag{12}$$

Allometric scaling is used in human risk assessment to predict biological responses in humans based on the corresponding responses in a laboratory animal. It would be of great benefit to test the reliability of this scaling procedure with respect to chemical susceptibility and disease development among different animal species, and among different groups of animals within the same species. These comparisons would provide "experimental confidence intervals," which would show to what degree risk scaling is appropriate in human or wildlife health assessment.

Extrapolation Models

Extrapolation models, in a sense, include scaling models as special cases. However, they reach beyond the simple allometric scaling in that they are designed to predict phenomena under unmeasured or unmeasurable conditions.

In many relevant situations of human health risk assessment, extrapolations are so uncertain that they are accounted for by increasing the risk estimate by some artificial factor. For instance, if an agent is inhaled under realistic conditions, but administered in the toxicological experiment by injection, the true response to the inhaled substance is unknown and instead is estimated by multiplying the response to injection with some "safety factor." Since these factors can dramatically inflate the resulting risk estimate, mathematical models should be explored that provide more precise estimates. One type of modeling that shows great potential for this purpose is physiologically based pharmacokinetic (PBPK) modeling (see next section).

A very important group of extrapolation models addresses responses to very low doses of an agent. As mentioned in an earlier section, only responses to relatively high doses are often measurable, and the task is to deduce responses to low, realistic doses. The problem is that any quantitative extrapolation requires knowledge of an underlying model. One might try to identify a model from the measured responses to high exposures, but such effort appears to be futile, since often several models capture the high range with comparable accuracy but differ dramatically when extrapolated to low-dose ranges. Quantitative black-box extrapolation without an underlying mechanistic, explanatory model (see next section) seems to be unreliable.

However complicated, some estimates of responses to concentrations in low doses are needed, and these are currently obtained by extrapolations of one of two types. Either the high-dose response data are modeled with some mathematical function and low-dose responses are estimated by extrapolating from this function, or it is argued, using Taylor series approximation, that the dose-response relationship is linear at low doses, and estimation is reduced to determining the slope of the linear relationship. In the latter case, it is explicitly or implicitly assumed that there is no "safe dose" and no response at zero dose.

Several models have been proposed to extrapolate high-dose response data. They are often variations on the concept that every individual in a population has a threshold for response to the chemical of interest, and that the thresholds occur within the population according to some distribution. For reasons of convenience, this distribution is assumed to be either normal, logistic, or Weibull in the logarithm of dose (cf., Hallenbeck, 1993). There is no reason not to specify a different type of distribution, except perhaps for simplicity of implementation. Judging by high-dose response data alone, any saturated or s-shaped cumulative distribution function could be a candidate model. The crux is that many models describe high-dose responses with comparable accuracy, but they may differ by orders of magnitude when extrapolated to very low doses, which are often in the most relevant environmental dose range.

The alternative to extrapolating *nonlinear* functions is extrapolation with a *linear* function which connects the lowest measured dose-response data of significance with the zero-dose/zero-response point. This great simplification is justified with arguments of conservatism and with the fact that all differentiable functions that go to zero are linear in a close vicinity of zero, according to Taylor's approximation theorem. In particular, the quasi-mechanistic multi-stage model of carcinogenesis (see below) is linear for low doses. While true in theory, the linear extrapolation is afflicted with great uncertainties which are very difficult to assess. In particular, it does not allow for the possibilities of a safe threshold dose or of hormesis, in which very small doses are actually beneficial.

There is serious doubt whether these types of extrapolation models in human or ecological assessment can be amended to become more reliable. One could use other distribution functions for response thresholds, or replace the linear low-dose extrapolation with a second-order polynomial, which would correspond to the next logical approximation of multistage or other models. However, given that data are not likely to be available for the low-dose range, the predictive power of these models is unknown, at least quantitatively. It appears that a better approach may lie in the use of mechanistic and dynamical models.

Models of Underlying Mechanisms and Dynamics

Mechanistic and dynamical models have been used in other disciplines for a long time but, with the exception of PBPK models and one type of cancer model, they are not often encountered in risk assessment. Prototype models in this group include

- Cancer models
- Metabolic models
- Pharmacokinetic models
- Ecosystem models

While these models may seem to be rather diverse, they have strong mathematical commonalties. In particular, they can be considered as compartment models, in which each compartment (size, number, or concentration) changes over time, according to a differential equation.

Cancer Models

The most prominent among the cancer models is the two-stage model (e.g., Moolgavkar and Venzon, 1979). It presumes that cells under normal conditions divide into healthy daughter cells, but that they can, in rare cases, divide to produce *initiated* cells, an event that is causally linked to some external agent, the *initiator*. If an initiated cell is exposed to a second agent, the *promoter*, it may develop into a tumor cell. A generalization is a multistage model, in which a cell line has to pass through more than two stages to become malignant.

The multistage concept can be translated into a mathematical model that can be used for different purposes. The full model is a compartment model, based on differential equations for cell numbers in different stages. This model can be employed to study the dynamics of the originally healthy cell population as it progresses toward tumor formation.

In most risk analyses, however, only one aspect of the multistage model is used, and that is the probability that at least one malignant cell is formed. This limitation in scope drastically simplifies analysis, ultimately leading to the probability of carcinogenesis as some exponential function of exposure. For low exposures, this function is approximately linear, and the only unknown of interest is its slope.

Theoretically (i.e., under the assumption that all necessary measurements are available), this *slope factor* can be estimated in two radically different ways. If all rates of cell division, cell death, initiation, and promotion are known, the linear approximation can be executed numerically, yielding the slope factor as model output. Alternately, if responses (tumor formation) could be experimentally measured for low doses, the slope factor could be measured directly from these data. Neither set of data is currently available, but the importance of this model lies in the potential of testability of results that are obtained from two intrinsically different sets of data.

The duality of data needs and results is not unique for multistage models, but is a criterion of quality for any systems model. Under ideal conditions, a good systems model is numerically identified with measurements of all involved parameters and all variables at some time point. The model then predicts the time courses of the system variables which, in turn, are compared to the corresponding observed time series. In this fashion, a good model bridges two levels of organization: the level of parameters, such as blood flow rates or Michaelis constants, and the output level of responses, such as the duration of time a drug spends in a particular organ.

Metabolic Models

Metabolic models address the transformation of chemical constituents in the body. They are differential equation models in which each metabolite (concentration) comprises a compartment and the processes are transport steps or enzyme-catalyzed reactions.

The transport steps are usually described as first-order kinetics. In other words, the transport rate, which is equivalent to the rate of change in concentration, is a linear function of the current concentration. For instance, the transport out of a compartment C, or the first-order decay of a substance C, is described as

$$\dot{C} = -kC \tag{13}$$

where the dotted variable on the left-hand side denotes the derivative with respect to time.

Enzyme-catalyzed reactions are traditionally described with Michaelis-Menten rate laws of the type

$$v = \frac{V_{max}S}{K_m + S} \tag{14}$$

where v is the rate of the process (i.e., the rate of conversion of a substrate into a product), S is the substrate concentration, and V_{max} and K_M are parameters. These rate laws and their generalizations have provided an enormously successful framework for understanding *in vitro* kinetics. Their disadvantages are that it is not clear whether these types of functions are appropriate *in vivo* (Hill et al., 1977; Savageau, 1992), and that even moderately large systems based on these functions become mathematically unwieldy to a point where analysis is unfeasible (e.g., Shiraishi and Savageau, 1992a). An alternative to these models will be given in the section *Canonical Modeling*.

Metabolic models offer great potential for the elucidation of toxicological mechanisms. If they can be successfully set up and parameterized, they can form a truly scientific basis for health assessments concerned with chemical hazards, especially at realistic dose ranges. Their role in human and ecological risk assessment is their ability to integrate kinetic information into models that describe metabolic or physiologic responses. In many cases, the results are transferable from one species or group to another, but it is also known that different species have sometimes developed different metabolic pathways that allow them to deal with chemical inputs in drastically different ways. A premier example is butadiene, which rats easily metabolize into harmless products, whereas the pathway in mice leads to fatally toxic epoxides (Henderson, 1995). One cannot exclude that such mechanistic differences may occur unpredictably for "new" toxicants, even if they are administered to two animal species that many times before had reacted to other compounds in a similar way.

Pharmacokinetic Models

PBPK models describe for chemicals the distribution and time spent in different organs or tissue compartments within an organism. They are formulated as sets of ordinary differential equations that each describe the temporal changes of concentration of the chemical in a given organ or compartment. The processes governing these models are typically either first-order kinetics, if they describe transport between compartments, or Michaelis-Menten rate laws, if the chemical is metabolized.

A great advantage of these models is that they are based on parameters that are often independent of the chemical in question. Examples are volumes of organs and blood flow rates in different tissues. Using these parameters and the original dose of the chemical agent, the PBPK model predicts concentrations and durations. For calibration purposes, these predictions can be compared with actual measurements obtained in the laboratory. PBPK models thus bridge two levels of organization.

The particularly interesting aspect for risk assessment is that PBPK models provide much sharper tools of extrapolation than simple allometric scaling. Most prominent may be interspecies extrapolation. The first step consists of designing, identifying, and validating a PBPK model in a laboratory species and for some chemical agent. In a second step, the necessary volumes, flow rates, and other parameters are measured in the target species, without involvement of the chemical agent. Substituting these values in the PBPK model and using chemical-specific parameters from the original model, the PBPK model for the target species is able to make predictions, based on dose, that are expected to have a reasonably high degree of reliability.

In human risk assessment, one usually extrapolates directly from a laboratory animal to humans. It appears that ecological risk assessment could both benefit from this experience and contribute new insights by analyzing more closely related species or animal groups. For instance, circumstantial evidence suggests that laboratory and free-living animals of the same species show rather different behaviors and may differ significantly in their physiological makeup. Comparative PBPK analyses of such groups of animals could greatly improve our intuition and experience with extrapolations from the laboratory to the wild, between different animal species, between animals and humans, and between different human populations.

PBPK models are also beginning to be used in evaluating the toxicity of mixtures. Combined with QSARs, this application could become very important in determining biological effects at environmental exposure levels.

Ecosystem Models

The simplest dynamical ecosystem models are based on linear differential equations of the form

$$\dot{X}_i = a_{i0} \pm a_{i1}X_1 \pm a_{i2}X_2 \pm \ldots - a_{i1}X_i \pm \ldots \pm a_{in}X_n \qquad (15)$$

The system variables represent compartments, and the differential equations describe the time courses of mass or energy fluxes into and out of compartments. The mathematical formulation implies first-order kinetics, which is typical for passive transport. Linear models of this type in the context of environmental risk assessment have addressed chemical uptake by wildlife (e.g., Clark et al., 1988) and the cycling of chemicals, including carbon and water, in the environment (e.g., Huggett, 1993). Linear models, while dynamic, are rather limited in their behaviors. For instance, they are unable to capture sustained, stable oscillations, as they are ubiquitous in nature.

Ecosystem models that are rich enough to represent biological systems necessarily require nonlinearities. The prototypes date back to the 1920s when Lotka (1924) and Volterra (1931) independently described the dynamics of coexisting species. These models are among the oldest mathematical structures in biology and, with subsequent generalizations, are still receiving enormous attention.

The typical Lotka-Volterra model for n species is again a compartment model based on ordinary differential equations. The crucial assumption is that the dynamics are governed by intrinsic growth and crowding within a species and interactions between each of two species. This assumption results directly in the sums of a growth term, a crowding term, and products of variables with positive coefficients a_{ij}:

$$\dot{X}_i = a_{i0}X_i \pm a_{i1}X_1X_i \pm a_{i2}X_2X_i . \pm \ldots - a_{i1}X_i^2 \pm \ldots \pm a_{in}X_nX_i \qquad (16)$$

In some cases, the intrinsic growth term $a_{i0}X_i$ is omitted. The coefficients in this equation are chosen to be positive since they indicate rates of the interaction processes. If they carry a plus sign, the species X_i gains from the interaction (e.g., as a predator does), and if they carry a minus sign, the species X_i is negatively affected by the interaction (as is, e.g., a prey).

If only one species is considered, the system (Equation 16) collapses to a simple logistic growth function:

$$\dot{X}_1 = a_{10}X_1 - a_{11}X_1^2 \qquad (17)$$

with the analytical solution

$$X_1(t) = \frac{K}{1 + \exp\left(\dfrac{r}{K} - rt\right)} \qquad (18)$$

where $r = a_{10}$ is the growth rate and $K = r/a_{11}$ is the carrying capacity of the population, which coincides with the maximum number of individuals that can permanently live under the given conditions. The mathematical form of this function was encountered before in the logistic hazard model.

The coefficients of the logistic growth model represent, in a lumped form, all effects contributing to growth (r, a_{10}) and all effects that retard or limit growth or contribute to death or decay (K, a_{11}), respectively. Adverse environmental conditions affect these coefficients and are reflected in either a decreased growth rate a_{10}, an increased decay rate a_{11}, or both. Analogous arguments hold for the full population system (Equation 16), where environmental conditions may affect any number of coefficients in a positive or negative way.

It is noted that division of the system equation (Equation 16) by X_i yields a linear system on the right-hand side, while the left-hand side describes a relative change (change per number of individuals) which is mathematically equivalent with the derivative of the logarithm of X_i. Thus, Equation 16 is equivalent with

$$\frac{d \ln(X_i)}{dt} = a_{i0} \pm a_{i1}X_1 \pm a_{i2}X_2 \pm \ldots - a_{ii}X_i \pm \ldots \pm a_{in}X_n \qquad (19)$$

It is noted that systems of the type (Equation 19) may look simple but that they can, in fact, model virtually any differentiable nonlinearity (cf., Peschel and Mende, 1986; Voit and Savageau, 1986; Savageau and Voit, 1987). The Lotka-Volterra models, obviously, are of some utility in ecological risk assessment.

"Meta-Models" for Evaluating Other Models

The risk assessment literature sometimes discusses Monte Carlo models, uncertainty models, or sensitivity models. In some sense, this is a misnomer, and one should refer to *methods* instead of *models*. The reason is that these methods of analysis are evaluations of models, which may have any of the abovementioned or totally different mathematical structures. For example, PBPK models can be evaluated with Monte Carlo simulations or subjected to sensitivity analyses.

Monte Carlo Methods

The famous casino on the Mediterranean Sea led to the name of this simulation method. The "gambling" here consists of (pseudo-)random determinations of parameter values that are used for solving a complicated model. The rationale for this approach is twofold. First, for many parameters of complicated models, precise numerical values are not known, and the only available information consists of possible or likely ranges within which the parameter values are assumed to fall. Given that a model may contain 50 or more parameters, the number of parameter value combinations becomes too large to permit systematic model evaluations. Second, if models are simple, the effect of uncertainties in parameter values can be assessed algebraically. However, for many realistic models, in particular, if ordinary or partial differential equations are involved, such analyses are not possible, and the only feasible alternative is repeated simulation of the models with different sets of parameter values.

A particular attraction of Monte Carlo simulations is the exploration of best case and worst case scenarios. Obviously, these are not often implemented experimentally, and good model simulations are the only feasible alternative for determining the range of possible outcomes, for instance, of a catastrophic chemical spill.

Monte Carlo simulations require efficient methods of numerical analysis. Two particular demands are the efficient determination of random numbers from different distributions and the efficient solution of ordinary and, if necessary, partial differential equations.

Uncertainty, Sensitivity, and Robustness Analysis

The design and implementation of a model is only the first step of a model analysis. Even if one is rather confident about the quality of the involved parameter values, it is necessary to analyze the model under perturbations. Obviously, environmental conditions change, and this lack of constancy is reflected in parameter values that differ from one situation to the next. Uncertainty, sensitivity, and robustness analyses explore the range of possible outcomes, often by means of Monte Carlo simulations.

Uncertainty analysis assesses the effects of known or presumed variations in parameter values and model inputs, as discussed above. Sensitivity analysis reveals how a model is affected if a parameter is slightly changed. The effects of such changes may be monotonic with the extent of change, or they may be "catastrophic," changing the model behavior qualitatively. An example is the shift from a stable steady state to a sustained oscillation. Robustness analysis assesses how strong a perturbation a system can tolerate before it is irreversibly damaged.

CANONICAL MODELING: A UNIFYING FRAMEWORK FOR ENVIRONMENTAL RISK ASSESSMENTS

It is quite evident that no single model would suffice for all purposes. Nonetheless, we propose the use of certain types of canonical models that have proven useful in analyses of other complex systems in biology. These models are called *canonical*, because they are always constructed according to the same general rules and always result in models of the same mathematical structure. This rigidity may give the impression of an overly inflexible straitjacket that would limit the scope of possible modeling, but it has been shown, and will be demonstrated later, that this first impression is wrong. In fact, we will show that most of the above models fall within this modeling framework either directly, in an approximate sense, or upon mathematical equivalence transformation.

The canonical models proposed here originally derived from the insights that linear models are not suitable to capture synergistic biological systems, and that nonlinear *ad hoc* models often do not possess enough mathematical tractability to be of practical use (e.g., Savageau, 1972). The combination of a convenient nonlinear structure with a set of strong mathematical tools for model design and analysis is the crucial feature of these models.

In the preface to his book *Biochemical Systems Analysis*, which deals entirely with canonical model analyses, Savageau (1976) acknowledged the great progress of the reductionist approach to molecular biology and the enormous amount of detailed knowledge gained. However, he added: "As a result of the molecular revolution a complementary synthetic approach has become not only possible, but necessary if we are to understand the more complex phenomena typical of higher levels of organization. It is the objective of this book to provide such an approach for the study of integrated biochemical phenomena." It is interesting to see analogous needs stated 20 years later in the area of risk assessment. Olden and Klein (1995) recently identified as an important task of environmental health risk assessment: "Along with understanding the basic building blocks of cellular and molecular systems and their interactions with environmental agents, we must integrate these single-system studies into multi-system studies. This effort is especially important for the study of environmental effects because environmental agents can affect different organs differently, effects can vary over time and with health status, and effects can be complicated by the multiple exposures that are typical of the human experience."

The formalism we propose has been described in the literature previously (e.g., Savageau, 1972; 1976; Voit, 1991; 1993), but since it appears that many risk assessors are unfamiliar with it, it seems useful to review pertinent key features, along with references that lead the interested reader to the original literature. The central tenet of biomathematical systems analysis is that the responses of natural systems are characterized by systems of differential equations (or difference equations) that equate the change in each variable to some function of potentially all dependent (state) variables, independent (external) variables, and various parameters. If spatial variation is important to consider, and it is in many exposure and transport models, the resulting equations are partial differential equations, but if spatial aspects can legitimately be ignored or coarsely discretized, the result is a system of much simpler ordinary differential equations. Also, as mentioned above, environmental transport models are often constructed from a combination of ordinary differential equation models, with spatial processes describing phenomena like advection and sorption (e.g., Rifai et al.,

1989). It is thus relevant, even for transport phenomena, to study systems of ordinary differential equations. The following will be limited to such systems.

The general form of a dynamical system equation is

$$\dot{X}_i = \sum_{j=1}^{k_i} F_{ij}(X_1, X_2, ..., X_n, X_{n+1}, ..., X_{n+m})$$ (20)

($i = 1, ..., n$), where the dotted variable on the left-hand side again denotes the derivative with respect to time, $X_1, ..., X_n$ are dependent variables, and $X_{n+1}, ..., X_{n+m}$ are independent variables, which are considered constant during the time period of interest. Parameters are explicitly omitted, but implicitly part of the functions F_{ij}.

In most situations, nothing really definite is known about these functions F_{ij}, but for simplicity of argument, we assume that they are sufficiently differentiable and positive-valued. If the latter is not the case, the problem can be reformulated in terms of additional functions to ensure positivity.

If nothing is known about these functions, it seems that the analysis would have to stop here. However, that is not the case. Approximation theory can be employed to capture the essence of the model (Equation 20), at least to some degree. For this purpose, an operating point is chosen, which consists of *nominal values* of all variables. A standard (but not necessary) choice is a steady-state point, which is a point at which the system, in an overall sense, does not change. At this operating point, the functions F_{ij} are approximated. The simplest variant is linearization, and the result is a system of linear differential equations. Such systems have been extremely successful in the engineering sciences.

For biological systems, linearization is not the best choice, since many biological phenomena simply are not linear. Instead, it has turned out that it is very useful to transport all variables X_i and all functions F_{ij} into logarithmic coordinates. Linearization is executed in this logarithmic space, and transporting back the variables and functions into Cartesian coordinates, the result is nonlinear. Specifically, the result of this procedure is that each function F_{ij} is ultimately approximated by a multivariate product of power-law functions:

$$F_{ij} = \gamma_{ij} \prod_{p=1}^{n+m} X_p^{q_{ijp}}$$ (21)

It is noted that this functional form is always the result of linearization in logarithmic coordinates, independent of the mathematical structure of the original function F_{ij}. The only information used is the operating point and the slopes at this point in logarithmic coordinates. Just as a system, under typical conditions, can always be approximated with a straight line, plane, or hyperplane, a system can always be approximated by simple power-law functions or their multivariate products.

If the original functions in Equation 20 are replaced with their power-law approximations, the result is a *Generalized Mass Action* (GMA) system of the form

$$\dot{X}_i = \gamma_{i1} \prod_{j=1}^{n+m} X_j^{g_{1ij}} \pm \gamma_{i2} \prod_{j=1}^{n+m} X_j^{g_{2ij}} \pm ... \pm \gamma_{ik} \prod_{j=1}^{n+m} X_j^{g_{kij}}$$ (22)

where the positive coefficients γ are *rate constants* of the associated processes and the exponents g are called *kinetic orders*.

As a variant of special interest, one can aggregate all functions F_{ij} that *increase* or *augment* the variable X_i into a single function F_i^+ and all functions that *decrease* or *degrade* X_i into a single function F_i^-. Subsequently, one performs the power-law approximation and obtains

$$\dot{X}_i = F_i^+ - F_i^-$$

$$\approx \alpha_i \prod_{j=1}^{n+m} X_j^{g_{ij}} - \beta_i \prod_{j=1}^{n+m} X_j^{h_{ij}} \qquad (23)$$

This particular form of system description is called an *S-system*, signifying its ability to model synergistic and saturable phenomena.

S-systems have numerous advantages. First, and maybe most significant, is the observation that they model biological systems with surprisingly high accuracy, even though they are by their very nature local approximations. It appears that many phenomena in nature actually exhibit responses that are similar to power-law functions. Second, it has been demonstrated mathematically (Savageau and Voit, 1987) that the mathematical structure of S-systems is rich enough to model virtually any differentiable nonlinearity, including different types of oscillations and chaos. Thus, in contrast to linear models, which, among other features, are unable to capture stable (limit cycle) oscillations, these S-systems are not limited in scope.

A third advantage is the fact that S-system models can be set up with a minimal amount of information. Their derivation reveals that a variable appears in a power-law term if and only if it has a direct influence on this particular term. Furthermore, if the influence is positive, the associated exponent is positive, and if the influence is negative, the exponent is negative. These simple rules render it possible to translate a list of direct influences immediately into a symbolic dynamical system model. Of course, quantitative information is needed to specify parameter values, but in many cases, even the symbolic model or a semiquantitative model, in which the parameter values are somewhat uncertain, can yield useful insights.

The fourth complex of advantages pertains to algebraic and numerical analysis. In contrast to most other nonlinear models, S-systems possess steady states that can usually be computed by analytical means. This is important, since it allows subsequent analyses of local and structural stability, sensitivities, and responses to changes in input variables. The algebraic nature of the steady states also facilitates questions of optimization.

Finally, the homogenous structure of GMA and S-systems is amenable to very efficient numerical evaluations, and specific software packages (e.g., Voit et al., 1989) are available for this purpose.

Relationships Between Canonical Models and Traditional Risk Assessment Models

Generalized Mass Action models and S-system models were developed without the specific intent of assessing health risks. Nonetheless, many of the traditional risk models fall naturally in this framework.

Risk and Exposure Models

The simple risk and exposure models are characterized by sums of products. They are expressed as algebraic equations, assuming that temporal changes in risk or exposure are negligible or can be averaged, or that the system is at steady state. Within the framework of canonical models, the assumption of temporal constancy corresponds to limiting the analysis to situations in steady state,

$$\dot{X}_i = 0 \tag{24}$$

Suppose an exposure model were constructed within the framework of GMA systems. The variables would include the concentration of the chemical in question as well as an assortment of physical features at the location of exposure. The full model would describe how the concentration and some of the physical features change over time. Restricting the exposure assessment to the steady state, the resulting equation consists of a sum of products of power-law functions, which includes sums of products as a special case.

The relationship between the full GMA model and a typical exposure equation becomes particularly obvious for the following scenario. Suppose the concentration of interest is coded as X_1 and that the variables $X_2, ..., X_n$ represent physical features that contribute to the dynamics of X_1. Suppose further that the only direct effect of X_1 on its own dynamics is a first-order decay. The GMA equation of X_1 in this situation reads

$$\dot{X}_1 = \gamma_{11} \prod_{j=2}^{n} X_j^{g11j} \pm \gamma_{12} \prod_{j=2}^{n} X_j^{g12j} \pm ... \pm \gamma_{1k} \prod_{j=2}^{n} X_j^{g1kj} - \gamma X_1 \tag{25}$$

Note that none of the products contains X_1, which reflects that the different contributions to exposure are independent of the concentration X_1 at the given point in time and space. This assumption may or may not be justified objectively, but is certainly in line with general assumptions of traditional exposure assessment.

At steady state, Equation 25 is set equal to zero and divided by the positive rate γ. Bringing X_1 to the left-hand side results in a steady-state exposure equation that consists of a sum of products of power-law functions. The same is true if the degradation term γX_1 is replaced with a full product of power-law functions. In this case, the steady-state equation is first divided by this product, except for X_1.

If all variables affect exposure in a linear fashion, all exponents g_{1ij} are either 0 or 1, and the steady-state exposure equation reduces to the traditional form. Thus, by simplifying the full GMA model in such a way that it reflects traditional assumptions, it naturally produces the well-known results, quasi as a special case. The full GMA model goes beyond the traditional model by accounting for nonlinear effects, synergisms, and feedbacks, and of course the overall dynamics of the exposure process.

Jiang and Succop's (1996) model describing the relationships between blood lead and environmental lead levels (Equation 6) is a direct special case of a canonical model. It is not so easy to see that the logistic model and Cox's proportional hazard model (Equations 2 and 3) would fall within the framework of canonical models, but that is indeed the case. It was shown recently that a dynamic S-system model under steady-state conditions, essentially without further assumptions, leads directly to these models (Voit and Knapp, 1997).

Allometric Scaling Models

It is not surprising that canonical power-law models are particularly amenable to allometric scaling (cf., Savageau, 1979b). If some or all variables of a GMA system are replaced with power-functions, the mathematical structure on the right-hand side is unaffected, and the derivative on the left-hand side simply picks up another power-law term. In fact, it has been shown elsewhere that canonical models are among very few mathematical structures that allow for complete allometry between any two system variables (Voit, 1992a).

Scaling of this type is also possible for time. It is becoming increasingly evident that environmental assessments are dependent on the appropriate time scale. This is particularly relevant in ecological assessments, where organisms with drastically different life expectancies are exposed to the same hazards but respond rather differently (e.g., Burger and Gochfield, 1992; Eakin, 1994). Allometric time scaling does not alter the structure of canonical models. For instance, if we make the original time scale in an S-system model explicit, by calling it X_0, and formally include it on the right-hand side, we obtain

$$\dot{X}_0 = 1$$

$$\dot{X}_i = \alpha_i \prod_{j=0}^{n} X_j^{g_{ij}} - \beta_i \prod_{j=0}^{n} X_j^{h_{ij}} \tag{26}$$

Introduction of a new time scale, $\tau = X_0^p$, application of the chain rule, and possibly renaming of the new time variable returns exactly the same mathematical form. In the simple case of proportional scaling (Eakin, 1994), all α and β coefficients are simply multiplied with the same scaling factor, while the exponents are unchanged.

The natural scalability of canonical power-law models has to be seen in contrast to alternative models. For example, consider a basic Michaelis-Menten model, as discussed before (see Equation 14). If the substrate is replaced with a power-law function $S = \alpha R^g$, the resulting rate law is

$$v = \frac{\tilde{V}_{max} R^g}{K_M + R^g} \tag{27}$$

which has the form of a Hill equation. This form is qualitatively different from a Michaelis-Menten rate law in that it is S-shaped and has a zero slope at zero, whereas the Michaelis-Menten law is simply saturated with a slope of 1 at 0.

Extrapolation Models

As discussed, extrapolation of response curves from high to low doses is often based on lognormal, logistic, or Weibull distributions, but there is no substantive reason not to use other types of distributions or s-shaped curves. A single S-system equation has enough flexibility to model essentially all univariate distribution functions with high accuracy (Voit, 1992b; Voit and Yu, 1994), and systems of one or two S-system equations were shown to contain all commonly used growth functions as exact special cases. This embedding of traditional functions in the framework of canonical models is not simply a curiosity, but can be derived from general approximation and simplification principles of biological systems (Savageau, 1979a). The fact that very simple power-law models are exact analogues of empirical mathematical descriptions suggests that these canonical models capture some of the essence of biological processes.

One advantage of using a simple S-system model instead of traditional growth or distribution functions is that the distinction between different models reduces from structural mathematical differences in the traditional cases to variations in parameter values. For instance, there are no obvious mathematical similarities between Weibull's and Bertalanffy's growth functions in their original, algebraic form, but the similarity is evident when they are embedded in the framework of S-systems (cf., Savageau, 1979a; 1980; Voit, 1991). The same is true for probability functions that are drastically different in mathematical form but become related when represented as S-systems (e.g., Savageau, 1982; Voit and Rust, 1992).

As an example, consider the Weibull, the Gompertz, and the logistic functions as possible candidates of an extrapolation model. The Weibull function takes the form

$$W(t) = a \times \exp[-b \times (t + c)^{-d}] \tag{28}$$

the Gompertz function reads

$$G(t) = p \times \exp[-q \times \exp(-\beta \times t)] \tag{29}$$

and the logistic function is given as

$$L(t) = \frac{k}{1 + \exp\left[m \times \left(\frac{1}{k} - t\right)\right]} \tag{30}$$

(cf., Savageau, 1980). All parameters in these functions are positive. A comparison of the three functions is *a priori* not easy, because their mathematical forms are different. Nonetheless, opportune choices of parameter values can produce graphs for all three that are rather similar, especially in the high range of the independent variable t. Savageau (1980) *recast* these and other functions as S-systems with one or two variables. In this form, which is mathematically equivalent to the original forms, the three functions become

$$\dot{W} = b \times d \times W \times Z^{d+1} \qquad \dot{Z} = -Z^2 \tag{31}$$

$$\dot{G} = G \times Y \qquad \dot{Y} = -\beta \times Y \tag{32}$$

and

$$\dot{L} = m \times L \times \left(1 - \frac{L}{k}\right) \tag{33}$$

where $Z = 1/(t+c)$ and $Y = q\, r \exp(-\beta\, t)$ are *auxiliary* variables that are introduced exclusively for mathematical purposes. For further simplified comparison, we define new auxiliary variables as $X = b \times d \times Z^{d+1}$ and $V = m \times (1 - L/k)$. Differentiation and substitution recasts the Weibull and the logistic functions (with mathematical equivalence) as

$$\dot{W} = W \times X \qquad \dot{X} = -\alpha \times X^{h_1} \tag{34}$$

and

$$\dot{L} = L \times V \qquad \dot{V} = -\gamma \times V \times L \tag{35}$$

where α, γ, and h are conglomerates of some of the former parameters. α and γ are positive, and $h = (d+2)/(d+1)$ is strictly between 1 and 2.

A comparison between the Weibull and the Gompertz functions in this form shows that they differ exclusively in the exponent of the equation for the auxiliary variable. Furthermore, a comparison between the Gompertz and the logistic functions identifies the factor L in the second equation as the only difference. Clearly, if a power g between 0 and 1 is associated with L in the second equation, the resulting system contains the Gompertz and the logistic functions as the special cases $g = 0$ and $g = 1$, and it also contains all "intermediate" functions ($0 < g < 1$), which presumably have no common name. If additionally V in the logistic system is taken to the power h, the Weibull function is included as a special case as well. A system that encompasses all three functions is thus

$$\dot{X}_1 = X_1 X_2 \qquad \dot{X}_2 = -\kappa X_1^g X_2^h \tag{36}$$

where κ is positive and $0 \le g \le 1$, $1 \le h \le 2$.

This system is not unique in its ability to categorize the above growth functions. Other equivalent formulations can be found that serve the same purpose but have a slightly different appearance. Whatever form is chosen, a consequence of this *embedding* is that no decision is required as to what functional form might be best suited. Moreover, simulations or sensitivity analyses can be used to determine the importance of those parameters that differ from one model to the next.

One should stress that the simplification in model identification that is made possible by the process of recasting does not alleviate the deeper underlying questions of high-dose to low-dose extrapolation. The fact that one or several models fit high-dose response data does not at all prove the quality of these models in the low-dose range. Indeed, one can rather easily construct mathematical situations in which two functions are close to equivalent in all measurable dose ranges, but differ very significantly in the unobservable low range. It is not unreasonable to assume that this situation may be relevant in reality.

Mechanistic and Dynamical Models

Canonical models were developed originally to describe biochemical and metabolic systems, and numerous applications, comparisons, and theoretical treatises have appeared in the biological and biomathematical literature (for references, see Voit, 1991). Until only a few years ago, canonical model analyses were limited to small numbers of variables, but recent advances in personal computers have made it feasible to investigate systems with dozens of components (e.g., Shiraishi and Savageau, 1992a–d; 1993; Ni and Savageau, 1996a,b; Curto et al., 1998a–c). In spite of the large number of variables, it was possible to perform systematic and exhaustive analyses, for instance, in order to determine the most sensitive components of the systems. Ni and Savageau (1996a,b) even implemented an algorithm to explore and compare systematically over a thousand possible network designs of human red blood cell metabolism in order to detect causes for weaknesses in the present models. Curto et al. (1998a–c) set up a series of models of purine metabolism, compared and diagnosed them in a systematic fashion, and developed a biochemical classification of associated diseases.

Shiraishi and Savageau (1992a–d; 1993) demonstrated quite clearly that large-scale model analyses are greatly facilitated by the use of canonical models. Comparable models based on traditional enzyme kinetics lack the analytical power, or at least efficiency, to execute analyses even of rather basic system features, such as stability and sensitivities. In fact, it was shown that the best previous efforts to model their particular system of interest, the tricarboxylic acid cycle, had failed, primarily because steady-state analyses were based on numerical simulations, which had been used in lieu of analytical methods.

Canonical models have not yet been used explicitly in risk assessment. Nonetheless, it appears that biochemical models and models of gene regulation, which have been the main applications of these types of models (e.g., Savageau, 1976; Hlavacek and Savageau, 1996), will be key components in understanding the metabolic, physiological, and clinical effects of environmental hazards. These types of models appear to have great potential for the characterization of regulatory patterns that control environmental protectors, such as cytochrome P450 enzymes and repair mechanisms, and they may ultimately provide a way around problems like low-dose extrapolation.

PBPK models typically use differential equations with first-order kinetics for transport and Michaelis-Menten functions for enzyme-catalyzed reactions within compartments. It has been shown that PBPK models can alternately be represented with power-law terms, either approximately or equivalently, thereby extending their flexibility and accuracy in data representation (Janszen, 1992).

As an example, consider a recent PBPK analysis of chemical mixtures by Krishnan et al. (1994). It was found that simple first-order or Michaelis-Menten kinetics did not always capture the observed dynamics of dichloroethylene in rats, and it was hypothesized that some of the involved enzymes were subject to inhibition. Indeed, the inclusion of inhibitory effects improved the data fit considerably. Accounting for inhibition in traditional enzyme kinetics requires knowledge or assumptions about the mechanism of inhibition, and the chosen mechanism dictates the mathematical format of the equation. For simple systems, this choice may not be difficult, but once PBPK models approach the level of sophistication of some of the more advanced models of biochemical systems, the use of generalized Michaelis-Menten models becomes very limiting (e.g., see models and comparative discussions in Shiraishi and Savageau, 1992a–d; 1993; Ni and Savageau, 1996a,b). An alternative is the use of canonical models, in which inhibitions and other regulatory signals are formulated in a streamlined fashion. The resulting power-law representations of metabolic fluxes or clearance terms are readily implemented as stand-alone models or within the shell of a PBPK model (e.g., Janszen, 1992).

Multistage models have not been approached with methods of canonical model analysis, but these models do have the form of simple GMA systems (e.g., Moolgavkar and Venzon, 1979; Serio, 1984).

Ecosystem models in the form of Lotka-Volterra models are direct special cases of GMA systems in which each product consists of one or two factors with exponents of 1. At the same time, GMA systems do *not* cover a larger domain of responses. In fact, it has been shown that, surprisingly, GMA systems, S-systems, Lotka-Volterra systems, and other canonical systems are one-to-one equivalent (e.g., Peschel and Mende, 1986; Voit and Savageau, 1986): every system that can be represented in one of these forms can also be represented exactly in any of the other forms. Furthermore, virtually any system of ordinary differential equations can be represented in any of these forms (e.g., Peschel and Mende, 1986; Savageau and Voit, 1987). One might add that all these forms are composed of a combination of differentiation, linear functions, and logarithmic functions. Lotka-Volterra systems describe the change in the logarithm of a variable as a linear function in all variables (as demonstrated in an earlier section), whereas GMA and S-systems result from linearization in logarithmic space.

Monte Carlo Simulations

As mentioned before, two ingredients are the key to efficient Monte Carlo simulations: an efficient solver for the underlying model, especially if it is a dynamical model, and an efficient method for obtaining random numbers to be selected from a predescribed statistical distribution. Both features are realized by canonical power-law models (cf., Voit et al., 1995).

Because the mathematical structure of GMA and S-systems is invariable, it has been possible to devise specific computer programs for their analysis. In particular, it was shown that derivatives of the right-hand sides can be computed recursively, which suggested the implementation of a very robust and efficient variable-order, variable-stepsize Taylor algorithm for integration (Irvine and Savageau, 1990). This algorithm was included in a comprehensive, interactive software package that executes numerical solutions, steady-state analyses, including stability, sensitivities, and gains, as well as data management, and graphing (Voit et al., 1989). Because of the homogeneous mathematical structure of the canonical models, this software program does not require compiling. This greatly facilitates repeated computations, as they are necessary for Monte Carlo simulations.

The obtainment of random numbers is greatly facilitated by the use of S-distributions of the form

$$\frac{dF}{dX} = \alpha(F^g - F^h) \tag{37}$$

where X is the random variable and F the cumulative distribution function (Voit, 1992b; Voit and Yu, 1994). These one-variable S-systems have been shown to model essentially all univariate distributional shapes with sufficient accuracy. Furthermore, it is easy to compute quantiles by inverting the system, i.e., by evaluating

$$\frac{dX}{dF} = \frac{1}{\alpha(F^g - F^h)} \tag{38}$$

This is important, since obtaining random numbers from a nonuniform distribution is most efficiently based upon the computation of quantiles.

This method of data analysis and Monte Carlo simulation based on S-systems is currently being applied in our laboratory to human exposure to mercury from eating fish.

Traditional Models vs. Canonical Models

Nature has not provided the biomathematician with instructions. There are uncounted processes and systems, but there are no guidelines for how to analyze them in the most appropriate way. This lack of objective guidance has naturally led to an "anything goes" mentality that pervades today's biomathematics and mathematical modeling. This is not all bad, since biomathematics is a relatively young enterprise, and a collection of good professional practices, which will eventually emerge, requires a comprehensive body of theoretical and practical experience, which is currently being accumulated.

The selection of a model for a given purpose traditionally uses criteria of former biological success and mathematical simplicity. Evaluation and quality control follow in the form of repeated application, successful or unsuccessful. If a model passes the muster of time, there is no reason to replace it.

This argument is certainly valid when we juxtapose traditional models and canonical models: If a traditional model has proven successful under relevant conditions, there is no need to explore alternate canonical models. The real advantage of the latter comes into play when information does not suffice to construct an *ad hoc* model, or when traditional models fail.

An analog is the use of linear models in engineering. There is no question that the percentage of linear phenomena among all processes is minimal. Nevertheless, "canonical" linear systems have been used in engineering probably more than any other mathematical structure, partly because of their mathematical elegance, and partly because no sufficient information was available to set up definite nonlinear models.

The same is true for canonical power-law models. If one really knows the underlying processes of a phenomenon and can represent them mathematically, there is no real need for alternatives. However, in most situations in biology, and even more so in environmental risk assessments, it is impossible to list all contributing factors, let alone quantify them sufficiently to allow mathematical modeling. In these situations, canonical modeling is very helpful (cf., Voit and Sands, 1996).

The models can be constructed from the topology of processes, which lists which system components affect each other. All parameters have a clearly defined meaning, and their values can either be estimated from data, or one can use the model in a manner that in engineering is sometimes called *order-of-magnitude modeling*. In the latter case, one acknowledges that parameter values are only known to be within certain ranges, sometimes as wide as one or two orders of magnitude, and the aim is to evaluate the model under this high degree of uncertainty. Since many features of canonical models, in particular S-system models, can be evaluated analytically, such coarse modeling goals can be pursued in an ideal fashion. Canonical models are not a panacea for insufficient data in environmental assessment, but they do provide advantages that are unmatched by most *ad hoc* models.

CONCLUSION

This chapter has attempted to show that biomathematical modeling may play an important role in integrating ecological and human risk assessments. Two areas were shown to be particularly amenable: extrapolation (including scaling) and systems analysis. Extrapolations are necessary at several crucial steps in human risk assessment, yet current knowledge about underlying mechanisms leaves much to be desired. It appears that much could be learned from extrapolations and scaling between closely related, as well as distant, animal species or groups. For instance, comparative analyses between laboratory and free-living individuals of the same animal species can be expected to provide valuable insights in physiological parameter variations and sensitivities, and thus in the importance of the physiological status of an organism.

Systems analysis could aid ecological and human risk assessments at various fronts. Biochemical and metabolic systems models have the potential of explaining in detail how agents affect an organism and why different species may respond differently to comparable exposures. Similar models are good candidates for quantifying and explaining the dynamics of biomarkers, whether they are used as diagnostic tools of exposure or in toxicity testing. Dynamic models are arguably the only hope if we are to understand the interactions within food webs. This is relevant not only in ecological risk assessment but also in human exposure assessment, especially if humans are top predators.

The use of canonical models was suggested here, especially for situations in which the governing processes of a system are not well understood. In contrast to typical *ad hoc* models, a general mass flow pattern, combined with overall regulatory information, is often sufficient in canonical modeling to gain new insights. The analysis of important characteristics like stability and sensitivities is facilitated, and the common structure of these models makes them uniquely suited for comparative analyses, whether they address human or ecological health.

REFERENCES

Adolph, E.F. Quantitative relations in the physiological constitutions of mammals. *Science* 109, pp. 579–585, 1949.

Arnold, S.F., D.M. Klotz, B.M. Collins, P.M. Vonier, L.J. Guilette Jr., and J.A. McLachlan. Synergistic activation of estrogen receptor with combinations of environmental chemicals. *Science* 272, pp. 1489–1491, 1996.

Bear, J. *Hydraulics of Groundwater*. McGraw Hill, New York, 1979.

Barnthouse, L.W. Issues in ecological risk assessment: the CRAM perspective. *Risk Anal.* 14, pp. 251–256, 1994.

Borden, R.C. and P.B. Bedient. Transport of dissolved hydrocarbons influenced by reaeration and oxygen-limited biodegradation: 1. Theoretical development. *Water Resour. Res.* 22, pp. 1973–1982, 1986.

Brown, K.G. and D.G. Hoel. Modeling time-to-time data: Analysis of the ED_{01} study. *Fundam. Appl. Toxicol.* 3, pp. 458–469, 1983.

Burger, J. How should success be measured in ecological risk assessment? The importance of predictive accuracy. *J. Toxicol. Environ. Health* 42, pp. 367–376, 1994.

Burger, J. and M. Gochfield. Temporal scales in ecological risk assessment. *Arch. Environ. Contam. Toxicol.* 23, pp. 484–488, 1992.

Burns, L.A. and G.L. Baugham. Fate modeling, in *Fundamentals of Aquatic Toxicology*, Rand, G.M. and S.R. Petrocelli, Eds., Hemisphere Publ., New York, 1985.

Cairns, J.C., Jr. The genesis of biomonitoring in aquatic ecosystems. *Environ. Prof.* 12, pp. 169–176, 1990.

Chen, C.W. and J.N. Blancato. Role of pharmacokinetic modeling in risk assessment: Perchloroethylene as an example, in *Pharmacokinetics in Risk Assessment. Drinking Water and Health*, Volume 8, NRC, National Academy Press, Washington, DC, pp. 369–384, 1987.

Clark, T., K. Clark, S. Paterson, D. Mackay, and R.J. Norstrom. Wildlife monitoring, modeling, and fugacity. *Environ. Sci. & Technol.* 22, pp. 120–127, 1988.

Colborn, T. The wildlife/human connection: Modernizing risk decisions. *Environ. Health Perspect.* 102 (Suppl. 12), pp. 55–59, 1994.

Cox, D.R. Regression models and life tables (with discussion). *J. R. Stat. Soc. B.* 34, pp. 187–220, 1972.

Curto, R., E.O. Voit, A. Sorribas, and M. Cascante. Mathematical models of purine metabolism in man. Submitted, 1998a.

Curto, R., E.O. Voit, A. Sorribas, and M. Cascante. Validation and steady-state analysis of a power-law model of purine metabolism. *Biochem. J.,* in press, 1998b.

Curto, R., E.O. Voit, A. Sorribas, and M. Cascante. Analysis of abnormalities in purine metabolism leading to gout and to neurological dysfunctions in man. *Biochem. J.,* in press, 1998c.

Eakin, T. Intrinsic time scaling in a survival analysis: Application to biological populations. *Bull. Math. Biol.* 56, pp. 1121–1141, 1994.

EPA (Environmental Protection Agency). *Assessment and Remediation of Contaminated Sediments Program: Risk Assessment and Modeling Overview Document.* Great Lakes National Program Office, Chicago, IL. EPA/905/R93/007, 1993a.

EPA. *Interim Report on Data and Methods for the Assessment of 2,3,7,7-Tetrachlorodibenzo-p-Dioxin (TCDD) Risks to Aquatic Organisms and Associated Wildlife.* EPA/600/R-93/055, 1993b.

EPA. *Ground-Water and Vadose Zone Models/Manuals.* Obtainable from: USEPA RSKERL/CSMoS, P.O. Box 1198, Ada, OK 74820, 1994.

Freeze, R.A. and R.B. Cherry. *Ground Water.* Prentice Hall, Englewood Cliffs, NJ, 1979.

Hallenbeck, W.H. *Quantitative Risk Assessment for Environmental and Occupational Health.* Lewis Publishers, Boca Raton, FL, 1993.

Harvey, T., K.R. Mahaffey, S. Velazquez, and M. Dourson. Holistic risk assessment: An emerging process for environmental decisions. *Regul. Toxicol. Pharmacol.* 22, pp. 110–117, 1995.

Hayton, W.L. Pharmacokinetic parameters for interspecies scaling using allometric techniques. *Health Phys.* 57, pp. 159–164, 1989.

Henderson, R. Species differences in metabolism of 1,3-butadiene. 5th International Symp. Biol. Reactive Intermediates, Munich, Germany, 1995.

Hemond, H.F. and F.J. Fechner. *Chemical Fate and Transport in the Environment.* Academic Press, San Diego, CA, 1994.

Hill, C.M., R.D. Waight, and W.G. Bardsley. Does any enzyme follow the Michaelis-Menten equation? *Mol. Cell. Biochem.* 15, pp. 173–178, 1977.

Hlavacek, W.S. and M.A. Savageau. Rules for coupled expression of regulator and effector genes in inducible circuits. *J. Mol. Biol.* 255, pp. 121–139, 1996.

Huggett, R.J. *Modelling the Human Impact on Nature. Systems Analysis of Environmental Problems.* Oxford University Press, Oxford, 1993.

Irvine, D.H. and M.A. Savageau. Efficient solution of nonlinear ordinary differential equations expressed in S-system canonical form. *SIAM J. Numer. Anal.* 27, pp. 704–735, 1990.

Janszen, D.B. APB-PK: An interactive physiologically-based pharmacokinetic modeling and simulation system," Doctoral Dissertation, Medical University of South Carolina, Charleston, SC, 1992.

Jiang, Q. and P.A. Succop. A study of the specification of the log-log and log-additive models for the relationship between blood lead and environmental lead. *J. Agric. Biol. Environ. Stat.* 1, pp. 426–434, 1996.

Karr, J.R. Biological integrity: A long-neglected aspect of water resource management. *Ecol. Appl.* 1, pp. 66–84, 1991.

Kelsey, J.L., W.D. Thompson, and A.S. Evans. *Methods in Observational Epidemiology.* Oxford University Press, New York, 1986.

Kooijman, S.A.L.M. A safety factor for LC_{50} values allowing for differences in sensitivity among species. *Water Res.* 21, pp. 269–276, 1987.

Krishnan, K., H.J. Clewell, III, and M.E. Andersen. Physiologically based pharmacokinetic analyses of simple mixtures. *Environ. Health Perspect.* 102 (Suppl. 9), pp. 151–155, 1994.

Lindstedt, S.L. Allometry: Body size constraints in animal design, in *Pharmacokinetics in Risk Assessment. Drinking Water and Health*, Volume 8, NRC, National Academy Press, Washington, DC, pp. 65–79, 1987.

Lotka, A.J. *Elements of Physical Biology.* Williams and Wilkins, Baltimore, 1924; reprinted as *Elements of Mathematical Biology*, Dover, NY, 1956, p. 465.

Mackay, D. and S. Paterson. Mathematical models of transport and fate, in *Ecological Risk Assessment*, Suter, G.W., III, Ed., Lewis Publishers, Boca Raton, FL, 1993.

Mill, T. Environmental chemistry, in *Ecological Risk Assessment*, Suter, G.W., II, Ed., Lewis Publishers, Boca Raton, FL, 1993.

Moolgavkar, S.H. and D.J. Venzon. Two event models for carcinogenesis: Incidence curves for childhood and adult tumors. *Math. Biosci.* 47, pp. 55–77, 1979.

Needham, J. *Biochemistry and Morphogenesis.* Cambridge University Press, Cambridge, 1942.

Ni, T.-C. and M.A. Savageau. Application of biochemical systems theory to metabolism in human red blood cells. *J. Biol. Chem.* 271, pp. 7927–7941, 1996a.

Ni, T.-C. and M.A. Savageau. Model assessment and refinement using strategies from biochemical systems theory: Application to metabolism in human red blood cells. *J. Theor. Biol.* 179, pp. 329–368, 1996b.

National Research Council (NRC). *Risk Assessment in the Federal Government: Managing the Process.* National Academy Press, Washington, DC, 1983.

Olden, K. and J.-L. Klein. Environmental health science research and human risk assessment. *Mol. Carcinog.* 14, pp. 2–9, 1995.

Peschel, M. and W. Mende. *The Predator-Prey Model: Do We Live in Volterra World?* Akademie-Verlag, Berlin, 1986.

Rifai, H.S., P.B. Bedient, and J.T. Wilson. BIOPLUME model for contaminant transport affected by oxygen limited biodegradation. USEPA Environmental Research Brief EPA/600/M-89/019, August, 1989.

Ryan, P.B. An overview of human exposure modeling. *J. Exp. Anal. Environ. Epidemiol.* 1, pp. 453–474, 1991.

Savageau, M.A. The behavior of intact biochemical control systems. *Curr. Top. Cell. Regul.* 6, pp. 63–129, 1972.

Savageau, M.A. *Biochemical Systems Analysis. A Study of Function and Design in Molecular Biology.* Addison-Wesley, Reading, Massachusetts, 1976.

Savageau, M.A. Growth of complex systems can be related to the properties of their underlying determinants. *Proc. Nat. Acad. Sci.* 76, pp. 5413–5417, 1979a.

Savageau, M.A. Allometric morphogenesis of complex systems: Derivation of the basic equations from first principles. *Proc. Nat. Acad. Sci. USA.* 76, pp. 6023–6025, 1979b.

Savageau, M.A. Growth equations: A general equation and a survey of special cases. *Math. Biosci.* 48, pp. 267–278, 1980.

Savageau, M.A. A suprasystem of probability distributions. *Biometrics* 24, pp. 323–330, 1982.

Savageau, M.A. Critique of the enzymologist's test tube, in *Fundamentals of Medical Cell Biology*, Vol. 3A, Bittar, E.E., Ed., JAI Press, Greenwich, CT, pp. 45–108, 1992.

Savageau, M.A. and E.O. Voit. Recasting nonlinear differential equations as S-systems: A canonical nonlinear form. *Math. Biosci.* 87, pp. 83–115, 1987.

Serio, G. Two-stage stochastic model for carcinogenesis with time-dependent parameters. *Stat. Probab. Lett.* 2, pp. 95–103, 1984.

Shiraishi, F. and M.A. Savageau. The tricarboxylic acid cycle in *Dictyostelium discoideum*. I. Formulation of alternative kinetic representations. *J. Biol. Chem.* 267, pp. 22912–22918, 1992a.

Shiraishi, F. and M.A. Savageau. The tricarboxylic acid cycle in *Dictyostelium discoideum*. II. Evaluation of model consistency and robustness. *J. Biol. Chem.* 267, pp. 22919–22925, 1992b.

Shiraishi, F. and M.A. Savageau. The tricarboxylic acid cycle in *Dictyostelium discoideum*. III. Analysis of steady state and dynamic behavior. *J. Biol. Chem.* 267, pp. 22926–22933, 1992c.

Shiraishi, F. and M.A. Savageau. The tricarboxylic acid cycle in *Dictyostelium discoideum*. IV. Resolution of discrepancies between alternative methods of analysis. *J. Biol. Chem.* 267, pp. 22934–22943, 1992d.

Shiraishi, F. and M.A. Savageau. The tricarboxylic acid cycle in *Dictyostelium discoideum*. Systemic effects of including protein turnover in the current model. *J. Biol. Chem.* 268, pp. 16917–16928, 1993.

SOT (Society of Toxicology) Task Force. Re-examination of the ED_{01} study. *Fundam. Appl. Toxicol.* 1, pp. 26–128, 1981.

Staffa, J.A. and M.A. Mehlman. *Innovations in Cancer Risk Assessment.* Pathotox Publ. Inc., Park Forest South, IL, 1979.

Stern, A.C., R.W. Boubel, D.B. Turner, and D.L. Fox. Transport and dispersion of air pollutants, in *Fundamentals of Air Pollution*, 3rd ed., Academic Press, Orlando, FL, 1994.

Suter, G.W., III. *Ecological Risk Assessment*, Lewis Publishers, Boca Raton, FL, 1993, p. 538.

van Veen, M.P. A general model for exposure and uptake from consumer products. *Risk Anal.* 16, pp. 331–338, 1996.

Voit, E.O. Dynamics of self-thinning plant stands. *Ann. Bot.* 62, pp. 67–78, 1988.

Voit, E.O., Ed. *Canonical Nonlinear Modeling. S-System Approach to Understanding Complexity.* Van Nostrand Reinhold, New York, 1991.

Voit, E.O. Selecting a model for integrated biomedical systems, in *Biomedical Modeling and Simulation*, Eisenfeld, J., M. Witten, and D.S. Levine, Eds., Elsevier Science Publishers, Amsterdam, 1992a.

Voit, E.O. The S-distribution. A tool for approximation and classification of univariate, unimodal probability distributions. *Biometrical J.* 34, pp. 855–878, 1992b.

Voit, E.O. S-system modelling of complex systems with chaotic input. *Environmetrics*, 4, pp. 153–186, 1993.

Voit, E.O., W.L. Balthis, and R.A. Holser. Hierarchical Monte Carlo modeling with S-distributions: Concepts and illustrative analysis of mercury contamination in king mackerel. *Environ. Int.* 21, pp. 627–635, 1995.

Voit, E.O., D.H. Irvine, and M.A. Savageau. *The User's Guide to ESSYNS.* Medical University of South Carolina, Charleston, 1989.

Voit, E.O. and R.G. Knapp. Derivation of the linear-logistic model and Cox's proportional hazard model from a canonical system description. *Statistics in Medicine,* 16, pp. 1705–1729, 1997.

Voit, E.O. and P.F. Rust: Tutorial. S-system analysis of continuous univariate probability distributions. *J. Stat. Comput. Simul.* 42, pp. 187–249, 1992.

Voit, E.O. and P.J. Sands. Modeling forest growth. I. Canonical approach. *Ecol. Modell.* 86, pp. 51–71, 1996.

Voit, E.O. and M.A. Savageau. Equivalence between S-systems and Volterra-systems. *Math. Biosci.* 78, pp. 47–55, 1986.

Voit, E.O. and S. Yu. The S-distribution. Approximation of discrete distributions. *Biometrical J.* 36, pp. 205–219, 1994.

Volterra, V. *Leçons sur la Théorie Mathématique de la Lutte pour la Vie.* Gauthier-Villars, Paris, 1931.

Weber, E.C., W.G. Deutsch, D.R. Bayne, and W.G. Seesock. Ecosystem-level testing of a synthetic-pyrethroid insecticide in aquatic mesocosms. *Environ. Toxicol. Chem.* 11, pp. 87–105, 1992.

White, J. The allometric interpretation of the self-thinning rule. *J. Theor. Biol.* 89, pp. 475–500, 1981.

Part 3
Measurement

7 ||| Estimating Human and Ecological Risks from Exposure to Radiation

Thomas G. Hinton

INTRODUCTION

The term *risk analysis* is now used routinely in the environmental sciences, with risk being defined as the probability of a deleterious effect. It is the quantitative probability component (e.g., a 1 in 10,000 chance of an event occurring) that separates a risk analysis from a more qualitative assessment of effects. Our abilities to conduct full-scale risk analyses differ greatly depending on the contaminants and organisms of concern. Indeed, most ecological risk analyses lack a probability statement and, therefore, are more accurately defined as assessments.

The primary goal of this chapter is to demonstrate some techniques for calculating both human and ecological risks from exposure to radiation. A few fundamental concepts about radiation are presented first, and then the reader is guided through two examples: a risk analysis for a hypothetical resident of a radionuclide-contaminated Superfund site, and a risk analysis of an exposed fish population. The scientific rigor inherent in the two risk analyses is compared and, although both examples are for radiation exposures, results of the comparison are not atypical for other environmental contaminants.

Exposure to radiation provides a good example because radiation is one of the most studied carcinogens. According to Eisenbud (1973), "...it is not unfair to say that more is known about this subject than is known about the effects of any other of the many noxious agents which man has introduced into his environment." A large database exists, due in part to the tremendous resources that were made available from the 1950s through the 1970s to study the environmental transport of radionuclides and their effects on humans and biota. The research was driven largely by the need to understand the effects of global fallout from nuclear weapons. Fortunately, the threat of nuclear war has greatly diminished. However, risks from nuclear accidents remain: the legacy of Chernobyl continues to be written as our knowledge of its effects expands. Risk analyses are also needed for making decisions about the cleanup of Department of Energy sites contaminated during the weapons production era. This chapter contrasts the scientific rigor, and thus the general level of confidence, inherent in analyses of human versus ecological risks from radiation exposure.

SOME KEY CONCEPTS RELATED TO RADIATION

Concentration, Activity, and Half-Life

Concentrations of radioactive contaminants are not expressed in the same manner as other contaminants. Nonradioactive contaminants are generally expressed on a mass basis (e.g., mg Hg kg^{-1} tissue), whereas concentrations of radioactive contaminants are expressed on an activity basis (e.g., Bq ^{90}Sr kg^{-1} tissue). The base unit of activity is the becquerel (Bq), equal to one disintegration per second.[1] A disintegration is the event in which a radioactive element releases photons or particles to gain stability.[2] Thus, activity is a measure of the rate at which a given quantity of radioactive material is emitting radiation.[3]

Radionuclides[4] have characteristic half-lives ($T_{1/2}$) and related decay constants (λ). The time required for one-half of the original material to decay is termed a half-life. At the end of two half-lives, 25% of the original activity is still present. Radioisotopes with very short half-lives are particularly unstable and decay quickly. The fraction of radioactive atoms (N) that will decay per unit time (t) is termed the decay constant (λ):

$$\frac{dN}{dt} = \lambda N \tag{1}$$

and is related to half-life by:

$$T_{1/2} = \frac{0.693}{\lambda} \tag{2}$$

Ionization Leads to Effects

The energetic rays and particles released during radioactive decay cause ionization[5] in the exposed biological material, resulting in some probability (i.e., risk) of molecular or genetic damage, either directly or through a multistep process. Part of the ionization pro-

---

[1] The traditional unit of activity (Curie) has been replaced in the International System of units with the Bq; 1 Ci = 3.7×10^{10} Bq.

[2] Radionuclides are unstable because they have an excess amount of energy. They gain stability by releasing their excess energy in the form of electromagnetic photons (X-rays or gamma rays) or particles (alpha, beta, or neutrons). The process of releasing energy is termed radioactive decay. All elements above bismuth (atomic number=83) are naturally unstable (radioactive), and a few of the lighter elements have one or more radioactive isotopes.

[3] Radiation is the electromagnetic photon or particle (alpha, beta, or neutron) emitted.

[4] Radionuclide, radioisotope, and radioactive contamination are used interchangeably and have similar meanings within the context of this chapter.

[5] If the radiation has sufficient energy to eject one or more orbital electrons from the atom or molecule that it interacts with, then the process is referred to as ionization. Ionizing radiation is characterized by a large release of energy (approximately 33 eV per event), which is more than enough to break strong chemical bonds; for example, only 4.9 eV are required to break a C=C bond (Hall, 1978).

cess often involves the formation of free radicals.[6] Free radicals are a principal cause of damage from exposure to radiation because they can easily break chemical bonds within DNA molecules. The OH• free radical is among the most common, as it is formed when cellular water is ionized.

Ionization causes biological damage, most of which is repaired. However, some probability exists, depending on the dose received, that deleterious effects can occur. For humans, the biological consequence upon which most risk calculations are based, following exposure to radiation, is cancer (NCRP, 1993). The health effects from cancer and genetic disorders are classified as stochastic. That is, the initiation of effects is probabilistic and the risk of incurring either is proportional, without threshold, to the dose in the relevant tissue. The severity of a stochastic health effect is independent of the dose. Nonstochastic effects (acute radiation syndrome, opacification of the eye lens, erythema of the skin and temporary impairment of fertility), in contrast, are dependent on the magnitude of the dose in excess of a threshold (ICRP, 1977).

Dosimetry

The derivation of risk factors—or slope factors if the Environmental Protection Agency (EPA) terminology is used—requires detailed knowledge about dose-response relationships. For nonradioactive contaminants, dose is expressed as a concentration (e.g., mg of Pb kg^{-1} tissue). For radioactive contaminants, however, dose goes beyond concentration and refers to the energy absorbed within biological tissues as a result of the radiation emitted during radioactive decay (i.e., 1 joule of energy kg^{-1} tissue = 1 gray).[7]

The advanced state of human radiation dosimetry is illustrated by the progressive improvement in methods to quantify dose. The fundamental unit, *absorbed dose rate* (Gy d^{-1}), can be calculated from a given radionuclide concentration using the general equation:

$$\begin{matrix} \text{Dose} \\ \text{Rate} \\ \text{(Gy d}^{-1}) \end{matrix} = \begin{matrix} \text{concentration} \\ \text{(Bq kg}^{-1}) \end{matrix} \times \begin{matrix} \text{Specific} \\ \text{eff. energy} \\ \text{(MeV dis}^{-1}) \end{matrix} \times \begin{matrix} 1.602 \times 10^{-13} \\ \text{(J MeV}^{-1}) \end{matrix}$$

$$\times \frac{\text{dis}}{\text{s Bq}} \times \frac{86{,}400\text{s}}{\text{d}}$$

(3)

---

[6] Free radicals are atoms or molecules with an unpaired or odd orbital electron. In addition to orbiting around the nucleus, electrons also spin, either clockwise or counterclockwise, about their own axis. In atoms or molecules with an even number of electrons, spins are matched; for every electron spinning clockwise, another is spinning counterclockwise. This leads to a high degree of chemical stability. Molecules or atoms with an odd number of electrons—free radicals—have an electron that is left with an unmatched spin, creating instability.

[7] Energy released from a radionuclide is measured in electron volts, often expressed as million electron volts (MeV). The energy absorbed by biological tissues is measured in grays. The traditional unit for absorbed dose was the rad. It has been replaced in the International System of units by the gray; 100 rad = 1 Gy.

where the radionuclide specific effective energy is found in reference tables (ICRP, 1983), and "dis" represents a disintegration. Equation 3 essentially converts the energy released during radioactive decay into energy absorbed by biological tissues. The energy absorbed is dependent on the concentration of the radionuclide and the energy that specific radionuclide emits upon disintegration.

The gray unit, however, is not entirely adequate to relate the amount of radiation absorbed to its effects. Specific characteristics of the exposed tissues and characteristics of the different types of radiation determine the effectiveness of absorbed energy at causing biological damage. For example, a 1 MeV alpha particle creates approximately 20 times more damage over a given distance than a 1 MeV gamma ray because more of the alpha particle's energy is transferred, per micron distance traveled, into surrounding tissues. This relative biological effectiveness of various types of radiation has been incorporated into quality factors that, if multiplied by absorbed dose, estimate *dose equivalent*, expressed in the SI units of sievert, Sv (ICRP, 1977; 1991).[8] Dose equivalent normalizes the different types of radiation to the same propensity for biological damage. Dose equivalent has no analog for chemical carcinogens. There is no common basis, similar to the quality factors used in radiation dosimetry, for relating biological effectiveness of one nonradioactive chemical to another.

A further improvement in radiation dosimetry was made in 1977 (ICRP) when the concept of *effective dose equivalent* was introduced. Effective dose equivalent weights the sum of dose equivalents to different organs or tissues, because a uniform irradiation of the whole body causes effects that differ from those initiated by a similar dose concentrated in specific tissues (as happens when some radionuclides are ingested, e.g., ^{131}I goes directly to the thyroid gland). Thus, the fractional contribution of organs and tissues to the total risk of stochastic health effects is normalized to when the entire body is uniformly irradiated (ICRP, 1991). Exposures with equal effective dose equivalents are assumed to result in equal risks, regardless of the distribution of the deposited energy among different body tissues.

The last consideration made when calculating dose involves the time period of the exposure. Because a one-time intake of radionuclides commits an individual to a future dose, dose rates are integrated over a 50 y period for adults and 70 y for children. The integration accounts for the time it takes for the contaminant to be removed from the body by biological elimination and radioactive decay.

The final product of all quality factors, organ weighting factors, and integration of dose over the individual's working lifetime results in the *committed effective dose* (CED). The CED is a rigorous attempt to quantify radiation exposure by not only specifying the amount of energy absorbed by tissues, but by accounting for numerous factors that influence the biological effectiveness of that energy at causing damage. From this perspective, radiation dosimetry is more quantitative and advanced than current methods used to determine dose from nonradioactive contaminants.

The CED is the fundamental unit by which risk factors are multiplied to estimate the probability of an individual human acquiring a fatal cancer from exposure to radiation (NCRP, 1993). There are numerous pathways for which a CED can be calculated, and these are presented in the next section.

---

[8] Dose equivalent was expressed traditionally in rem (Roentgen Equivalent Man), 100 rem = 1 Sv.

RISKS TO HUMANS FROM EXPOSURE TO RADIATION

Example: Health Risks to a Hypothetical Resident of a Contaminated Site

The multistep process of calculating risks to humans is perhaps best illustrated with an example. Whicker et al. (1993) provide an illustrative one. Their data concern a 10.5 km² cooling reservoir (Par Pond) on the U.S. Department of Energy's Savannah River Site (SRS) near Aiken, South Carolina. The reservoir was built in 1958 as the primary component of a thermal cooling loop for two nuclear production reactors. Accidental releases from the reactors contaminated Par Pond with fission products in the late '50s and '60s. In the summer of 1991, the water level of Par Pond was lowered 5.8 m so that repairs could be made to the dam. The drawdown exposed 5.3 km² of contaminated sediments and the exposed lake bed was declared a Superfund site by the Environmental Protection Agency. Whicker et al. (1993) calculated the radiation exposure and carcinogenic risk to hypothetical residents on the contaminated lake bed. The general procedure requires determining the dose received by the hypothetical resident from various pathways of exposure (e.g., ingestion of contaminated food, inhalation of contamination, and external irradiation). The cumulative dose from all pathways was determined, and when multiplied by a risk factor derived from accepted dose-response relationships, predicted the lifetime probability of a resident acquiring a fatal cancer.

Whicker et al. (1993) based much of their analyses on two years of site-specific data collected from the CERCLA site, including the uptake of radionuclides in vegetable crops grown directly on the exposed lake bed. Assumptions for the hypothetical resident of the lake bed (such as quantities of water, meat, milk, and vegetables consumed) were taken from the EPA guidance document for Superfund sites (EPA, 1991). This document provides "reasonable maximum exposure" default recommendations for residential and agricultural scenarios in terms of residence times, annual occupancy, and ingestion and inhalation rates. Calculations based on these assumptions provided conservative yet reasonable dose estimates. Whicker et al. (1993) state that *maximum worst case* exposures would be higher than the "reasonable maximum exposure," in part because the authors chose to use mean concentrations of radionuclides in soil, water, and dietary components within their calculations. Had they chosen the 95% percentile for input parameters, the predicted doses would have been correspondingly greater. The authors bound their calculations by performing an additional analysis using scenario assumptions less conservative than the "reasonable maximum exposure" recommended by the EPA. The authors termed these predictions "reasonable average exposures." Inputs for this scenario were based subjectively on the authors' judgment, and were believed to better represent the long-term, self-sufficient resident farmer. The general steps used by the authors to calculate risks are systematically presented below.

Ingestion

As a first step, mean radionuclide concentrations were determined in numerous environmental components within the lake and on the exposed sediments, including concentrations in garden crops grown directly on the contaminated lake bed (Table 7.1). The committed effective dose from consuming such contaminated food was then calculated:

Table 7.1. Mean Concentrations of Radionuclides Within the Par Pond Ecosystem (adapted from Whicker et al., 1993).

Component	^{137}Cs	^{90}Sr	^{239}Pu
Water (Bq L^{-1})	0.13	0.02	4×10^{-6}
Sediments (Bq cm^{-3})	2.0	0.17	0.02
Macrophytes (Bq kg^{-1} dry)	700	28	0.009
Largemouth bass (Bq kg^{-1} wet tissue)	470	6.3	0.002
Native terrestrial plants (Bq kg^{-1} dry)	2,300	24	0.004
Garden vegetables (Bq kg^{-1} dry)[a]	2,300	24	0.004
Garden vegetables (Bq kg^{-1} dry)[b]	970	10	0.0004
Feral pig muscle (Bq kg^{-1} wet)[c]	930	[d]	[d]
Whitetailed deer muscle (Bq kg^{-1} wet)[c]	300	[d]	[d]

[a] No K-Ca fertilizer applied.
[b] K-Ca fertilizer applied and subsequent uptake in vegetables was reduced.
[c] Data obtained from Dr. I.L. Brisbin, Savannah River Ecology Laboratory.
[d] Not analyzed.

$$
\begin{array}{l}
\text{Ingestion} \\
\text{CED} \\
\text{(mSv)}
\end{array}
=
\begin{array}{l}
\text{Concentration} \\
\text{in food} \\
(\text{Bq kg}^{-1})
\end{array}
\times
\begin{array}{l}
\text{Ingestion} \\
\text{rate} \\
(\text{kg d}^{-1})
\end{array}
\times
$$

$$
\begin{array}{l}
\text{Exposure} \\
\text{frequency} \\
(\text{d y}^{-1})
\end{array}
\times
\begin{array}{l}
\text{Effective} \\
\text{exposure} \\
\text{duration} \\
(\text{y})
\end{array}
\times
\begin{array}{l}
\text{Ingestion} \\
\text{dose factor} \\
(\text{mSv Bq}^{-1})
\end{array}
\qquad (4)
$$

Ingestion rates used by Whicker et al. (1993) in Equation 4 for the "reasonable maximum" and "reasonable average" exposures, as well as other standard assumptions, are listed in Table 7.2. Risk calculations generally require standardized data for numerous assumptions related to dietary consumption patterns, breathing rates, exposure frequency, and exposure duration for various levels of occupancy. These are located in guidance documents for Superfund sites (EPA, 1991), and other pertinent references (NCRP, 1989; NCRP, 1984; Pao et al., 1982; Yang and Nelson, 1984).

The *effective exposure duration* within Equation 4 accounts for radioactive decay by:

$$
\int_0^t e^{\lambda t}\, dt \qquad (5)
$$

where t is the residence time, and λ is the radioactive decay constant.

The ingestion *dose conversion factor* (Equation 4) converts the ingested activity to a committed effective dose per unit intake, and accounts for the fractional uptake from the small intestine to blood of the common chemical forms of the radionuclides, as well as tissue distribution and effective retention time. Dose conversion factors for each radionuclide are found in Eckerman et al. (1988) and Eckerman and Ryman (1993) for the path-

Table 7.2. Assumptions Used to Calculate Dose for a Hypothetical Resident on the Contaminated Lake Bed (adapted from Whicker et al., 1993).

Parameter	Reasonable Maximum Exposure[a]	Reasonable Average Exposure[b]
Water intake (L d^{-1})	2	0
Inadvertent soil intake (mg d^{-1})	100	50
Inhalation rate (m^3 d^{-1})	20	15
Vegetable and fruit intake (g d^{-1})	122	60
Meat intake (g d^{-1})	75	40
Milk intake (L d^{-1})	0.3	0.15
Fish intake (g d^{-1})	132	33
Annual occupancy (d)	350	335
Time outdoors (h d^{-1})	12	4
Length of residence (y)	30	15
Effective length of residence (y)[c]	22	13
Use of K/Ca fertilizer for garden	no	yes

[a] Data from EPA (1991).
[b] Data based on the authors' judgment of a less conservative scenario.
[c] Accounts for physical decay of ^{137}Cs and ^{90}Sr.

ways of ingestion, inhalation, submersion in a contaminated plume, submersion in contaminated water, and external irradiation from contaminated soils. These authors have taken the primary guides[9] for radiation protection, issued by the International Commission on Radiological Protection (ICRP, 1977) and EPA (1987), and derived dose conversion factors (Eckerman et al., 1988).

Inhalation

Dose from inhaling contaminated soil was calculated by Whicker et al. (1993) using a resuspension factor of 1×10^{-5} in Equation 6.

$$
\begin{array}{l}
\begin{matrix} \text{Inhalation} \\ \text{CED} \\ \text{(mSv)} \end{matrix}
=
\begin{matrix} \text{Inventory} \\ \text{in soil} \\ \text{(Bq m}^{-2}) \end{matrix}
\times
\begin{matrix} \text{Resuspension} \\ \text{factor} \\ \text{(m}^{-1}) \end{matrix}
\times \\[2em]
\begin{matrix} \text{Annual} \\ \text{inhalation} \\ \text{(m}^3 \text{ y}^{-1}) \end{matrix}
\times
\begin{matrix} \text{Effective} \\ \text{exposure} \\ \text{duration} \\ \text{(y)} \end{matrix}
\times
\begin{matrix} \text{Inhalation} \\ \text{dose} \\ \text{factor} \\ \text{(mSv Bq}^{-1}) \end{matrix}
\end{array}
\qquad (6)
$$

---

[9] Primary guides are the current recommendations for radiological protection (e.g., the committed effective dose for a radiation worker in a given year should not exceed 50 mSv).

External Irradiation

In addition to the pathways of contaminant transfer already discussed, external irradiation has to be considered with most forms of radioactive contamination. This pathway is generally not needed for nonradioactive contamination or that involving alpha emitters. Whicker et al. (1993) actually measured external irradiation using sophisticated field instruments. In the absence of site-specific field data, calculating the external dose to humans exposed to radioactively contaminated soil, water column, or plume of air is possible using external dose coefficients provided in Eckerman and Ryman (1993). Equation 7 could be used in situations involving contaminated soil.

$$
\begin{array}{c}
\text{External} \\ \text{CED} \\ (\text{mSv})
\end{array} =
\begin{array}{c}
\text{Activity} \\ \text{in soil} \\ (\text{Bq m}^{-3})
\end{array} \times
\begin{array}{c}
\text{Exposure} \\ \text{frequency} \\ (\text{s y}^{-1})
\end{array} \times
$$

$$
\begin{array}{c}
\text{Effective} \\ \text{exposure} \\ \text{duration} \\ (\text{y})
\end{array} \times
\begin{array}{c}
\text{Indoor} \\ \text{shielding} \\ \text{factor} \\ (\text{unitless})
\end{array} \times
\begin{array}{c}
\text{Soil dose} \\ \text{coefficient} \\ (\text{mSv} - \text{m}^3 / \text{Bq} - \text{s})
\end{array} \tag{7}
$$

Whicker et al. (1993) used an indoor shielding factor of 0.5 to account for time spent indoors, where the irradiation from the contaminated soil would be reduced.

Similar calculations can be performed for exposure from a contaminated plume or immersion in contaminated water. Neither pathway had significant source terms in this scenario and thus they were not considered by Whicker et al. (1993).

Additional data were needed by Whicker et al. (1993) because the calculations were for a hypothetical self-sufficient resident. For example, parameters were needed to estimate the concentration of contaminants in milk and meat consumed by humans. Data in Table 7.3 were used to derive such information.

Each of the calculations presented above results in an estimate of the committed effective dose for that particular pathway of contaminant exposure. Effective dose equivalents for all pathways, and all radioactive contaminants, were summed to yield a total CED of 51 mSv (Table 7.4) for an adult with a 30 y residence time on the contaminated sediments of the exposed lake bed. The adult is assumed to be self-sufficient and produces sustenance directly from the contaminated area—the reasonable maximum exposure scenario. The less conservative, "reasonable average exposure" scenario resulted in a total CED of 12 mSv (Table 7.4). Recall that the CED accounts for the different types of radiation emitted from the various radionuclides, and normalizes the dose for numerous factors mentioned earlier in the dosimetry section, including accounting for physical and biological decay of the contaminants.

A dose distribution table (Table 7.4) reveals differences among the radionuclide contaminants. Strontium contributed least to the total dose. This was primarily due to the much lower concentration of ^{90}Sr within the Par Pond ecosystem when compared to ^{137}Cs (Table 7.1). Activities of ^{137}Cs and ^{90}Sr were released in similar amounts, but now—35 years later—^{137}Cs is 10 to 100 times more abundant because of strontium's greater mobility within the environment; much of the ^{90}Sr has passed through the Par Pond system, whereas

Table 7.3. Additional Data Used by Whicker et al. (1993) in Estimating Dose to a Hypothetical Resident of a Radioactively Contaminated Site.

| Parameter | Radionuclide Specific | | | Generic |
	^{137}Cs	^{90}Sr	^{239}Pu	
Resuspendable soil activity (Bq m^{-2})	2900	25	5.2	
Milk transfer coefficient (d L^{-1})[a]	8x10^{-3}	1x10^{-3}	1x10^{-7}	
Meat transfer coefficient (d kg^{-1})[a]	2x10^{-2}	1x10^{-3}	1x10^{-5}	
Feeding rate, milk cow (dry kg d^{-1})[b]				16
Feeding rate, beef cow (dry kg d^{-1})[b]				12
Dose factor, ingestion (mSv Bq^{-1})[c]	1.4x10^{-5}	3.9x10^{-5}	9.6x10^{-4}	
Dose factor, inhalation (mSv Bq^{-1})[c]	8.6x10^{-6}	3.5x10^{-4}	1.2x10^{-1}	
Indoor shielding factor				0.5

[a] Transfer coefficients estimate the amount of activity incorporated into milk or edible meat per unit intake of contamination in the animal feed. Multiplying the transfer coefficient (d L^{-1}) times the ingestion rate (kg d^{-1}) and activity in the animal feed (Bq kg^{-1}) yields the contamination in the human consumable milk or meat product (Bq kg^{-1} meat or Bq L^{-1}).
[b] Standardized value from NCRP (1989).
[c] Eckerman et al. (1988).

Table 7.4. Committed Effective Dose (mSv) for Hypothetical, Self-Sufficient Resident on the Par Pond Superfund Site (adapted from Whicker et al., 1993).

Exposure Pathway	^{137}Cs	^{90}Sr	^{239}Pu	Totals
Reasonable Maximum Exposure				
External gamma	22.22	nc[a]	nc	22.22
Inhalation	0.038	0.014	1.31	1.36
Ingestion				
Soil	0.008	< 0.001	0.004	0.01
Water	0.028	0.012	< 0.001	0.05
Fish	6.69	0.25	0.003	6.94
Vegetables	4.04	0.12	< 0.001	4.16
Meat	6.69	0.007	< 0.001	6.70
Milk	9.51	0.035	< 0.001	9.54
Total Ingestion	<u>26.97</u>	<u>0.42</u>	<u>0.007</u>	<u>27.40</u>
Total All Pathways	49.23	0.43	1.32	50.99
Reasonable Average Exposure				
External gamma	4.88	nc[a]	nc	4.88
Inhalation	0.016	0.006	0.470	0.49
Ingestion				
Soil	0.002	< 0.001	0.001	0.003
Water	0.0	0.0	0.0	0.0
Fish	0.95	0.035	< 0.001	0.98
Vegetables	0.48	0.0014	< 0.001	0.49
Meat	2.02	0.002	< 0.001	2.02
Milk	2.69	0.010	< 0.001	2.70
Total Ingestion	<u>6.14</u>	<u>0.061</u>	<u>0.001</u>	<u>6.20</u>
Total All Pathways	11.04	0.07	0.47	11.58

[a] Not considered because of negligible contribution.

^{137}Cs has a greater tendency to bind to clay particles in the sediments and remain closer to the point of discharge.

The negligible contribution of ^{90}Sr and ^{239}Pu to external irradiation (Table 7.4) is due to the type of radiation each predominantly releases: ^{90}Sr is a beta emitter and ^{239}Pu emits alpha particles, neither of which are capable of appreciably penetrating the skin. The gamma emissions from ^{137}Cs, however, are a major contributor to external dose (22 mSv), providing almost one-half of the dose from all pathways and all radionuclides combined (51 mSv).

The relative importance of the radiation types is switched when inhalation is of concern: alpha emissions from ^{239}Pu dominate the dose contribution (1.3 mSv), while the emissions from ^{137}Cs and ^{90}Sr are almost inconsequential (0.04 and 0.01 mSv, respectively). This is because alpha particles are very effective at causing biological damage within a distance of a few microns. If inhaled and deposited on sensitive lung tissues, alpha particles are by far the most destructive. In the Par Pond scenario, however, the inhalation pathway contributes least to total dose (1.36 mSv), even with the contribution from ^{239}Pu.

Having obtained a dose, the authors were then confronted with determining its relevance. Are doses of 51 and 12 mSv dangerous to the resident? Is some type of control or remediation of the contaminated site called for? Relevance can be gauged through risk analysis. Thus, a final computation was needed in which the total CED is multiplied by an appropriate risk factor to obtain the probability of harmful effects. Derivation of those risk factors is the subject of the next section.

Derivation of Risk Factors

Human epidemiological data form the basis for determining the probability of deleterious effects from exposure to radionuclides. In contrast, risks from exposure to most chemical carcinogens are largely extrapolated from laboratory experiments using nonhuman subjects (EPA, 1989a). The latter involves a greater degree of uncertainty.

Groups of exposed individuals from which the effects of radiation have been observed include uranium miners occupationally exposed to radon gas; individuals medically treated with acute partial-body X-rays to counter the effects of various illnesses; and painters of luminous dials on clocks and watches. The latter inadvertently ingested radium-laden paint when they orally moistened their brushes to a fine point. By far, however, the most informative dose-response data come from studies of the Japanese atomic-bomb survivors. A very meticulous reconstruction of each individual's physical position has been performed relative to ground zero, taking into consideration the presence of buildings and other structures that shielded the individual from the radiation at the time of the blast. This has allowed dose estimates to be made for individuals that are now being followed for lifetime health consequences.

The average dose to the Japanese cohort of 75,991 individuals was 0.12 Gy.[10] The ongoing study of the Japanese population has revealed that 344 of the 5,936 cancer deaths have

---

[10] Epidemiological studies use the base unit of absorbed dose (Gy) because they are interested in accurately quantifying the dose to specific individuals, from which dose-response relationships can be obtained. The additional uncertainties associated with quality and weighting factors inherent in the unit of dose equivalent (Sv) would decrease the precision of the dose estimate. In

been attributed to radiation. A significant dose-related increase in genetic effects has not been observed in the Japanese population (NCRP, 1993). Such epidemiological studies have resulted in today's consensus that induction of cancer is the predominant risk from exposure to radiation rather than genetic effects (NCRP, 1993). EPA (1989b) concurs that the risk of cancer may be used as the sole basis for assessing the radiation-related human health risks for a site contaminated with radionuclides.

Currently Accepted Risk Factors

The recommended risk factors, furnished by the International Commission on Radiological Protection, are derived from the dose-response relationships obtained in the epidemiological studies presented above. The ICRP (1991) has quantified risks and provided risk factors for four categories:

Detriment	Risk Factor (mSv^{-1})
Fatal cancer	5.0×10^{-5}
Severe genetic effects	1.3×10^{-5}
Nonfatal cancer	1.0×10^{-5}
Total Detriment	7.3×10^{-5}

Recall that cancer incidence is the accepted endpoint for human risk analyses of exposure to radiation. EPA slope factors differ slightly from the ICRP recommendations, primarily in that the incidence of cancer is the endpoint of interest in the EPA slope factors, rather than the probability of fatality from a cancer as used by the ICRP (morbidity vs. mortality).

Returning to the Par Pond example, and using:

$$Risk = Committed\ Effective\ Dose \times Risk\ Factor \qquad (8)$$

Whicker et al. (1993) quantitatively estimated the lifetime probability that a subsistence resident would acquire a fatal cancer from the radiation exposure (using the "reasonable maximum exposure"):

$$Risk = 51\ mSv \times 5{\times}10^{-5}\ mSv^{-1} = 0.00255$$

That is, a probability of 2.6 in 1000, or odds of 1 in 385. Thus, a village of self-sufficient farmers on the exposed lake bed with 3850 residents would be expected to experience 10 excess lifetime cancer fatalities due to the contamination. Under the "reasonable average exposure" the probability decreases to 0.0006 (6 additional cancers per 10,000 residents).

Such quantitative analyses can now be used to help manage the contaminated site. Risk managers would compare the estimated risks to those deemed acceptable by the EPA (i.e., 1 in 10,000; Figure 7.1). The need for some type of remediation action is apparent if self-sufficient farming is the chosen long-term land use for the contaminated Par Pond sediments. Using the tools of risk analysis, risk managers could alter the parameters used in the

---

contrast, risk analyses and calculations involving radiation protection are concerned with estimating doses to groups of people in general terms, rather than actual doses received by individuals from real intakes of radionuclides.

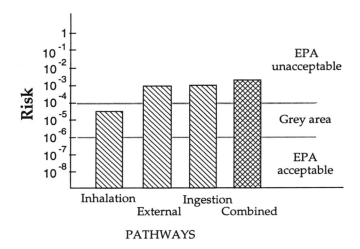

Figure 7.1. Risk of developing fatal cancer to a 30 y self-sufficient resident of the contaminated Par Pond lake bed (adapted from Whicker et al., 1993).

risk scenario (such as residency time, dietary consumption patterns, etc.) and determine risks for situations other than self-sufficient residency. Risk analyses could also be used to choose remediation strategies, and to evaluate the success of decontamination efforts. The strength of risk analysis as a management tool is in being able to quantify the probability of specific events, and importantly, that the probability component is derived from rigorous scientific data.

ECOLOGICAL RISK ANALYSIS

Example: Risk to a Fish Population from Radioactive Releases

Let's now contrast an ecological risk analysis to what was presented above. Again, an example will be illustrative. The example is particularly pertinent because risk to a fish population was estimated using the same contaminant source term and within the same drainage system used above to illustrate a human risk analysis. The data come from Hinton and Whicker (1997), where they estimated the probability of harmful effects to fish from historic releases of radionuclides on the SRS. Their goals were to (1) examine past radionuclide releases into aquatic systems on the SRS, (2) use a screening model to conservatively estimate the absorbed dose rate fish might have received from chronic exposure to such releases, and (3) estimate the probability of damage to the fish population as a result of that exposure.

Hinton and Whicker (1997) adapted a screening model approach to accomplish their goals. A screening model uses conservatively chosen input parameters that cause the dose estimate to be unrealistically high. If dose levels estimated using this approach were inadequate to cause concern, then it is probable that the actual doses received by the fish were inconsequential, because the real doses were less than the conservatively obtained model predictions. If, however, doses approach levels at which injuries have been documented, then a reevaluation and more precise calculation may be necessary.

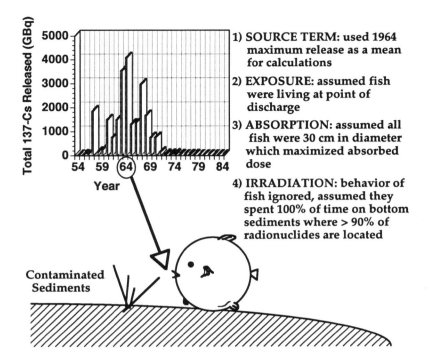

Figure 7.2. Conservative assumptions used in the screening model to assure that the predicted dose to fish was larger than the real dose they experienced (adapted from Hinton and Whicker, 1997).

The authors used four conservative assumptions in their screening model to maximize the dose estimate to the fish population (Figure 7.2): (1) Maximum contaminant concentrations were used. Initial calculations indicated that the largest contributor to fish dose was exposure from ^{137}Cs. Maximum ^{137}Cs discharges occurred in 1964, when a failed fuel rod resulted in the release of 3.1 TBq (83 Ci) into the drainage of Lower Three Runs Creek, Pond B, and the previously mentioned Par Pond (Carlton et al., 1992). Hinton and Whicker (1997) used the 1964 ^{137}Cs releases, combined with those for ^{3}H, ^{60}Co, ^{90}Sr, ^{131}I, and ^{239}Pu, to maximize the source term and to estimate a cumulative impact. (2) Fish were assumed to live at the point of discharge. (3) The biology of the fish was ignored, and Hinton and Whicker (1997) assumed that the fish lived 100% of the time on the sediment-water interface. This maximized the fish dose, because over 90% of the contaminants become attached to the sediments, and living on the sediment would substantially increase external irradiation. (4) Fish were assumed to be large, round spheres of 30 cm diameter. This maximized the probability that the incoming gamma ray would interact within the body of the fish, thereby maximizing the dose. A higher percentage of gamma photons would pass through thin, fusiform-shaped fish without causing damage.

To estimate dose to the fish, Hinton and Whicker (1997) chose BIORAD, a model recently used by the National Council on Radiation Protection and Measurements (NCRP, 1991) and the International Atomic Energy Agency (IAEA, 1992) to analyze the effects of ionizing radiation on aquatic organisms. BIORAD requires an equilibrium concentration of radionuclides in water (Bq L^{-1}) as the starting point for subsequent calculations. To esti-

Table 7.5. 1964 Radionuclide Releases into the Lower Three Runs Drainage. Initial Concentrations are Based on 1964 Water Discharges Added to Normal Stream Flow (3.2 x 10^10 L). Dividing these Initial Concentrations by the Inventory Ratios (IR) from Pond B Produced Estimates of 1964 Water Concentrations after Equilibration with Sediments. The Equilibrated Values were Used in Subsequent Dose Calculations (adapted from Hinton and Whicker, 1997).

Radionuclide	Annual Release (Bq)	Initial Conc. (Bq L^{-1})	IR	Conc. @ Equil. (Bq L^{-1})
^{3}H[a]	6.0 x 10^{14}	1.8 x 10^4	0.0	1.8 x 10^4
^{60}Co[b]	8.8 x 10^{10}	2.7 x 10^0	141.0	0.02
^{90}Sr	2.2 x 10^{11}	6.9 x 10^0	6.9	1.00
^{131}I[c]	—	—	—	3.77
^{137}Cs	3.1 x 10^{12}	9.7 x 10^1	150.0	0.65
^{239}Pu[d]	—	—	—	0.01

[a] Tritium is not sorbed to soil particles and remains soluble (Murphy et al., 1991).

[b] Data from Pond B were not available, the IR for ^{60}Co was estimated from ^{137}Cs's K_d and IR relationship. $[IR]_{Co} = ([K_d]_{Co} / [K_d]_{Cs}) [IR]_{Cs}$

[c] A maximum concentration measured in 1962 from Steel Creek on the SRS was used (Kantelo et al., 1991) because 1964 release data were not available.

[d] A maximum concentration reported for Four Mile Creek on the SRS was used (Holloway and Hayes, 1981) because early '60s data were not available and only gross alpha was reported.

mate equilibrium concentrations, Hinton and Whicker (1997) used the 1964 release rates, adjusting them to account for the high sorption affinity of specific radionuclides to sediments. The sorption correction was derived from existing inventory ratios (IR) in Pond B, a well-studied aquatic system within the same drainage as the 1964 release (Whicker et al., 1990). The IR is the total activity in the sediments divided by the total activity within the water column. Data for IRs and the resulting sediment-equilibrated water concentrations ($C_{w(equil)}$) are presented in Table 7.5.

Dose to Fish from Ingestion

To estimate the dose to fish from ingesting contaminated food, the BIORAD model applies a concentration factor to the radionuclide levels in the water column and then accounts for the effective absorbed energy per disintegration for each radionuclide (Turbey and Kaye, 1973). The calculations performed by the BIORAD model have been reduced to dose conversion factors by the IAEA (1992; Table 7.6), and the resulting dose from ingestion is found in Table 7.7.

Dose to Fish from External Exposure (water)

The dose rate to fish from external irradiation (D_E) within the water column is derived assuming an effectively infinite, uniformly contaminated source (NCRP, 1991).

$$D_E(\mu Gy\ h^{-1} / Bq\ L^{-1}) = 5.76 \times 10^{-4} \times E_m \qquad (9)$$

Table 7.6. Distribution Coefficients (K_d) and Decay Energies of Radionuclides used in Estimating External Exposure to Fish from Contaminated Sediments. K_d for ^{90}Sr, ^{137}Cs and ^{239}Pu are from Whicker et al. (1990). Dose Conversion Factors are also Given for the Internal and External Exposure of Fish (mGy y^{-1} per Bq L^{-1}; Turbey and Kaye, 1973; IAEA, 1992; adapted from Hinton and Whicker, 1997).

| | | | | Dose Conversion Factors | |
Radionuclide	K_d (L kg^{-1})	β Energy (MeV)	γ Energy (MeV)	Internal Exposure	External Water (β + γ)
^{3}H	0	5.7×10^{-3}	0.0	5.1×10^{-5}	1.6×10^{-5}
^{60}Co	30,000	9.7×10^{-2}	2.5×10^0	1.5×10^{-1}	1.3×10^{-2}
^{90}Sr	1,200	1.1×10^0	0.0	2.7×10^{-2} (a)	2.7×10^{-3}
^{131}I	200	1.9×10^{-1}	3.8×10^{-1}	3.2×10^{-2}	2.5×10^{-3}
^{137}Cs	32,000	2.5×10^{-1}	5.6×10^{-1}	1.2×10^0 (a)	3.2×10^{-3}
^{239}Pu	5.9×10^6	6.7×10^{-3}	8.0×10^{-4}	9.5×10^{1}(a)	2.2×10^{-5}

[a] These generic dose conversion factors were multiplied by 150, 21, and 0.34 for ^{90}Sr, ^{137}Cs, and ^{239}Pu, respectively, to incorporate site-specific concentration factors from Pond B (Whicker et al., 1990).

Table 7.7. Total Dose (mGy y^{-1}) to Fish from Internal Exposure Due to the Ingestion of Contaminated Food and Water, Plus External Irradiation from the Water and Sediments (adapted from Hinton and Whicker, 1997).

Radionuclide	Internal Exposure	External Exp. Water (β + γ)	Ext. Exp. Sediment (β)	Ext. Exp. Sediment (γ)	Total
^{3}H	9.97	3.13	0.0	0.0	13.10
^{60}Co	0.003	0.0003	0.15	3.8	3.95
^{90}Sr	4.05	0.0027	3.3	0.0	7.35
^{131}I	0.12	0.01	0.36	0.73	1.22
^{137}Cs	16.13	0.002	12.9	28.9	57.93
^{239}Pu	0.32	2.2×10^{-7}	1.0	0.12	1.44
Combined	30.59	3.14	17.71	33.55	84.99
Grand Total = 84.99 mGy y^{-1} (**0.23 mGy d^{-1}** or 0.023 rad d^{-1})					

where E_m is the mean energy of all radiation emitted (MeV dis^{-1}). Due to their limited range, alpha particles were assumed not to contribute to external dose, and the D_E for beta particles was reduced by 50%. Conversion factors are presented in Table 7.6, with resultant doses in Table 7.7.

Dose to Fish from External Exposure (Sediments)

The BIORAD model does not estimate exposure of fish via irradiation from contaminated sediments. Hinton and Whicker (1997) followed the approach used by the NCRP (1991) and the IAEA (1992) and assumed that the concentration of the radionuclides in the sediments could be determined using their distribution coefficients (K_d), a measure of the partitioning of an element between the sediments and overlying water column. They used site-specific K_d from Pond B when possible (Whicker et al., 1990). The dose rate at the sediment-water interface was estimated by:

$$D_S (\text{mGy y}^{-1}) = 2.52 \times 10^{-3} \times C_{w(equil)} \times K_d \times E_{total} \qquad (10)$$

where $E_{(total)}$ is the total energy of the decay modes from the radionuclides. Values for $C_{w(equil)}$ are found in Table 7.5; K_d and $E_{(total)}$ data are in Table 7.7.

Dose Rate Summary

Table 7.7 summarizes the dose rates received by fish. Combining the six radionuclides considered, across all pathways of exposure, resulted in a dose rate of 0.23 mGy d^{-1}. The major contributor to internal exposure was ^{137}Cs (16 mGy y^{-1}), followed by ^{3}H (10 mGy y^{-1}). Tritium was also the primary source of external irradiation from contaminants in the water column, contributing three orders of magnitude more than those radionuclides that tend to sorb to the sediments. ^{137}Cs was the largest contributor to external irradiation from the sediments (40 mGy y^{-1}), an order of magnitude greater than the next larger contributor (^{60}Co). If dose rates from all radionuclides are combined, the contributions from internal contamination and external exposure from the sediments are similar. Recall, however, that the behavior of the fish is not considered in this conservative approach and the model assumes that the fish spends 100% of its time on the sediment surface where irradiation would be maximum (Figure 7.2).

It was interesting that the use of site-specific concentration factors from Pond B increased the dose rate to the fish by a factor of 150 for ^{90}Sr and 21 for ^{137}Cs (footnote Table 7.6). The increased availability of the radionuclides is due largely to the low K and Ca concentrations in the water (Whicker et al., 1990). The abundance of K and Ca, chemical analogs of ^{137}Cs, and ^{90}Sr, respectively, affects the radionuclides' bioavailability. The resulting change in dose rates also shifted the relative importance of the radionuclides away from ^{3}H and toward ^{137}Cs. The data of Hinton and Whicker (1997) demonstrate the unusual case where a generic—falsely assumed to be conservative—model underestimated dose and points to the importance of collecting site-specific data.

Risk Computation

Next, Hinton and Whicker (1997) needed to determine the significance of a 0.23 mGy d^{-1} dose rate to the fish population. Had this been a human risk analysis, they would have simply applied a risk factor to the estimated dose and obtained the probability of the human acquiring a fatal cancer. To my knowledge, however, there are no such risk factors for nonhumans. Lacking risk factors, the authors evaluated the significance of their predicted 0.23 mGy d^{-1} dose rate to studies in the literature, including reviews by the NCRP (1991) and the IAEA (1992) that documented specific effects.

Hinton and Whicker (1997) considered two endpoints: fish mortality and physiological effects, such as decrease in response of the immune system or reduced testicular size. The latter are sublethal endpoints that may nevertheless have an impact on fish populations by altering mortality or natality rates. Their comparisons are summarized in Figure 7.3 for the two endpoints. The dose rates received by fish in the SRS watershed, even with the very conservative assumptions used, seem to be at least an order of magnitude lower than those documented to have caused mortality (Figure 7.3a). Most important, because of their overly conservative assumptions, the predicted dose was much greater than the actual doses received by the fish. Therefore, Hinton and Whicker (1997) were confident that the actual doses were insufficient to cause fish mortality.

These findings corroborate the NCRP (1991) conclusion that "deleterious effects of chronic irradiation have not been observed in natural populations at dose rates ≤ 10 mGy d^{-1} over the entire history of exposure to ionizing radiation." Similarly, the IAEA (1992) recommended that limiting the dose rate to the maximally exposed individual in an aquatic population to ≤ 10 mGy d^{-1} would provide adequate protection to the population. That level of exposure is considerably greater than the dose rate calculated for fish on the SRS (0.23 mGy d^{-1}); thus there is a very low probability that fish mortality occurred from historic releases of radionuclides, since the conservatively maximized dose rate was well below 10 mGy per day.

The dose rate to SRS fish was also below those observed to have caused physiological damage, the second endpoint considered (Figure 7.3b). However, unlike the mortality endpoint, the studies referenced for these effects were still documenting damage at the lowest range of exposures tested, which did not extend down to the low rate encountered by the SRS fish. Thus, data were inconclusive as to whether SRS fish might have experienced harmful physiological effects.

Our ability to judge the significance of radionuclide exposures to nonhuman organisms is currently limited by the lack of data relating dose to the probability of harmful effects. The absence of risk factors is a fundamental problem in ecological risk analyses. Until those data are available, a screening model approach may be of value. The approach, however, has limitations. A screening model is only viable when the predicted dose is considerably less than doses at which harmful effects have been documented. This was the case in Hinton and Whicker's (1997) study for the endpoint of mortality (Figure 7.3a). However, the screening model approach failed when the conservatively predicted dose approached doses at which effects had been observed, as was the case in their study for the endpoint of physiological damage. They could not say with certainty that physiological effects had not occurred. This example illustrates three fundamental problems of quantitatively estimating the probability of deleterious effects to nonhuman species exposed to radioactive contaminants: (1) the difficulty in accurately determining dose, (2) problems of interpreting data when sublethal endpoints are used, and (3) the lack of risk factors makes it difficult to make quantitative statements concerning the probability of specific effects. The latter two are not unique to radioactive contaminants, but are generic to ecological risk analyses. Indeed, it is questionable if the term "risk" should even be used, as the science is not sufficiently developed to attach a probability statement to the analysis; "ecological assessment" may be a better description.

RELATIVE UNCERTAINTIES IN HUMAN AND ECOLOGICAL RISK ANALYSES

Quantifying uncertainty in environmental dose and risk calculations is essential to the decisionmaking process (NCRP, 1996). Uncertainty analysis quantifies the level of confi-

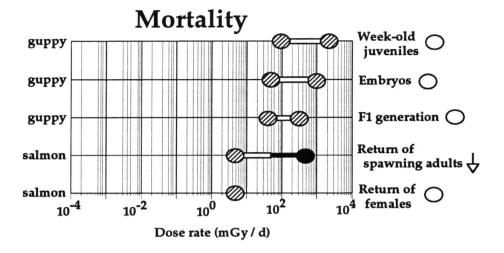

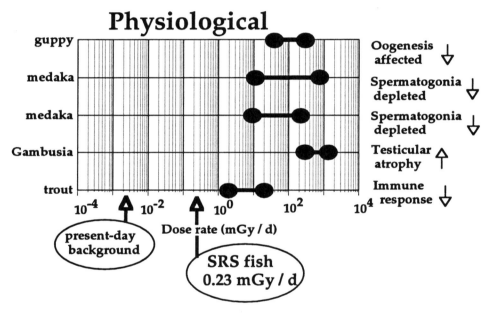

Figure 7.3. Documented mortality and physiological effects from chronic exposure to radionuclide contamination (NCRP, 1991; IAEA, 1992). The left side of the y-axis lists the organisms studied, with the corresponding endpoint on the right side. Dose rate comprises the logarithmic-scaled x-axis. The connecting circles show the range of dose rates examined for each study. The dose rate at which an effect was observed is highlighted in black, compared to the hatched circles indicative of dose rates at which effects were not observed. This is reinforced by the symbols following the endpoints on the right side of the y-axis; an oval indicates that effects were not found over the range of doses studied, and arrows indicate the direction of response from observed effects. Additionally, the dose rate estimated for fish exposed to the 1964 SRS release is indicated, as well as the present average background (adapted from Hinton and Whicker, 1997).

dence about the prediction of dose or risk. The amount of confidence in a risk analysis can be estimated from a number of different techniques (Hoffman and Gardner, 1983; Iman and Helton, 1988; IAEA, 1989; Ferson and Kuhn, 1992; NRC, 1995; NCRP, 1996). A rather simple means of estimating uncertainty is to bound the true but unknown risks within a range generated from screening level calculations (NCRP, 1996), as was done by Whicker et al. (1993) for the hypothetical resident on the exposed sediments of Par Pond. They calculated a risk based on two scenarios that ranged in conservatism from a "reasonable maximum exposure" to a "reasonable average exposure." Their approach estimated a range of possible risk values. Because the results indicated that some kind of remedial action was required (the risk estimates were greater than those accepted by the EPA; Figure 7.1), and the remediation alternatives could cost several millions of dollars, a more detailed uncertainty analysis was warranted.

A common method of acquiring detailed uncertainty estimates is through Monte Carlo simulations (Rubinstein, 1981). The technique requires estimating the range and probability distribution for each input parameter. For example, rather than simply using a fish consumption parameter of 132 g d^{-1} (Table 7.2), Whicker et al. (1993) would have to determine the minimum and maximum fish consumption rates, as well as the type of distribution the data fall into (e.g., normal, lognormal, triangular; NCRP, 1996). This would have to be done for each input parameter. The model (e.g., the model used to estimate ingestion dose, Equation 4) would be run numerous times, each time with a suite of different input parameters chosen from their possible distributions. The Monte Carlo simulation is the driver that chooses the individual parameter values for each model run. At the end of several hundred runs a suite of possible answers exists, from which the most probable one, with an associated confidence level or uncertainty, can be obtained. Uncertainty analysis allows one to express results using confidence statements such as "I am 95% confident that the true risk falls within the range of 3.2×10^{-5} to 4.7×10^{-4}." Achieving such quantitative results is expensive and time-consuming, but possible with human risk analyses because ranges and distributions can be determined for the entire spectrum of needed input parameters, including the dose conversion and risk factors. The confidence intervals are subjective, however, because extensive use of judgment is required to quantify uncertainty in both the parameters and the structure of the model (NCRP, 1996).

Uncertainties in human risk analyses arise from incomplete knowledge about epidemiology, dosimetry, transferring results gathered from one population (Japanese) to another (U.S.) and extrapolation of results from high-dose to low-dose regions (Sinclair, 1993). Cumulative uncertainty of a factor of 10 to 20 would not be unreasonable in a risk analysis conducted for human exposure to radiation.

As complex as human risk analyses are, determining risks to other environmental components is even more so, and entails much greater uncertainty. Problems in ecological risk analyses that generally contribute to a larger uncertainty than exists in human risk analyses are enumerated below:

1. *Measuring the dose received by most nonhuman species is not a trivial task.* Dose measurement instruments can be placed on animals, but recapturing the same animal and recovering the instrument is problematic. Mathematical models that predict dose to free-ranging organisms are poorly developed and have not been as well validated as have those used for human dosimetry.

2. *Tables of dose conversion factors that transform a unit concentration of contaminant to a dose received by the organism exist for only a few nonhuman species.* Notice that the fish doses presented by Hinton and Whicker (1997) were given in units of Gy rather than the committed effective dose units of Sv, as was used for the human risk analysis presented earlier. Recall that the CED normalizes the dose based on the distribution of the exposure within the human body, accounts for the relative biological effectiveness of different types of radiation, and integrates the body burden over time to account for physical and biological elimination rates. Dosimetry estimates for humans are derived from rigorous metabolic and physiological models based on chemical properties of the radionuclides, solubility factors, and clearance rates for specific organs. Dosimetry has not advanced to the same degree for nonhumans, thus the fundamental unit of absorbed dose (Gy) was used. Recently a compilation of dose conversion factors for nonhuman biota has emerged (Amiro, 1997). The work is a significant contribution to the data needed for ecological risk analyses. Amiro (1997) states, however, that the dose conversion factors "apply to generic target organisms and are most useful as conservative indicators of radiological dose; they are not designed to give exact dose estimates to specific species or body parts." Unlike human dose conversion factors, those presented by Amiro (1997) for nonhumans (1) do not account for the relative biological effectiveness of different radiation types (gamma vs alpha emissions); (2) conservatively assume that all energies emitted by radionuclides from within the organism are also absorbed by the organism, resulting in an overestimation of dose for smaller organisms; and (3) do not define the geometry of the organism. Thus, the same dose conversion factors are used on all organisms, whose diverse geometries "could range from an ant to an elephant, or from a lichen to a tree" (Amiro, 1997). The dose conversion factors are for a "generic organism," not accounting for differences in physical size, physiology, or behavior among species. The dose conversion factors presented by Amiro (1997) are useful, but it is apparent that when the disparities in detail and parameters considered are accounted for, they are far less rigorous than those calculated for humans.

3. *There are difficulties in interpreting the effects of sublethal endpoints.* Indeed, there is considerable debate over what should be the pertinent endpoint for an ecological risk analysis. The critical endpoint for humans (cancer incidence in individuals) is not considered relevant for most populations of plants and animals. A basic premise is that the main concern for nonhuman species is the population (a self-sustaining unit of a particular species) rather than specific individuals (IAEA, 1992). Ecologists are not normally as concerned that a single fish or single tree survives, but rather that the populations of these organisms remain viable. Thus, there can be effects in individuals, which, if there are no consequences at the population level, may be considered acceptable (NCRP, 1991).

Contaminants can exert their effects on all levels of biological organization, from molecules to ecosystems. Most ecotoxicological research concentrates on effects at lower levels of organization, often because of the complexity of documenting effects at the levels of population, community, and ecosystem. Also, many of the mechanisms that connect effects between individuals and populations are poorly understood; yet it is these upper population level effects that are of concern for nonhuman species. Documentation of cellular/molecular damage, for example, generally represents a sublethal endpoint that may provide good early warning signs of potential contaminant impact, but the relationships between cellular/molecular damage and consequences at upper levels of biological organization have not been made (Underwood and Peterson, 1988; Clements and Kiffney, 1994). Current environmental risk analyses that use cellular/molecular effects as endpoints are of questionable value because of the large uncertainties when speculations about the relevance to higher levels of biological organization are presented. Many cellular and molecular effects are repairable within the organism or can be removed from the population through natural selection.

4. *Most important, it appears that there are no risk factors established for nonhumans.* Thus, even if a probabilistic analysis is performed within an ecological risk analysis, as was expertly done by MacIntosh et al. (1994) for PCB and mercury contamination in mink and great blue herons, the ability to equate the body burden or dose to a probabilistic risk is severely limited. Dose-response data pertinent to nonhumans are generally inadequate to quantify the probability of deleterious effects (i.e., risks).

The problems listed above substantially increase the uncertainties in most ecological risk analyses, and illustrate the additional data required for the science to reach the rigor currently associated with human risk analyses.

CONCLUSIONS

This author believes that risk, when rigorously defined, should include a probability statement concerning the incidence of a deleterious event (i.e., a 1 in x chance of some effect occurring). The science supporting the determination of hazards to humans from radiation exposures is sufficiently developed to meet that rigorous definition, albeit, with considerable uncertainties attached. Documentation of effects from human exposures to radiation began with the discovery of X-rays in the late 1890s and has since developed into a science that crosses disciplines of nuclear physics, human physiology, chemistry, biology, and the medical professions. The uncertainties associated with risks to humans from exposure to radiation are probably less than those from many nonradioactive contaminants, and most important, they are quantifiable. This is due, in part, to the availability of human epidemiological data from which risk factors can be derived and the agreement upon a specific, well-defined endpoint (the probability of an individual developing cancer).

In contrast, the science of estimating ecological risks from radioactive contamination is much less developed. Particularly lacking are quantitative dosimetry for nonhuman organisms, calibrated endpoints, and risk factors. Without the latter, developing a quantitative statement of the probability of harm is problematic, particularly for chronic, low-level exposure conditions where sublethal effects dominate.

Risk factors are powerful tools that help managers make informed decisions about contaminated sites. They provide a common measure in which comparisons can be made. Most important, they are generated from scientific input. Many decisions regarding the management of contaminated sites will be influenced by political motivations and the public's perception of the hazard. Risk analyses are the primary mechanism by which science can enter into the decisionmaking process. It is therefore critical for risk analyses to be conducted with the utmost professionalism and scientific rigor. Anything less will result in the risk analysis process losing credibility and, thereby, the loss of scientific input into public and environmental health issues.

ACKNOWLEDGMENTS

This work was supported by Financial Assistance Award Number DE-FC09-96SR18546 from the U.S. Department of Energy to the University of Georgia Research Foundation. The critical reviews by C. Bell, J. Graves, J. Joyner, M. Malak, L. Marsh, F.W. Whicker, and an anonymous reviewer were much appreciated.

REFERENCES

Amiro, B.D. Radiological dose conversion factors for generic non-human biota used for screening potential ecological impacts. *J. Environ. Radioact.* 35, pp. 37–51, 1997.

Carlton, W.H., L.R. Bauer, A.G. Evans, L.A. Geary, C.E. Murphy, Jr., J.E. Pinder, III, and R.N. Strom. *Cesium in the Savannah River Site Environment.* Report No. WSRC-RP-92-250, Westinghouse Savannah River Company, Savannah River Laboratory, Aiken, SC, 1992, p. 97.

Clements, W.H. and P.M. Kiffney. Assessing contaminant effects at higher levels of biological organization. *Environ. Toxicol. Chem.* 13, pp. 357–359, 1994.

Eckerman, K.F. and J.C. Ryman. *External Exposure to Radionuclides in Air, Water and Soil.* Federal Guidance Report No. 12, EPA 402-R-93-081, U.S. Environmental Protection Agency, 1993, p. 235.

Eckerman, K.F., A.B. Wolbrast, and A.C.B. Richardson. *Limiting Values of Radionuclide Intake and Air Concentration and Dose Conversion Factors for Inhalation, Submersion, and Ingestion.* Federal Guidance Report No. 11, EPA-520/1-88-020, U.S. Environmental Protection Agency, 1988, p. 225.

Eisenbud, M. *Environmental Radioactivity.* Academic Press, New York, 1973, p. 542.

EPA. Radiation Protection Guidance to Federal Agencies for Occupational Exposure. *Federal Register* 52, Issue 17, pp. 2822, 1987.

EPA. *Risk Assessment Guidance for Superfund, Volume II: Environmental Evaluation Manual.* EPA 540 1-89 001, NTIS, Springfield, VA, 1989a.

EPA. *Risk Assessment Guidance for Superfund, Volume I: Human Health Evaluation Manual (Part A).* EPA 540 1-89 002, NTIS, Springfield, VA, 1989b.

EPA. *Risk Assessment Guidance for Superfund, Volume I: Human Health Evaluation Manual/ Supplemental Guidance-Standard Default Exposure Factors.* OSWER Directive 9285.6-03, Springfield, VA, 1991, p. 28.

Ferson, S. and R. Kuhn. Propagating uncertainty in ecological risk analysis using interval and fuzzy arithmetic, in *Computer Techniques in Environmental Studies IV*, Zannetti, P., Ed., Elsevier Applied Science, London, pp. 387–402, 1992.

Hall, E.J. *Radiobiology for the Radiologist.* Harper & Row Publishers, San Francisco, CA, 1978, p. 460.

Hinton, T.G. and F.W. Whicker. A screening model approach to determine probable impacts to fish from historic releases of radionuclides, in *Freshwater and Estuarine Radioecology,* Desmet, G. et al., Eds. Elsevier, London, pp. 441–448, 1997.

Hoffman, F.O. and R.H. Gardner. Evaluation of uncertainties in radiological assessment models, in *Radiological Assessment: A Textbook on Environmental Dose Analyses*, Till, J. and H.R. Meyer, Eds., U.S. NRC, Washington, DC, pp. 11.1–11.55, 1983.

Holloway, R.W. and D.W. Hayes. *Transuranics in a Stream Near a Nuclear Fuel Chemical Separations Plant.* Report No. DP-1592, E. I. du Pont de Nemours and Company, Savannah River Technology Center, Aiken, SC, 1981, p. 15.

Iman, R.L. and J.C. Helton. An investigation of uncertainty and sensitivity analysis techniques for computer models. *Risk Anal.* 8, pp. 71–90, 1988.

IAEA (International Atomic Energy Agency). *Evaluating the Reliability of Predictions Made Using Environmental Transfer Model.* IAEA Safety Series 1000, IAEA, Vienna, Austria, 1989, p. 106.

IAEA. *Effects of Ionizing Radiation on Plants and Animals at Levels Implied by Current Radiation Protection Standards.* Technical Reports Series No. 332, Vienna, Austria, 1992, p. 73.

ICRP (International Commission on Radiological Protection). *Recommendations of the International Commission on Radiological Protection.* ICRP Publication 26, Annals of the ICRP 1, Pergamon Press, New York, 1977, p. 53.

ICRP. *Radionuclide Transformations: Energy and Intensity of Emissions.* ICRP Publication 38, Annals of the ICRP, Vols. 11–13, Pergamon Press, New York, 1983, p. 1250.

ICRP. *1990 Recommendations of the International Commission on Radiological Protection.* ICRP Publication 60, Annals of the ICRP 21, Pergamon Press, New York, 1991, p. 211.

Kantelo, M.V., L.R. Bauer, W.L. Marter, C.E. Murphy, Jr., and C.C. Ziegler. *Radioiodine in the Savannah River Site Environment.* Report No. WSRC-RP-90-424-2, Westinghouse Savannah River Company, Savannah River Laboratory, Aiken, SC, 1991, p. 121.

MacIntosh, D.L., G.W. Suter, II, and F.O. Hoffman. Uses of probabilistic exposure models in ecological risk assessments of contaminated sites. *Risk Anal.* 14, pp. 405–419, 1994.

Murphy, C.E., L.R. Bauer, D.W. Hayes, W.L. Marter, C.C. Ziegler, E.E. Stephenson, D.D. Hoel, and D.M. Hamby. *Tritium in the Savannah River Site Environment.* Report No. WSRC-RO-90-424-1, Westinghouse Savannah River Company, Savannah River Laboratory, Aiken, SC, 1991, p. 172.

NCRP (National Council on Radiation Protection and Measurements). *Radiological Assessment: Predicting the Transport, Bioaccumulation, and Uptake by Man of Radionuclides Released to the Environment.* NCRP Report No. 76, Bethesda, MD, 1984, p. 304.

NCRP. *Screening Techniques for Determining Compliance with Environmental Standards.* NCRP Commentary No. 3, Bethesda, MD, 1989, p. 134.

NCRP. *Effects of Ionizing Radiation on Aquatic Organisms.* NCRP Report No. 109, Bethesda, MD, 1991, p. 148.

NCRP. *Risk Estimates for Radiation Protection.* NCRP Report No. 115, Bethesda, MD, 1993, p. 148.

NCRP. *A Guide for Uncertainty Analysis in Dose and Risk Assessments Related to Environmental Contamination.* NCRP Commentary No. 14, Bethesda, MD, 1996, p. 54.

NRC (Nuclear Regulatory Commission). *Probabilistic Accident Consequence Uncertainty Analysis.* Volumes 1–3. NUREG/CR-6244, NTIS, Springfield, VA, 1995, p. 512.

Pao, E.M., K.H. Fleming, P.M. Gueuther, and S.J. Mickle. *Food Commonly Eaten by Individuals: Amount Per Day and Per Eating Occasion.* U.S. Department of Agriculture, Springville, VA, 1982, p. 431.

Rubinstein, R.Y. *Simulation and the Monte Carlo Method.* John Wiley & Sons, NY, 1981, p. 278.

Sinclair, W.K. *Science, Radiation Protection and the NCRP.* (Lauriston S. Taylor lectures in Radiation Protection and Measurements, Lecture 17) NCRP, Bethesda, MD, 1993, p. 56.

Turbey, D.K. and S.V. Kaye. *The EXREM III Computer Code for Estimating External Radiation Doses to Populations from Environmental Release.* Report ORNL-TM-4322, Oak Ridge National Laboratory, TN, 1973, p. 82.

Underwood, A.J. and C.H. Peterson. Towards an ecological framework for investigating pollution. *Mar. Ecol. Prog. Ser.* 46, pp. 227–234, 1988.

Whicker, F.W., J.E. Pinder, III, J.W. Bowling, J.J. Alberts, and I.L. Brisbin, Jr. Distribution of long-lived radionuclides in an abandoned reactor cooling reservoir. *Ecol. Monogr.* 60, pp. 471–496, 1990.

Whicker, F.W., T.G. Hinton, and D.J. Niquette. Health risks to hypothetical residents of a radioactively contaminated lake bed, in *Proceedings of the Environmental Remediation Conference, Augusta, Georgia, 24–28 October, 1993.* U.S. Department of Energy, Savannah River Site, Aiken, SC, pp. 619–624, 1993.

Yang, Y. and C.B. Nelson. *An Estimation of the Daily Average Food Intake by Age and Sex for Use in Assessing the Radionuclide Intake of Individuals in the General Population.* EPA Report No. 520/1-84-021, Office of Radiation Programs, U.S. Environmental Protection Agency, Washington, DC, 1984, p. 30.

8 ||| Ecotoxicological Effects Extrapolation Models[1]

Glenn W. Suter II

INTRODUCTION

One of the central problems of ecological risk assessment is modeling the relationship between test endpoints (numerical summaries of the results of toxicity tests) and assessment endpoints (formal expressions of the properties of the environment that are to be protected) (Suter, 1993a; EPA, 1992). For example, one may wish to estimate the reduction in species richness of fishes in a stream reach exposed to an effluent and have only a fathead minnow 96 hr LC_{50} as an effects metric. The problem is to extrapolate from what is known (the fathead minnow LC_{50}) to what matters to the decisionmaker (the loss of fish species). Models used for this purpose may be termed effects extrapolation models (EEMs) or activity-activity relationships (AARs), by analogy to structure-activity relationships (SARs).

These models have been previously reviewed in Chapters 7 and 9 of Suter (1993a) and by an OECD workshop (OECD, 1992). This chapter updates those reviews and attempts to further clarify the issues involved in the development and use of EEMs. Although there is some overlap, this chapter does not repeat those reviews and the reader is referred to them for a more complete historical perspective, and for treatment of additional extrapolation issues.

This class of models is defined as being limited to empirical models and does not include mechanistic mathematical simulation models, which are treated in other chapters of this volume as well as in Bartell et al. (1992) and Suter (1993a). Although these empirical models are simple and often imprecise, models of this sort have in general proved to be more predictive of ecological responses than models with a more theoretical basis (Peters, 1991). However, there is a continuum from models that are purely descriptive to those that are purely theoretical. Although all of the models discussed below are derived from toxicological data by expert judgement or statistical techniques, their basis in theory ranges from none (e.g., factors), to minimal (e.g., species sensitivity distributions), to explicit (e.g.,

---

[1] Publication No. 4718, Environmental Sciences Divisions, ORNL. (Oak Ridge National Laboratory managed by Lockheed Martin Energy Research Corp. for the U.S. Department of Energy under contract number DE-AC05-96OR22464.)

allometric scaling models). Although association with theory is reassuring, it does not mean that a statistically fitted model is more true or more useful. All models should stand or fall based on their predictive power.

It would be a mistake to portray these empirically based and statistically derived models as being in competition with theoretically based and mathematically derived models. Ecological risk assessments are best conducted using multiple lines of evidence, each with its own strengths and weaknesses. Therefore, one model need not drive out another if both are credible and appropriate. In addition, empirical models may be used to derive parameter values for theoretical models (Barnthouse et al., 1990).

This chapter begins by describing different techniques for extrapolation modeling and illustrates them by showing how they have been used to extrapolate between species. Interspecies extrapolations are considered much more often than other extrapolations, and a name has even been proposed for them—quantitative species sensitivity relationships (QSSRs) (Notenboom et al., 1995). The chapter then presents brief discussions of some other extrapolations and concludes with some predictions and recommendations concerning future directions in the field.

TECHNIQUES FOR EXTRAPOLATION

Classification

The simplest way to extrapolate test endpoints is to assume that the observed responses are representative of some class of responses. For example, if the LC_{50} for fathead minnows in a particular test is x_t then we may assume that half of all fish will die at that concentration, or half of all cyprinids, or simply half of all fathead minnows, depending on how broadly we define the classes. The model is Ea = Et + e, where Ea and Et are the assessment and test endpoints, respectively.

Classification is the inevitable starting point of extrapolation modeling. That is, we must begin by deciding what differences are important enough to be modeled and which are not. If only this method is used, then we must conclude either that the equality applies and the test endpoint approximately equals the assessment endpoint, or that the species, life stage, and other specifics of the test endpoint are sufficiently different from the assessment endpoint that they belong to different classes and therefore there is no useful relationship between them.

Species may be classified on various bases. For example, plants are often classified by growth form and the EPA classifies freshwater fish as warm water and cold water species (Stephan et al., 1985). However, the most common system of species classification is taxonomy. Usually, vertebrates are effectively classified to the class level. That is, birds are lumped but separated from mammals and bony fishes are lumped but separated from amphibians. However, all invertebrates and all plants are typically lumped in assessments of both aquatic and terrestrial contaminants. In calculation of U.S. water quality criteria, LC_{50} for all members of a genus are lumped by taking their geometric mean and carrying that genus mean value through the calculations (Stephan et al., 1985). Studies based on correlations of LC_{50} of species at different taxonomic distances indicated that for both freshwater and marine fishes and arthropods, species within genera and genera within families tended to be relatively similar, which suggests that they could be lumped (Suter, 1993a; Suter and

Rosen, 1988; Suter et al., 1983). The same conclusion was reached by the same method for terrestrial vascular plants (Fletcher et al., 1990). Application of analysis of variance to acute toxicity data sets for aquatic organisms classified to the class level (Hoekstra et al., 1994) and for birds classified to family (Baril et al., 1994) found taxonomy to be a statistically significant variable.

Multivariate ordination techniques would seem to be logical for identifying clusters of species with similar sensitivities to a set of chemicals. However, a research program at the Dutch National Institute for Public Health and Environmental Protection (RIVM) that applied principal components analysis (PCA) to acute toxicity data for birds and mammals and to acute and chronic data sets for aquatic species did not find clear clusters, except for daphnids in the aquatic acute data set (Vaal et al., 1994; Van der Wal et al., 1995a; Van der Wal et al., 1995b). A PCA of avian LD_{50} for pesticides showed distinct clusters of Icteridae and Phasianidae, but not other taxa (Baril et al., 1994); however, only eight species were included.

Factors

If the test endpoint is judged not to belong to the same class as the assessment endpoint, the simplest and most common way to extrapolate between them is to multiply by a factor. The model is Ea = fEt, where f is a factor. Factors may be applied to account for uncertainties concerning the nature of the relationship between assessment and test endpoints or for biases in test endpoints as estimators of assessment endpoints. For example, if the assessment endpoint is defined in terms of some sublethal response to chronic exposures and the test endpoint is an acute LC_{50}, then a factor may be applied to account for the bias associated with the fact that the chronic sublethal responses usually occur at lower concentrations. Factors may be derived by analysis of data, but more often they are based on professional judgment. Most often they are factors of ten, which makes their imprecision apparent and aids computation. Probably the most commonly used set of factors are those developed by the EPA for use in regulation of industrial chemicals and adapted by an OECD workshop (Table 8.1).

Distributions

One may assume that the assessment endpoint is a random variate represented by the distribution of test endpoints (Figure 8.1). The most common form of this extrapolation technique is the species sensitivity distribution. One simply assumes that for a particular chemical the distribution with respect to concentration (or dose) of test endpoints for different species is an estimate of the distribution of concentrations (or doses) at which species exposed in the field will display the endpoint response. For example, if the distribution of 96 hr LC_{50} values for fish exposed to a chemical is normally distributed (m_t, s_t) then half of fish species in the field would be expected to experience mass mortality after exposure to concentration m_t for 96 hours. This approach was developed by the U.S. EPA for the calculation of water quality criteria (Stephan et al., 1985). Distributions used have included the log triangular (Stephan et al., 1985), log normal (Wagner and Lokke, 1991), and log logistic (Kooijman, 1987).

Table 8.1. Factors for Estimating Environmental Concern Levels from Different Eco-toxicological Data Sets for the U.S. EPA and OECD (OECD, 1992; Zeeman, 1995).

U.S. EPA Data Sets	OECD Data Sets	Assessment/Extrapolation Factor
Limited (e.g., only one acute LC_{50} via SAR/QSAR)	Lowest acute LC_{50}, EC_{50}, or QSAR estimate for 1–2 aquatic species	1000
Base set acute toxicity (e.g., fish and daphnid LC_{50}s and algal EC_{50})	Lowest acute LC_{50}, EC_{50}, or QSAR estimate in a set of at minimum algae, crustaceans, and fish	100
Chronic toxicity MATCs	Lowest chronic NOEC value or QSAR estimate in a set comprising at minimum algae, crustaceans, and fish	10
Field test data		1

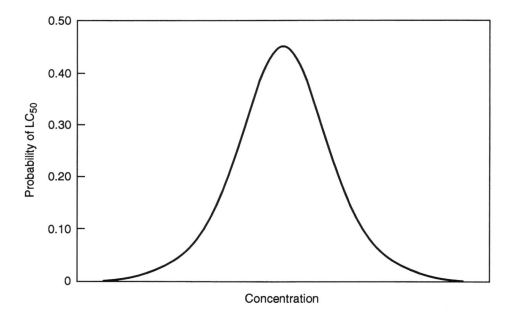

Figure 8.1. Illustration of a species sensitivity distribution for LC_{50}.

The lower fifth percentile of such distributions has been used as the basis for environmental quality standards (Figure 8.2). Because of this use, species sensitivity distributions have received much attention and criticism (OECD, 1992; Smith and Cairns, 1993). Among those who use these models, there is no controversy about the use of log transformation of exposures or the choice of the fifth percentile (no other number fills scientists with as much confidence and comfort as 5%), but differences of opinion occur on the following points.

Distribution function—In the absence of evidence that one function is better than another, functions have been chosen largely on the basis of convenience and personal prefer-

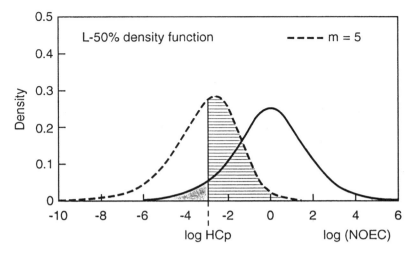

Figure 8.2. The estimated species sensitivity distribution (solid line) and the proportion of species not protected by a given HCp (shaded area for p = 5%). The probability distribution of HCp (dashed line) and the likelihood of protecting less than 9% of species at HCp (line-hatched area) when the number of species tested is five. Redrawn from Aldenberg and Slob (1993).

ence. However, the use by the U.S. EPA of the triangular distribution with its lower limit to sensitivity has been criticized by the Dutch and others who prefer distributions with potentially unlimited sensitivity (OECD, 1992).

Minimum data set—The EPA requires eight acute values; others require as few as three or as many as 20 for establishing distributions (Hoekstra et al., 1994; OECD, 1992). Also, certain species or representatives of certain taxa or groups (e.g., daphnids or cold water fish) may be required.

Inclusiveness—The EPA includes multicellular aquatic animals (Stephan et al., 1985) but others include algae and other plants as well (Aldenberg and Slob, 1993; Wagner and Lokke, 1991). Although inclusiveness seems desirable, it strains the assumption that the species are drawn from a single unimodal distribution.

Confidence in the fifth percentile—The EPA uses the most likely (median) estimate of the fifth percentile, but others have required 95% confidence of not exceeding the fifth percentile (i.e., the lower 95% tolerance limit) (Aldenberg and Slob, 1993).

Distributions may be fit and percentiles calculated by any statistical software. However, convenient software is available for this purpose, including calculation of both HC5 and its lower, one-tailed 95% confidence limit for log normal, log logistic, and log triangular distributions (Aldenberg, 1993).

Regression

If the assessment endpoint can be approximated by some test endpoint and if values of that test endpoint and the test endpoint from which we wish to extrapolate are available for several chemicals, then we may create an extrapolation model by regressing the one endpoint against the other. The linear version of the model is $Ea = a + bEt + e$. For example, if the assessment endpoint is acute lethality in rainbow trout and a fathead minnow LC_{50} is available, then a regression of rainbow trout LC_{50} against fathead minnow LC_{50} for chemicals that were tested on both species under similar conditions could be used to perform the extrapolation (Kenaga, 1978).

In that simple form, regression models can be used only to extrapolate among commonly tested species. However, clustering species in the taxonomic hierarchy permits one to predict the responses of untested as well as tested species (Suter, 1993a; Suter and Rosen, 1988; Suter et al., 1983). If the test and endpoint species are in the same genus then a species-to-species model is used; if the species belong to different genera within a family, then a regression of those two genera is used, etc. (Figure 8.3). For example, if one wishes to predict the response of largemouth bass from rainbow trout, the lowest taxonomic level that they have in common is the class *Osteichthyes,* so the extrapolation is between the orders *Perciformes* and *Salmoniformes* (Figure 8.4). In performing an interfamily extrapolation, one assumes that the relative sensitivities of the two species are represented by the relative sensitivities of the families to which they belong and that the uncertainty in that assumption is represented by the variance in the model. For North American aquatic species, most of the needed equations are available, although many possible combinations of taxa are missing (Suter, 1993a). This is because most of the missing equations are for low-level extrapolations (species within genera) while most extrapolations required for assessments are relatively high level (orders within classes). For those that are missing, it is not possible to estimate relative sensitivity or extrapolation-specific variance. However, one may assume that the weighted mean of prediction intervals from extrapolations performed at a particular taxonomic level are estimates of the uncertainty concerning any particular extrapolation at that level (Calabrese and Baldwin, 1994; Suter et al., 1987). The prediction interval is the appropriate basis for these uncertainty factors because we are interested in intervals that contain a pair of species at that taxonomic level with 95% confidence (or some other level) (Calabrese and Baldwin, 1994; Suter et al., 1987).

A third regression-based approach to taxonomic extrapolation is regression of all members of a taxon against a standard test species (Barnthouse and Suter, 1986; Holcombe et al., 1988; Suter and Rosen, 1988; Suter et al., 1987). This approach is less precise than the taxonomic hierarchy approach and does not work when the Et is for nonstandard test species, but it does estimate the sensitivity of standard species relative to larger sets of species and the uncertainty concerning predicted effects on a randomly chosen species (Suter, 1993a).

Scaling

Extrapolations can be made by assuming that the differences in sensitivity among organisms and species are differences of physical scale. The simplest and most common example of this is the expression of doses to wildlife as dose per unit mass (mg/kg), which amounts to an assumption that toxicity is a function of the dilution of the toxicant in the mass of the organism. The model in extrapolation by scaling is Ea = Et + e when both the endpoint species and test species are appropriately scaled. The formal analysis of the consequences of organism size in physiology, ecology, pharmacology, and other branches of biology is termed allometry.

The most commonly used allometric model is a power function of weight, $Ex = a\,W^b$. This form has been adopted by toxicologists because various physiological processes, including metabolism and excretion of drugs and other chemicals, are approximated by that

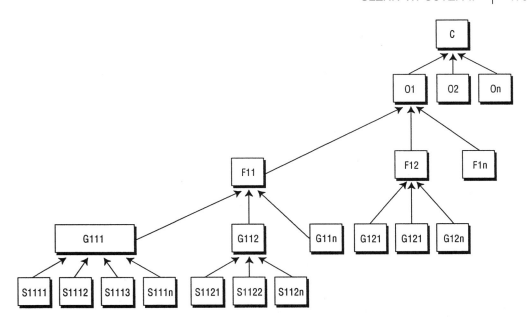

Figure 8.3. A diagrammatic representation of the use of phylogenetic relatedness to extrapolate between species. If species (Sx) are within the same genus (Gx), the model extrapolates directly between the two species. If the species are in different genera in the same family (Fx), data for all members of the two genera are aggregated and the model developed to extrapolate between genera. If they are in different families in the same order (Ox), then toxicity data are aggregated within families, etc.

form (Davidson et al. 1986; Peters, 1983). Exponents for various processes range from 0.6 to 0.8. Some ecotoxicologists followed the EPA's (EPA, 1986) practice of using 2/3 in human health risk assessments for wildlife (Opresko et al., 1993). This practice is conservative for humans and mammalian wildlife in that it makes large species such as deer more sensitive than the small rodents that are typically used in mammalian toxicity testing while making small wild species approximately equal in sensitivity. More recently, the EPA has used the less conservative 3/4 power for piscivorous wildlife (EPA, 1993b), and others have followed their lead (Sample et al., 1996). Acute mammalian toxicity data sets yield exponents that are closer to 3/4 than 2/3 on average, but are consistent with either value (Goddard and Krewski, 1992; Travis and Morris, 1992; Watanabe et al., 1992).

Little attention has been paid to allometric models for avian toxicology. However, use of the same models for birds as mammals with the same exponents was supported by allometric models of avian physiology (Peters, 1983) and pharmacology (Pokras et al., 1993). In fact, Pokras et al. (1993) present models for the extrapolation of effective doses of drugs from mammals to birds based on a common exponent of 3/4 but with a higher a value for birds. However, Mineau et al. (1996) performed allometric regression analyses on 37 pesticides with between 6 and 33 species of birds. They found that for 78% of chemicals the exponent was greater than 1, with a range of 0.63 to 1.55 and a mean of 1.1 (Figure 8.5). If a chemical has an exponent greater than 1 for birds, that implies that most avian species would be more sensitive than standard test species, such as mallard ducks and bobwhite

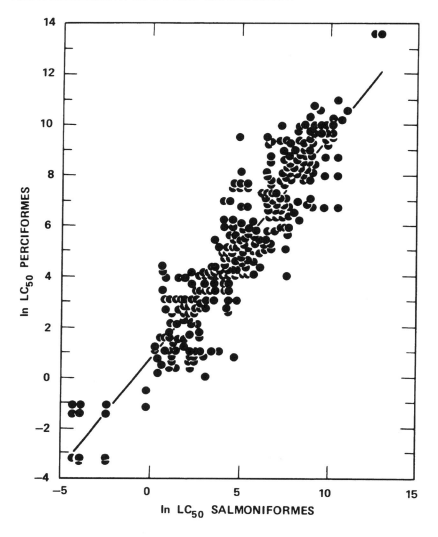

Figure 8.4. Regression of log LC_{50} values for members of the order *Perciformes* against LC_{50} values for the same chemicals for members of the order *Salmoniformes.* From Suter et al. (1983).

quail. It is not clear why these toxicological results would differ from the pharmacological experience. Use of an exponent of 1 is consistent with the mean exponent from the only available allometric study of avian toxicology (Mineau et al., 1996), and, because it reduces to scaling to weight, it is parsimonious, making neither small nor large birds more sensitive to a given dose (Sample et al., 1996).

Fractional exponents of weight would be expected when effects are a function of some time integral of internal dose. This is because the mechanism controlling integral internal exposure is drug or toxicant clearance, which is more rapid in small species due to their more rapid metabolism. However, if effects are a function of peak internal dose, as may be the case for lethal effects of cholinesterase inhibitors, dilution in the mass of the organism (exponent of one) would be the expected exposure model. These sorts of toxicokinetic and

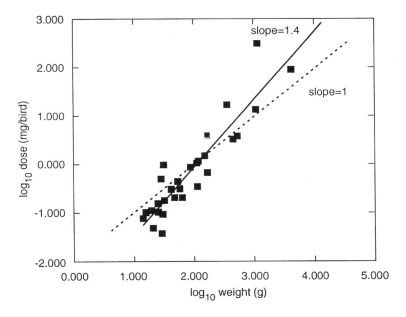

Figure 8.5. Regression of log LD_{50} against weight of the test species for the carbamate pesticide methiocarb (n = 32). Redrawn from Mineau et al. (1996).

toxicodynamic considerations could provide the basis for scaling models that are specific to chemical classes.

The scaling approach has not been used for aquatic ecological risk assessments, although Patin (1982) argued that sensitivity of aquatic species is a function of size. The reasonableness of this proposition is supported by observations that chemical uptake, elimination, and body burden are power functions of weight (Newman and Mitz, 1988; Newman and Heagler, 1991) and that intraspecies variation in sensitivity (time to death) of mosquitofish to NaCl was a function of weight as well as concentration (Newman et al., 1994). However, interspecific extrapolation based on allometry has apparently not been investigated for aquatic species beyond Patin's (1982) initial effort.

An alternative proposed for fish is scaling toxicity of lipophilic chemicals to lipid content (Geyer et al., 1994). This approach is based on the theory of "survival of the fattest" which proposes that tissues that are the site of toxic action are protected by partitioning chemicals to lipids (Lassiter and Hallam, 1990). Greyer et al. (1994) found that acute toxicity of lindane in 16 fish species was a linear function of lipid content (%) above 5% lipid and curvilinear below that level (Figure 8.6). This relationship is likely to be more complex for chronic exposures in which lipid stores fluctuate with season and reproductive cycle and toxic effects of the chemicals may include effects on lipid accumulation or metabolism.

ACUTE-CHRONIC EXTRAPOLATIONS

After extrapolation between species, the extrapolation that is most often addressed is that between acute and chronic toxicity. Conventionally, acute toxicity test endpoints are median lethal concentrations or doses (LC_{50} or LD_{50}) from tests of a few days duration, and

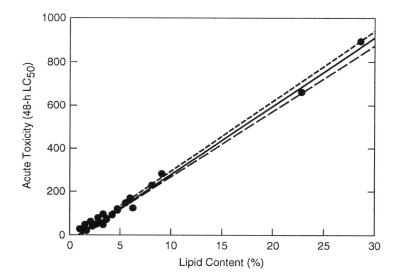

Figure 8.6. Regression of 48-h LC_{50} values against percent lipid for various fish species exposed to lindane. Dashed lines delimit 95% confidence intervals. Redrawn from Geyer et al. (1994).

chronic test endpoints are concentrations or doses that constitute thresholds for statistically significant differences from controls in any of a variety of lethal and sublethal responses (NOELs or LOELs) in tests extending over more than 10% of an organism's lifespan. Therefore, extrapolation between acute and chronic toxicity involves extrapolation between short and long durations, between lethality and various responses, and between a biological effect level and a statistically significant difference. The acute-chronic extrapolation is typically performed using a generic factor (e.g., 10) or a chemical-specific factor. In the latter case, acute and chronic tests are performed for a particular species-chemical combination, and the ratio of the resulting test endpoints is then applied to acute test endpoints for other species to estimate the chronic test endpoint (Mount and Stephan, 1967).

Because the acute-chronic distinction is such a hodgepodge, it has been recommended that it be abandoned, that exposure duration be treated as a continuous variable, and that the individual response parameters be distinguished (Suter et al., 1987). Extrapolation between response parameters (e.g., between mortality and fecundity) has been addressed using factors (Mayer et al., 1986) and regression models (Suter et al., 1987).

Similarly, extrapolation between different exposure durations may be performed using ratios of response levels at different durations (chronicity factors) or regression-based techniques. One may simply assume a concentration-duration function and fit it to test endpoints for different durations (e.g., 24, 48, 72, and 96 hour LC_{50}). One may then use the model to extrapolate to other durations, including indefinitely long exposures (Green, 1965; Mayer et al., 1994).

ORGANISM TO COMMUNITY/ECOSYSTEM

If one assumes that responses of communities or ecosystems are simply the aggregate responses of the organism-level responses of individual species, the interspecies extrapolations described above can be reinterpreted as extrapolations from organismal test endpoints to community or ecosystem endpoints. That is, if fathead minnows are assumed to represent all other individual fish species they may also represent the aquatic community as a whole. This interpretation is commonly applied to species sensitivity distributions. That is, rather than assuming that one is protecting species with 95% confidence, it is assumed that one is protecting the community by protecting 95% of species (Stephan et al., 1985). The same assumptions can be applied to regressions of all species against standard test species (Suter, 1993a).

Alternatively, one may extrapolate from organism-level tests to tests that are assumed to represent ecosystems (i.e., microcosms or mesocosms). Microcosm and mesocosm NOECs for individual chemicals have been regressed against the lowest reported laboratory LC_{50} and NOEC values for those chemicals (Sloof et al., 1986). These models have similar precision to regressions between orders of fish.

A direct approach to extrapolation from laboratory species to the field is to regress field measurements of the assessment endpoint against the test endpoint. Hartwell et al. regressed fish diversity (Margalef's index) and IBI in a number of streams against a combined score for tests of stream water and sediment with multiple test species (Hartwell et al., 1995a; 1995b) (Figure 8.7). They found that toxicity of water and sediment was correlated with diversity but not IBI. These models provide the most complete and realistic extrapolation.

BETWEEN MEDIA/COMMUNITIES

Far more ecotoxicological data exist for freshwater aquatic species than for communities inhabiting other media, and it would be desirable to extrapolate from freshwater toxicity to toxicity in other media. In some cases the media are not qualitatively different and one may simply use factors or regression models to extrapolate between media. For example, regressions of standard chronic test endpoints for the most commonly tested saltwater fish and crustacean against the most commonly tested freshwater species resulted in equations with slopes approximately equal to one and intercepts of approximately zero (Figure 8.8) (Suter and Rosen, 1988).

The lack of data is particularly problematical for organisms in soil and sediment. One solution is to extrapolate from the aquatic organisms to soil and sediment organisms. This is done by assuming that the organisms are exposed to the aqueous phase of soil or sediment, that the aqueous concentration can be estimated, and that the sensitivity of organisms to aqueous phase exposures is the same in all media. The U.S. EPA calculates sediment quality criteria for neutral organic chemicals by assuming that the chemicals are at equilibrium and partition between the aqueous phase and the organic fraction of the solid phase (EPA, 1993a). The EPA then uses aquatic toxicity data to estimate the effects of that exposure on sediment organisms. Lokke (1994) has tentatively proposed that an extrapolation from aquatic to soil organisms can be made by assuming that soil exposures are only to the soil pore water, that concentration in pore water can be estimated from soil concentrations using the distribution coefficient for bulk soil K_d, and the sensitivity of soil and sediment organisms to the aqueous phase chemical is the same as the sensitivity of aquatic organisms.

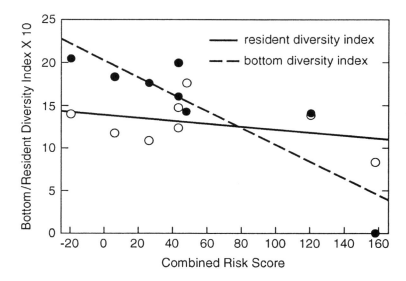

Figure 8.7. Regression of Margalef's diversity for bottom-dwelling fish and all resident fish species against the combined risk score, a value that integrates the results of multiple tests of the toxicity of ambient waters. Redrawn from Hartwell et al. (1995a).

EXTRAPOLATION OF MODE OF ACTION

Although the extrapolation of mode of action between species is a critical issue in human health risk assessment (e.g., Is a rat carcinogen also a human carcinogen?), the issue is seldom considered in ecological risk assessments. However, ecological risk assessments are concerned with estimating reproductive effects, which are relatively costly and difficult to measure. Therefore, there is considerable need for models to predict whether a chemical has a specific mode of action on the reproductive system or for some other reason causes reproductive effects at low exposures relative to adult lethality. This issue has gained some urgency with the recent emphasis on chemicals that affect the reproductive system by acting as agonists or antagonists to endocrine hormones. Further, since multigenerational tests of pesticides are nearly always performed on rats, it would be useful to extrapolate from rats to other vertebrates. The fact that only 44% of chemicals that were found to be avian reproductive toxicants were also reproductive toxicants in rats (Mineau et al., 1994), suggests that this extrapolation will not be simple.

MULTIPLE EXTRAPOLATIONS

Although extrapolation models are usually used singly in ecotoxicological assessments, it is often apparent that one extrapolation does not incorporate all of the differences between the measurement and assessment endpoints. For example, if one is assessing risks to fish species richness from a fathead minnow LC_{50}, one must extrapolate from one species to many, from lethality to all responses that influence viability, and from individual responses to population responses. The simplest approach is factor chains: Ea = (f1 f2 f3 ... fn)Et + e, where each factor accounts for a particular difference between the measurement and

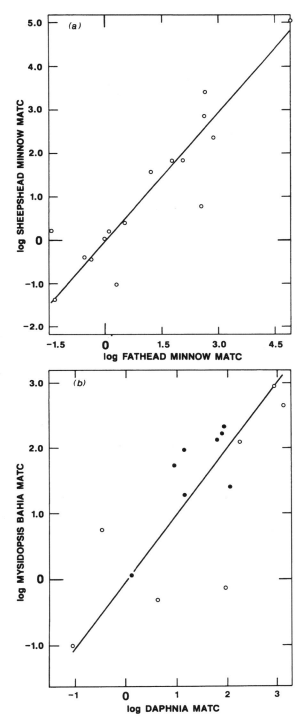

Figure 8.8. Regression of Maximum Acceptable Toxicant Concentrations (MATCs) for a standard saltwater test fish and crustacean species against MATCs for standard freshwater test species. Dark points are metals; all others are organic chemicals. From Suter and Rosen (1988).

assessment endpoint. This approach is very conservative in practice because it amounts to an assumption that all of the differences between the endpoints are simultaneously extreme (Suter, 1993a; National Research Council, 1994). For example, the assessment endpoint species is extremely sensitive; the endpoint life stage is extremely sensitive; and the effects of differences in duration are large, simultaneously.

Regression models lend themselves to multiple extrapolations. The output of one regression model (e.g., cyprinid LC_{50} to salmonid LC_{50}) may be used as input to another (e.g., LC_{50} to EC_{25} for weight of juveniles per egg) to generate an extrapolation from a fathead minnow LC_{50} to brook trout early life stage effects (Suter, 1993a; Suter et al., 1983). The principal difficulty in this approach is correct propagation of uncertainty through the chain of models.

Multiple extrapolations may be performed using multiple techniques. For example, in calculation of chronic water quality criteria, the EPA uses species sensitivity distributions to estimate a fifth percentile acute value and then applies an acute/chronic factor to estimate a fifth percentile chronic value (Stephan et al., 1985).

All of these multiple extrapolations depend on the assumption that the individual extrapolations are concordant. For example, it is assumed in calculating chronic water quality criteria and in the example of multiple regression models that relative sensitivities of species are the same in acute and chronic exposures. That assumption is unlikely to be perfectly true. For example, there are more modes of action involved in the various chronic effects than in acute lethality, which is likely to result in increased range in sensitivities among species. On the other hand, if the mode of action is the same for acute and chronic toxicity, then interspecies variance may be lower in the chronic tests because differences in response rate would be less important.

UNCERTAINTIES

These empirical models deal adequately with uncertainty estimable from the data used, but not the uncertainty inherent in the selection of data. Although this is true to some extent of all models based on samples of data, it is a particular problem for these models due to the following issues.

Representative species—Extrapolation models—such as species sensitivity distributions—that use sets of test species to represent endpoint species depend on the assumption that test species are representative. Although test species have not been randomly or systematically selected, there is no reason in general to believe that there has been a bias in their selection. However, some taxa—such as fishes—are clearly overrepresented in data sets relative to their abundance in nature.

Representative life stages—Most of the data used to develop extrapolation models are for a single life stage. The absence of life stages that may be more sensitive than tested stages is particularly problematical when extrapolation models are limited to interspecies extrapolations, and acute or subchronic data are used. Some sensitive life stages—such as reproducing adults—are seldom represented (Suter et al., 1987).

Representative chemicals—Extrapolation models such as factors and regression models that use multiple chemicals depend on the assumption that tested chemicals represent all chemicals of concern. However, chemicals are not randomly chosen for testing. For example, because ecotoxicological data sets are dominated by pesticides, they can have an inordinate influence. Partitioning data sets into chemical classes reduces but does not eliminate this uncertainty.

Small data sets—Many extrapolation models are derived from small data sets, which has obvious implications for their reliability.

Data quality and consistency—Poor quality data obviously can cause bad predictions and inconsistent data obscures relationships and increases variance. These issues must be carefully considered and balanced when choosing data sets. For example, the very large data set generated at the former Columbia National Fisheries Research Laboratory would be considered low quality because of the use of static rather than flow-through exposures (Mayer and Ellersieck, 1986). However, it is unlikely that the bias associated with static testing would be significantly different in different species of fish. Therefore, the large size and consistency of this data set makes it useful for extrapolation modeling.

Model extrapolation—Often one type of data is used to develop a model that is then used for extrapolation of another type. For example, interspecies extrapolation models developed with acute lethality data are used to perform interspecies extrapolations of chronic data. This extrapolation of the model to other uses is usually unacknowledged but may be important.

Incomplete extrapolation—In most assessments, one or a few extrapolations are modeled and the others are assumed to contribute negligibly to uncertainty. For example, it is common to explicitly extrapolate between taxa but not from the laboratory to the field (OECD, 1992; Smith and Cairns, 1993).

The importance of these considerations depends on their relative contribution to total uncertainty, but the analysis of relative uncertainty must be made in the proper context. For example, many studies have pointed out that variance among species is small relative to variance among chemicals. Such comparisons tend to reassure the reader that uncertainties in the extrapolation models are relatively insignificant. However, these comparisons are based on analyses of sets of heterogeneous chemicals. Real assessments are likely to be concerned with estimating the risks of a set of alternative cholinesterase-inhibiting pesticides rather than comparing a pesticide to ethane. In that context, differences in species and life stage are relatively large contributors to uncertainty in the results of the assessment.

CONCLUSIONS AND RECOMMENDATIONS

Although extrapolation models have been an important tool for ecotoxicological assessments for decades, they have been neglected as a research topic. Far more effort has been devoted to developing new toxicity tests than to developing models to relate the test results to relevant endpoints. In part, this is because of inertia. For example, the EPA Office of Pollution Prevention and Toxics (OPPT) developed a set of factors for their assessments of new industrial chemicals based on a short, unpublished, and unreviewed literature review (Branch, 1984) (Table 8.1). OPPT continues to use and defend these simple factors because more complex models have not been demonstrated to be more appropriate (Zeeman, 1995). However, the choice of extrapolation method can clearly make a large difference in the results of assessments. A comparison of HC_5 values derived by the van Straalen and Denneman model and concern levels derived using the OPPT factors applied to sets of three LC_{50} for eight chemicals found that the results differed by factors of 2.6 to 1020 (Okkerman et al., 1991). Similarly, a comparison of the results of applying seven published extrapolation models to a chemical with a fathead minnow LC_{50} of 100 mg/L found a factor of 19,200 difference in an estimated effects threshold (Suter, 1993b).

The claim that no other extrapolation method is better than simple factors of 10, 100, and 1000 can be made because there is no good basis for the comparison. That is, there is no agreement about what the assessment endpoints should be (i.e., what the models should predict) and no set of data from the field that is agreed to represent the responses of those assessment endpoints. Therefore, no validation of extrapolation models has been possible. One possible standard endpoint is mass mortalities (Suter, 1993b). Nearly everyone agrees that streams littered with dead fish or a park littered with dead geese is undesirable, and the ability to estimate concentrations causing mass mortalities is a minimal goal for assessment models. An appropriate and more challenging standard assessment endpoint for validation purposes would be a 5% reduction in species richness. Its advantages include the following:

1. The fifth percentile of species sensitivity distributions is used for regulation of chemicals and effluents in the United States, the Netherlands, and elsewhere.
2. Species richness is sensitive to toxic effects relative to other ecosystem or community level endpoints (Dickson et al., 1992; Hartwell et al., 1995b).
3. The concern with preservation of biodiversity implies that species richness is a societally valued endpoint.
4. Species richness data are likely to be available for most field studies of toxic effects.
5. A 5% reduction in species richness is likely to be detectable in many field studies of toxic effects.

A validation study for extrapolation models is long overdue, and one based on these assessment endpoints is quite feasible.

REFERENCES

Aldenberg, T. *ETX 1.3a, a Program to Calculate Confidence Limits for Hazardous Concentrations Based on Small Samples of Toxicity Data*. RIVM, Bilthoven, The Netherlands, 1993.

Aldenberg, T. and W. Slob. Confidence limits for hazardous concentrations based on logistically distributed NOEC toxicity data. *Ecotoxicol. Environ. Saf.* 25, pp. 48–63, 1993.

Baril, A., B. Jobin, P. Mineau, and B.T. Collins. (1994). *A Consideration of Inter-species Variability in the Use of the Median Lethal Dose (LD$_{50}$) in Avian Risk Assessment*. Report No. 216, Canadian Wildlife Service, Hull, PQ.

Barnthouse, L.W. and G.W. Suter, II, Eds. *User's Manual for Ecological Risk Assessment*. Report ORNL-6251, Oak Ridge National Laboratory, 1986.

Barnthouse, L.W., G.W. Suter, II, and A.E. Rosen. Risks of toxic contaminants to exploited fish populations: Influence of life history, data uncertainty, and exploitation intensity. *Environ. Toxicol. Chem.* 9, pp. 297–311, 1990.

Bartell, S.M., R.H. Gardner, and R.V. O'Neill. *Ecological Risk Estimation*. Lewis Publishers, Boca Raton, FL, 1992, p. 252.

Branch, E.E. *Estimating "Concern Levels" for Concentrations of Chemical Substances in the Environment*. U.S. Environmental Protection Agency, Washington, DC, 1984.

Calabrese, E.J. and L.A. Baldwin. A toxicological basis to derive a generic interspecies uncertainty factor. *Environ. Health Perspect.* 102, pp. 14–17, 1994.

Davidson, I.W.F., J.C. Parker, and R.P. Beliles. Biological basis for extrapolation across mammalian species. *Regul. Toxicol. Pharmacol.* 6, pp. 211–237, 1986.

Dickson, K.L., W.T. Waller, J.H. Kennedy, and L.P. Ammann. Assessing the relationship between ambient toxicity and instream biological response. *Environ. Toxicol. Chem.* 11, pp. 1307–1322, 1992.

EPA. Guidelines for carcinogenic risk assessment. *Federal Register* 51, pp. 33992-34003, 1986.

EPA. *Framework for Ecological Risk Assessment.* Report No. EPA/630/R-92/001. Risk Assessment Forum, Washington, DC, 1992.

EPA. *Technical Basis for Deriving Sediment Quality Criteria for Nonionic Organic Contaminants for the Protection of Benthic Organisms by Using Equilibrium Partitioning.* Report No. EPA-822-R-93-001. Office of Water, U.S. Environmental Protection Agency, Washington, DC, 1993a.

EPA. *Wildlife Criteria Portions of the Proposed Water Quality Criteria for the Great Lakes System.* Report No. EPA/822/R-93/006. Office of Science and Technology, U.S. Environmental Protection Agency, Washington, DC, 1993b.

Fletcher, J.S., F.L. Johnson, and J.C. McFarlane. Influence of greenhouse versus field testing and taxonomic differences on plant sensitivity to chemical treatment. *Environ. Toxicol. Chem.* 9, pp. 769–776, 1990.

Geyer, H.J., I. Scheunert, R. Bruggemann, M. Matthies, C.E.W. Steinberg, V. Zitko, A. Kettrup, and W. Garrison. The relevance of aquatic organisms' lipid content to the toxicity of lipophilic chemicals: toxicity of lindane to different fish species. *Ecotoxicol. Environ. Saf.* 28, pp. 53–70, 1994.

Goddard, M.J. and D. Krewski. Interspecies extrapolation of toxicity data. *Risk Anal.* 12, pp. 315–317, 1992.

Green, R.H. Estimation of tolerance over an indefinite time period. *Ecology.* 46, pp. 887, 1965.

Hartwell, S.I., C.E. Dawson, E.Q. Durell, and D.H. Jordahl. *Integrated Measures of Ambient Toxicity and Fish Community Diversity.* Report No. CBRM-TX-95-2. Maryland Department of Natural Resources, Chesapeake Bay Research and Monitoring Division, Annapolis, MD, 1995a.

Hartwell, S.I., C.E. Dawson, D.H. Jordahl, and E.Q. Durell. *Demonstrating a Method to Correlate Measures of Ambient Toxicity and Fish Community Diversity.* Report No. CBRM-TX-95-1. Maryland Department of Natural Resources, Chesapeake Bay Research and Monitoring Division, Annapolis, MD, 1995b.

Hoekstra, J.A., M.A. Vaal, J. Notenboom, and W. Sloof. Variation in the sensitivity of aquatic species to toxicants. *Bull. Environ. Contam. Toxicol.* 53, pp. 98–105, 1994.

Holcombe, G.W., G.L. Phipps, and G.D. Veith. Use of aquatic lethality tests to estimate safe chronic concentrations of chemicals in initial ecological risk assessments, in *Aquatic Toxicity and Hazard Assessment: Eleventh Symposium,* Suter, G.W. and M. Lewis, Eds., American Society for Testing and Materials, Philadelphia, PA, pp. 442–467, 1988.

Kenaga, E.E. Test organisms and methods useful for early assessment of acute toxicity of chemicals. *Environ. Sci. & Technol.* 12, pp. 1322–1329, 1978.

Kooijman, S.A.L.M. A safety factor for LC_{50} values allowing for differences in sensitivity among species. *Water Res.* 21, pp. 269–276, 1987.

Lassiter, R.R. and T.G. Hallam. Survival of the fattest: implications for acute effects of lipophilic chemicals on aquatic populations. *Environ. Toxicol. Chem.* 9, pp. 585–595, 1990.

Lokke, H. Ecotoxicological extrapolation: tool or toy, in *Ecotoxicology of Soil Organisms,* Donker, M., H. Eijsackers, and F. Heimbach, Eds., Lewis Publishers, Boca Raton, FL, 1994.

Mayer, F.L.J. and M.R. Ellersieck. *Manual of Acute Toxicity: Interpretation and Data Base for 410 Chemicals and 66 Species of Freshwater Animals.* Report Resource Pub. 160. U.S. Fish and Wildlife Service, Washington, DC, 1986.

Mayer, F.L., K.S. Mayer, and M.R. Ellersieck. Relationship of survival to other endpoints in chronic toxicity tests with fish. *Environ. Toxicol. Chem.* 5, pp. 737–748, 1986.

Mayer, F.L., G.F. Krause, D.R. Buckler, M.R. Ellersieck, and G. Lee. Predicting chronic lethality of chemicals to fishes from acute toxicity test data: concepts and linear regression analysis. *Environ. Toxicol. Chem.* 13, pp. 671–678, 1994.

Mineau, P., D.C. Boersma, and B. Collins. An analysis of avian reproduction studies submitted for pesticide registration. *Ecotoxicol. Environ. Saf.* 29, pp. 304–329, 1994.

Mineau, P., B.T. Collins, and A. Baril. On the use of scaling factors to improve interspecies extrapolation of acute toxicity in birds. *Regul. Toxicol. Pharmacol.* 24, pp. 24–29, 1996.

Mount, D.I. and C.E. Stephan. A method for establishing acceptable toxicant limits for fish— malathion and butoxyethanol ester of 2,4-D. *Trans. Am. Fish. Soc.* 96, pp. 185–193, 1967.

National Research Council. *Science and Judgement in Risk Assessment.* National Academy Press, Washington, DC, 1994, p. 651.

Newman, M. and S.V. Mitz. Size dependence of zinc elimination and uptake from water by mosquitofish *Gambusia affinis* (Baird and Girard). *Aquat. Toxicol.* 12, pp. 17–32, 1988.

Newman, M.C. and M.G. Heagler. Allometry of metal bioaccumulation and toxicity, in *Metal Ecotoxicology, Concepts and Applications*, Newman, M.C. and A.W. McIntosh, Eds., Lewis Publishers, Boca Raton, FL, pp. 91–130, 1991.

Newman, M.C., M.M. Keklak, and M.S. Doggett. Quantifying animal size effects in toxicity: a general approach. *Aquat. Toxicol.* 28, pp. 1–12, 1994.

Notenboom, J., M.A. Vaal, and J.A. Hoekstra. Using comparative ecotoxicology to develop quantitative species sensitivity relationships (QSSR). *Environ. Sci. & Pollut. Res.* 2, pp. 242–243, 1995.

OECD. *Report of the OECD Workshop on the Extrapolation of Laboratory Aquatic Toxicity Data to the Real Environment.* Report No. OECD/GD(92)169. Organization for Economic Cooperation and Development, Paris, 1992.

Okkerman, P.C., V.D. Plassche, W. Sloof, C.J. Van Leeuwen, and J.H. Canton. Ecotoxicological effects assessment: A comparison of several extrapolation procedures. *Ecotoxicol. Environ. Saf.* 21, pp. 182–193, 1991.

Opresko, D.M., B.E. Sample, and G.W. Suter, II. (1993). *Toxicological Benchmarks for Wildlife.* Report ES/ER/TM-86. Oak Ridge National Laboratory, Oak Ridge, TN.

Patin, S.A. *Pollution and the Biological Resources of the Oceans.* Butterworth Scientific, London, 1982, p. 287.

Peters, R.H. *The Ecological Implications of Body Size.* Cambridge University Press, Cambridge, 1983, p. 329.

Peters, R.H. *A Critique for Ecology.* Cambridge University Press, Cambridge, 1991, p. 366.

Pokras, M.A., A.M. Karas, J.K. Kirkwood, and C.J. Sedgewick. An introduction to allometric scaling and its uses in raptor medicine, in *Raptor Biomedicine*, Redig, P.T., J.E. Cooper, J.D. Remple, and D.B. Hunter , Eds., U. Minnesota Press, Minneapolis, MN, 1993.

Sample, B.E., D.M. Opresko, and G.W. Suter, II. (1996). *Toxicological Benchmarks for Wildlife.* Report No. ES/ER/TM-86/R3. Oak Ridge National Laboratory, Oak Ridge, TN.

Sloof, W., J.A.M. van Oers, and D. de Zwart. Margins of uncertainty in ecotoxicological hazard assessment. *Environ. Toxicol. Chem.* 5, pp. 841–852, 1986.

Smith, E.P. and J. Cairns, Jr. Extrapolation methods for setting ecological standards for water quality: statistical and ecological concerns. *Ecotoxicology* 2, pp. 203–219, 1993.

Stephan, C.E., D.I. Mount, D.J. Hanson, J.H. Gentile, G.A. Chapman, and W.A. Brungs. (1985). *Guidelines for Deriving Numeric National Water Quality Criteria for the Protection of Aquatic Organisms and Their Uses.* Report No. PB85-227049. U.S. Environmental Protection Agency, Washington, DC.

Suter, G.W., II. *Ecological Risk Assessment.* Lewis Publishers, Boca Raton, FL, 1993a, p. 538.

Suter, G.W., II. New concepts in the ecological aspects of stress: the problem of extrapolation. *Sci. Total Environ. Supp.* 1993b, pp. 63–76, 1993b.

Suter, G.W., II and A.E. Rosen. Comparative toxicology for risk assessment of marine fishes and crustaceans. *Environ. Sci. & Technol.* 22, pp. 548–556, 1988.

Suter, G.W., II, D.S. Vaughan, and R.H. Gardner. Risk assessment by analysis of extrapolation error: a demonstration for effects of pollutants on fish. *Environ. Toxicol. Chem.* 2, pp. 369–378, 1983.

Suter, G.W., II, A.E. Rosen, E. Linder, and D.F. Parkhurst. Endpoints for responses of fish to chronic toxic exposures. *Environ. Toxicol. Chem.* 6, pp. 793–809, 1987.

Travis, C.C. and J.M. Morris. On the use of 0.75 as an interspecies scaling factor. *Risk Analysis.* 12, pp. 311–313, 1992.

Vaal, M.A., J.T. Van der Wal, and J.A. Hoekstra. *Ordering Aquatic Species by Their Sensitivity to Chemical Compounds: A Principal Component Analysis of Acute Toxicity Data.* Report No. 719102 028. National Institute for Public Health and Environmental Protection, Bilthoven, The Netherlands, 1994.

Van der Wal, J.T., R. Luttik, M.A. Vaal, and J.A. Hoekstra. *Ordering Birds and Mammals by Their Sensitivity to Chemical Compounds: A Principal Component Analysis.* Report No. 719102 036. National Institute for Public Health and Environmental Protection, Bilthoven, The Netherlands, 1995a.

Van der Wal, J.T., M.A. Vaal, J.A. Hoekstra, and J.L.M. Hermens. *Ordering Chemical Compounds by Their Chronic Toxicity to Aquatic Species: A Principal Component Analysis.* Report No. 719102 041. National Institute for Public Health and Environmental Protection, Bilthoven, The Netherlands, 1995b.

Wagner, C. and H. Lokke. Estimation of ecotoxicological protection levels from NOEC toxicity data. *Water Res.* 25, pp. 1237–1242, 1991.

Watanabe, K., F.Y. Bois, and L. Zeise. Interspecies extrapolation: a reexamination of acute toxicity data. *Risk Analysis* 12, pp. 301–310, 1992.

Zeeman, M.G. Ecotoxicity testing and estimation methods developed under section 5 of the Toxic Substances Control Act (TSCA), in *Fundamentals of Aquatic Toxicology: Effects, Environmental Fate, and Risk Assessment,* Rand, G., Ed., Taylor and Francis, Washington, DC, pp. 703–715, 1995.

9 ||| Dynamic Measures for Ecotoxicity[1]

S.A.L.M. Kooijman, J.J.M. Bedaux, A.A.M. Gerritsen, H. Oldersma, and A.O. Hanstveit

OVERVIEW

There are three required components of dynamic models for toxic effects: toxicokinetics, effects on a target parameter coupled to the internal concentration, and the physiological component. The Dynamic Energy Budget (DEB) model, which is used to model the latter component, relates a change in a target parameter of a particular physiological process, such as the specific costs for growth, to an output variable, such as the cumulative number of offspring. We compare the logit/probit and the DEB-based models conceptually and numerically, and conclude that the DEB-based model is more effective as an effect model. The DEB-based model solves the problem of estimating the No-Effect Concentration and provides the required information to evaluate the consequences of effects on individuals for population dynamics.

INTRODUCTION

In environmental risk assessments Predicted No Effect Concentrations (PNECs), which are derived from No Observed Effect Concentrations (NOECs) in standard single-species toxicity tests, are compared with Predicted Environmental Concentrations (PECs), derived from production volumes, use patterns, and transport in the environment. The purpose of the toxicity tests, together with their designs and experimental protocols, evolved gradually toward sophistication. The analysis of the results of these tests, however, did not catch up with these changes for a long time. This chapter introduces a process-based approach for the analysis of toxicity tests, which has a firm rooting in biology.

The purpose of this chapter is to identify the aim of toxicity tests and to present the DEB-based effect model as a method to achieve this aim. The static and the dynamic methods are compared, conceptually and numerically. We tried to refrain from technical details in this

---

[1] Background Document for the OECD Workshop on Analysis of Data from Aquatic Ecotoxicity Test, Braunschweig, Oct. 1996.

chapter and refer to Kooijman and Bedaux (1996) for a full discussion of the DEB-based model for toxic effects.

Aims of Toxicity Tests

The primary purpose of standard toxicity tests is to provide information about

- The maximum concentration that gives no effect on a response variable (such as survival, body growth, reproduction, and population growth). This information is used to derive a level in the environment that can be considered as "safe."
- What effects are to be expected in the environment if these levels are exceeded (by a small amount).
- The requirement for further in-depth studies with respect to the ecotoxicity of the chemical. The priority is high if expected levels in the environment are likely to approach or exceed the level that is considered "safe" and, in that case, the effect is likely to be substantial. The actual decision regarding further research depends on financial possibilities for research and socioeconomic factors in industrial activity.

The second and third application of toxicity tests imply a quantification of effects as a function of the concentration. They also imply an extrapolation of the effects as observed during the tests to effects at long-term exposure and a translation to effects in the environment.

Classification of Approaches

The analyses of toxicity tests can be classified into three approaches: the ANOVA method, static methods, and dynamic methods.

- The ANOVA method aims to identify the highest tested concentration with a response that does not deviate from the blank on the basis of a statistical procedure: the NOEC.
- The static method quantifies effects at a standardized exposure time on the basis of concentration-response models, such as the logit, probit, or Weibull model, which are all very similar. The toxic effect is quantified by EC_{50} (LC_{50}) or an EC_x value for some small value of x. The method is called static because information about the rate at which effects build up during exposure is not used. Fixed extrapolation factors are used to predict effects at long exposure.
- The dynamic method quantifies effects as functions of the concentration *and* the exposure time. The Dynamic Energy Budget (DEB)-based model is an example of this approach. The toxic effect is quantified by two parameters: the No Effect Concentration (NEC), and the killing rate (for survival) or the tolerance concentration (for other endpoints) as a measure of the effect if the NEC is exceeded.

THE DYNAMIC APPROACH

The dynamic approach for the analysis of standard toxicity tests characterizes toxic effects with a No Effect Concentration (NEC), a tolerance concentration (or a killing rate in the case of effects on survival), and an elimination rate. These parameters directly relate to long-term exposure, so no extrapolation factor is required here. They relate to changes in processes. From these three parameters, it is possible to calculate the static parameters, including the EC_0 at the end of the test (therefore comparable to the NOEC), as well as the full EC_x-time behavior. It is not possible to calculate the dynamic parameters from the static ones, which shows that the dynamic parameters contain more information than the static ones. The dynamic approach uses information about the rate at which effects build up during exposure. It is important to realize that static and dynamic approaches are alternative analyses for the *same* toxicity tests. The approaches only differ in the type of information that is extracted from the data. Given the choice for the analysis of the data, we can try to optimize the experimental setup of the test in terms of efficiency. This is another issue that is beyond the scope of this chapter.

The Components of Dynamic Effect Models

The three components that any dynamic model for effect should have are the kinetics, the effect, and the physiological component, which are discussed in the next subsections. The implication is that responses are modeled as a function of the concentration and exposure time. We have a response *surface*, rather than a concentration-response relationship. For a response such as the cumulative number of offspring, this means that we include any delay of the start of the reproduction into the analysis.

Kinetics

The kinetics component links internal concentrations to external concentrations. The first-order kinetics (also called the one-compartment model) is the simplest choice. It assumes that uptake is proportional to the external concentration and elimination is proportional to the internal concentration. If the organism grows during exposure, dilution by growth results in a deviation from first-order kinetics that should be taken into account. The DEB-based model takes uptake and elimination rates proportional to the surface area, while the DEB component (see below) specifies the growth process.

Actual measurements of time profiles for the internal concentration sometimes show deviations from first-order kinetics. More-compartment models are frequently used to improve the fit, because they have more parameters. A more detailed modeling of the various uptake rates (via water and/or food), elimination rates (via water, reproduction, and/or respiration), changes in fat content, and metabolic transformations, is frequently more realistic than using multicompartment models. Some chemicals are not taken up at all, but have their effect on the surface of the organism. Multicompartment models should only be used if the compartments are identified and their concentrations of toxicant measured. The latter requirement applies to all kinetics models that are more complex than the first-order model. Reconstructions of complex kinetics from observed effects, rather than from measured internal concentrations, run into problems rather easily.

Since internal concentrations are not measured in standard toxicity tests, application of more advanced models for kinetics in these toxicity tests is not feasible. The alternative—to refrain from dynamic modeling and apply an arbitrary extrapolation factor to arrive at predicted long-term effects—is certainly not a better alternative. We consider the application of a safety factor to compensate for possibly inappropriate use of first-order kinetics to be a more promising alternative. This problem is inherent in the use of standard toxicity tests for environmental risk assessments; an in-depth scientific study for long-term effects is the only sensible alternative. Because of the financial costs of such a study, this will only be possible for a limited number of chemical compounds.

Effects

The effect component links effects of a target parameter—such as the volume-specific costs for maintenance or growth—to the internal concentration. The linear effect model is the simplest choice: the change in the target parameter is proportional to the internal concentration that exceeds the internal No Effect Concentration (NEC). Translation of this concept into formulae gives the following relationships for sublethal and lethal effects of a target parameter: if int. conc(t) $\geq$ int. NEC at exposure time t

$$\text{par}_c(t) = \text{par}_0 \left(1 + \frac{\text{int. conc(t)} - \text{int. NEC}}{\text{int. tolerance conc.}} \right)$$

$$\text{haz}_c(t) = \text{haz}_0 + \text{killing rate} \frac{\text{int. conc(t)} - \text{int. NEC}}{\text{BCF}}$$

where BCF stands for the BioConcentration Factor (the ratio of the internal and the external concentration after long-term exposure) and haz stands for the hazard rate, i.e., the instantaneous death rate. As long as the internal concentration is less than the internal NEC, we have that $\text{par}_c = \text{par}_0$ and $\text{haz}_c = \text{haz}_0$. If we divide the internal concentrations by BCF, we get external concentrations. The result is for conc (t) $\geq$ NEC

$$\text{par}_c(t) = \text{par}_0 \left(1 + \frac{\text{conc(t)} - \text{NEC}}{\text{tolerance conc.}} \right)$$

$$\text{haz}_c(t) = \text{haz}_0 + \text{killing rate} (\text{conc(t)} - \text{NEC})$$

where conc(t) has the dimensions of an external concentration but is proportional to the internal one. Note that the NEC refers to long-term exposure. The NOEC for a particular toxicity test depends on the maximum observed exposure time t. It is therefore conceptually more or less comparable with the $\text{EC}_{0,t}$, while the NEC equals the $\text{EC}_{0,\infty}$.

The abovementioned formulae define the killing rate and the tolerance concentration; these parameters occur as proportionality coefficients in the description of effects. The killing rate has dimension "per environmental concentration per time" and is a measure of the toxicity of a chemical if it exceeds the NEC—a toxicity measure that is independent of the exposure time. The tolerance concentration has the dimension "environmental concentration." The more toxic the chemical, the lower its value. It is essential to specify the target parameter to which the tolerance concentration relates.

Five target parameters are distinguished for effects on reproduction: two for direct and three for indirect effects. Direct effects on reproduction are defined as effects on what happens with the investment in reproduction, not on the size of the investment itself. One mode of action increases the energetic costs of each offspring, the other affects the survival of each ovum during a short sensitive period. Indirect effects on reproduction affect the investment in reproduction, not the conversion of this investment into offspring. Indirect effects increase the costs of growth or maintenance, or decrease the assimilative input. These three indirect effects not only reduce the reproduction, but also delay the start of the reproduction.

Consistent with the five target parameters for effects on reproduction, there are three target parameters for effects on body growth: one direct effect (the increase of the specific costs for growth), and two indirect ones—the increase of maintenance costs and the decrease of the assimilative input.

For effects on population growth of algae, we distinguish a direct effect on cell growth (by increasing the specific costs for growth), and two effects on survival (so the target parameter is the hazard rate): one effect lasts during population growth and the other only operates during a very short period after inoculation. In the latter case, effects are supposed to occur only during the transition from the culture to the experimental conditions. The sensitivity of the cells here is supposed to relate to the cell cycle. The overall effect is a delay of population growth, rather than a reduction. The uptake/elimination kinetics are assumed to be fast relative to the population growth process, so that the internal concentration equals the product of the BCF and the external concentration.

The linear relationships between the internal concentration and the target parameter follow from two arguments: effective molecules operate independently, and it is a simple approximation for small changes in the target parameter. The improvement of goodness of fit for large changes in the target parameter probably does not balance extra parameters. Moreover, at higher concentrations, more physiological processes are likely to be affected simultaneously. This means that many parameters have to be introduced to capture large effects. We consider such improvements counterproductive in standard toxicity tests. The argument implies that if the goodness of fit of the model to the responses is less than excellent, a higher weight should be given to responses at low concentrations.

Note that a linear relationship between the target parameter and the internal concentration does not imply a linear relationship between the response (i.e., output variable) and the external concentration. These relationships work out to be sigmoid.

Blank Physiology

For acute lethal effects, we do not need a model for the blank physiology. For effects on algal population, such a model can be very simple. For effects on body growth and reproduction, however, we need a physiological component that links output variable(s) to the target parameter. The output variable is the variable that is measured, such as body length, cumulative number of offspring, number of surviving individuals, etc. The Dynamic Energy Budget (DEB) model is the simplest choice that links all essential processes: feeding, digestion, respiration, maintenance, growth, development, reproduction, and aging. Many other models for the uptake and use of food that typically involve many parameters have been described in the literature. A comparative discussion of these models is beyond the

Table 9.1. Effects of PCP on the Survival of the Fathead Minnow *Pimephales promelas*.

Survival, Hazard Model		ASD	Correlation Coefficients		
Blank mortality rate	2.296e–011 d^{-1}	0.000			
NEC	190.2 $\mu g\ L^{-1}$	0.783	0.000		
Killing rate	0.009304 $L\ \mu g^{-1}\ d^{-1}$	0.001	0.000	0.081	
Elimination rate	7.019 d^{-1}	2.510	–0.000	0.134	–0.700
Deviance	31.41				

scope of this chapter. Since the physiological component only relates to the ecophysiology of the test species involved, and not to the toxicant, its applicability does not need to be studied for each individual toxicity test. It needs to be studied only once per species. The physiological component in this respect differs fundamentally from the other two components: the kinetics and the effect component.

Some chemicals, such as endocrine disrupters, might have an effect at the molecular level that does not directly relate to the energetics of the organism, but other effects will eventually translate into effects on the energetics; it is the choice of output variable (e.g., growth or reproduction) that directly relates to the energetics, not the molecular mode of action of the chemical. If a chemical has neither direct nor indirect effects on energetics—such as a chemical that only affects behavior—it will have no effects in toxicity tests for effects on growth or reproduction.

Effects on survival directly affect the hazard rate, which can be studied without detailed reference to energy budgets. It is perhaps better to call the model a hazard-based model, rather than a DEB-based one; the rich structure of the DEB model only comes into play for effects on growth and, especially, reproduction. The fact that changes in the hazard rate of toxicants links up beautifully with the effect of the aging process in the DEB model can be considered as a happy coincidence that has little relevance for most standard toxicity tests. It becomes more relevant if the length of the toxicity test is not short with respect to the lifespan of the test organism. Evaluation of the consequences of effects on survival for population dynamics does involve the complete structure of the DEB model.

Effects on the population growth rate for dividing organisms such as algae only involve certain simple aspects of the DEB model, as is explained in Kooijman (1993). This is because the surface area/volume ratio only changes within a very restricted range during the cell cycle. This does not hold for animals such as water fleas and fish.

EXAMPLES OF APPLICATION OF THE DEB-BASED MODEL

Examples of the application of the dynamic approach for toxicity tests for effects on survival, body growth, reproduction, and population growth are given in Tables 9.1, 9.2, 9.3, and 9.4, and Figures 9.1, 9.2, 9.3, and 9.4. All figures and tables are composed from output files of the software package DEBtox, as provided in Kooijman and Bedaux (1996). It fits the response surface to all available data simultaneously, so the different curves in the concentration and exposure time profiles are linked and are not fitted independently.

Since the use of profile ln likelihood functions for the identifications of confidence intervals (here for the NEC) is not standard (probably because of the substantial amount of

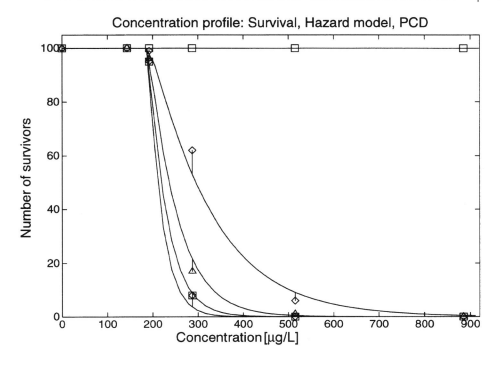

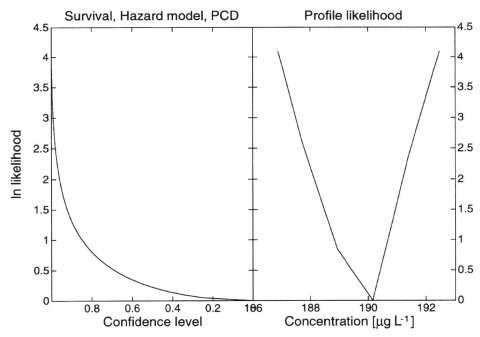

Figure 9.1a,b. Effects of PCP on the survival of the fathead minnow *Pimephales promelas*. The profile ln likelihood function for the NEC, given in the graph on the lower right, has a nontypical shape, because DEBtox used few evaluation points. The reason is that the confidence interval of the NEC here is very small with respect to its value.

Table 9.2. Effects of DCA on the Body Growth of the Rainbow Trout *Oncorhynchus mykiss.* **Data from the OECD Ring Test 1988/9. Costs of Maintenance has been Selected as Target Parameter.**

Body Growth, Maintenance Model		ASD	Correlation Coefficients			
NEC	111.3 $\mu g\ L^{-1}$	30.988				
Blank ultimate length	3.065 $g^{1/3}$	0.026	−0.298			
Tolerance concentration	546.8 $\mu g\ L^{-1}$	188.734	0.403	0.028		
Elimination rate	0.1059 d^{-1}	0.092	0.602	0.014	0.956	
Initial length	1.56 $g^{1/3}$					
Von Bertalanffy growth rate	0.01 d^{-1}					
Energy investment ratio	1					
Mean deviation	0.01425 $g^{1/3}$					

calculation that is involved), some guidance might be appropriate. The idea is that a confidence level is first selected on the horizontal axis in the left panel; the threshold value for the ln likelihood function is then read off on the vertical axis, and the graph in the right panel is (mentally) intersected at this level. The intersection points represent the boundaries of the appropriate confidence interval. More than one confidence interval might result, because the ln likelihood function for the NEC can deviate substantially from a simple parabola, which is the shape that it should have if the large-sample theory for parameter estimation applied. This is why this way of obtaining confidence sets is much more reliable than making use of the Asymptotic Standard Deviation (ASD, given in the parameter tables). The profile ln likelihood functions for the NEC in the toxicity tests on fish body growth and algae population growth are close to the expected parabola in these examples. DEBtox can also produce the numerical values of the interval estimates. Note that, generally, the NEC cannot easily be guessed from the concentration profiles, because this threshold corresponds with the $EC_{0,t}$, while we have that $NEC = EC_{0,\infty}$.

The confidence intervals for the NEC in Figure 9.3 are not small. The main reason is the uncertainty about the value of the elimination rate. If other information about this rate could be supplied (for instance from direct measurements, comparison with other toxicity tests, or quantitative structure activity relationships), the confidence interval for the NEC could be reduced substantially.

DEBtox can also be used to test parameter values statistically and to extract all kinds of information about the effect surface, such as EC_x values. This is not illustrated here.

DYNAMIC VERSUS STATIC APPROACHES

Toxicity Comparisons

Apart from comparing different models for the same data, we can and should compare data from different toxicity tests (different species of test organisms, different test chemicals, different endpoints). These comparisons yield some arguments that play a role in the comparison of the different models to the same data, and are therefore presented first.

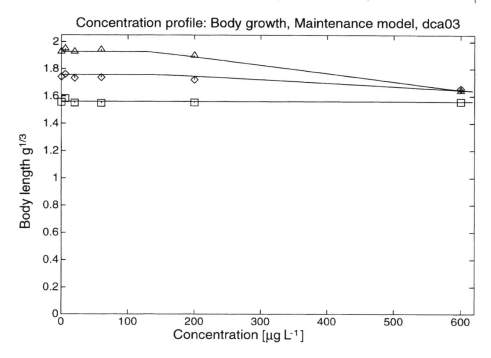

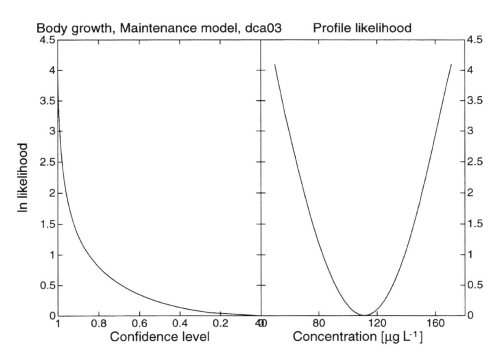

Figure 9.2a,b. Effects of DCA on the body growth of the rainbow trout *Oncorhynchus mykiss*. Data from the OECD ring test 1988/9. Cost of maintenance has been selected as the target parameter. The profile ln likelihood function for the NEC, given in the right of the upper graph, is close to its large sample shape: the parabola.

Table 9.3. Effects of Phenol on the Reproduction of the Water Flea *Daphnia magna*. Data from the OECD Ring Test 1994/5. Cost of Growth has been Selected as the Target Parameter.

Reproduction, Growth Model			ASD	Correlation Coefficients		
NEC	2.157	mg L^{-1}	0.117			
Tolerance concentration	0.702	mg L^{-1}	0.133	−0.193		
Maximal reproduction rate	30.94	No. d^{-1}	0.359	−0.189		
Elimination rate	0.6689	d^{-1}	0.124	0.122	0.950	−0.086
Von Bertalanffy growth rate	0.1	d^{-1}				
Scaled length at birth	0.13					
Scaled length at puberty	0.42					
Energy investment ratio	1					
Mean deviation	10.95					

Solubility in Fat

Solubility in fat is a property that is very relevant to the toxicity of a chemical. The linear effect model (i.e., the effect component of the DEB-based model), assumes that effective molecules operate independently. Since the ultimate number of molecules in an organism is directly proportional to the octanol/water partition coefficient, P_{ow}, the tolerance concentration and the NEC should be proportional to P_{ow}^{-1}, and the killing rate to P_{ow}. The latter relationship directly follows from the argument that the inverse of the killing rate is proportional to a tolerance concentration, as is obvious from the dimensions of the killing rate.

The symmetry argument states that the uptake flux depends on P_{ow} in the same way as the elimination flux depends on P_{wo}, while $P_{wo} = P_{ow}^{-1}$. This argument directly results in the expectation that the uptake rate is proportional to $\sqrt{P_{ow}}$ and the elimination rate is proportional to $1/\sqrt{P_{ow}}$. This logic is easily seen when we realize that the BCF equals the ratio of the uptake and the elimination rate, while it is proportional with P_{ow}.

The strength of these simple relationships becomes obvious if we have the results of a toxicity test with a chemical with a P_{ow} of P_1 available and wonder about the toxicity of another test chemical with a P_{ow} of P_2. The relationships tell us to multiply the elimination rate with $\sqrt{P_1/P_2}$, and the NEC and tolerance concentration with P_1/P_2. These three statistics define the complete $EC_{x,t}$ behavior of the second chemical. These expectations can help a lot in choosing the concentrations that are to be used in a toxicity test for the second chemical. This increase in efficiency helps to reduce the financial costs of a toxicity test.

Figure 9.5 presents the NEC, killing rate, and elimination rate as functions of the octanol/water partition coefficients for alkyl benzenes. The theoretical predictions seem to apply, but the scatter, particularly for the elimination rate, is substantial.

The simplicity of the relationships of the DEB-based model parameters with the P_{ow} also helps to detect patterns in toxicity, comparing many test chemicals.

The EC_{50} of the logit model depends on the P_{ow} in a much more complex way, due to the fact that it depends on the standardized exposure time. Chemicals with a sufficiently small P_{ow} reveal their toxic properties fully during the toxicity test, but chemicals with a large P_{ow} do not, and the apparent toxicity will increase if the test is extended. This can be understood on the basis of the relationship between the elimination rate and the P_{ow}, as mentioned

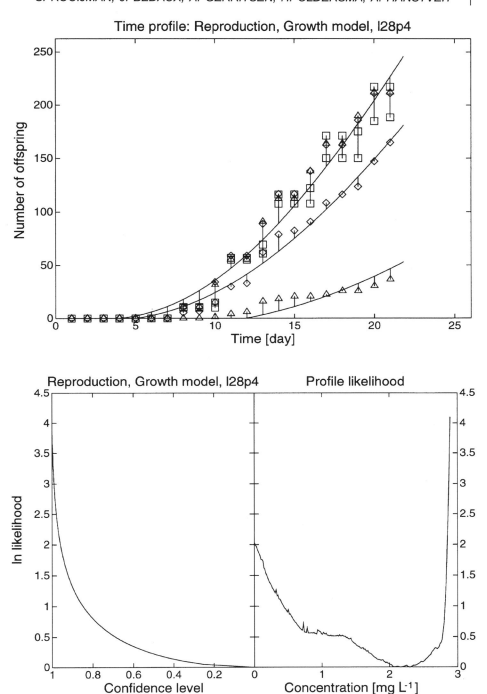

Figure 9.3a,b. Effects of phenol on the reproduction of the water flea *Daphnia magna*. Data from the OECD ring test 1994/5. Cost of growth has been selected as target parameter. The little "teeth" in the profile ln likelihood function for the NEC, given in the graph on the left, are artifacts that resulted from numerical integrations to obtain the reproductive output. They are not typical; their occurrence depends on parameter values.

Table 9.4. Effects of $K_2Cr_2O_7$ on the Population Growth Rate of *Skeletonema costatum*. Data are from the ISO Ring Test. Cost of Growth has been Selected as the Target Parameter.

Population growth, Growth Model		ASD	Correlation Coefficients			
Inoculum size	5.663 $\cdot 10^3$ cells mL^{-1}	0.685				
Population growth rate	2.1 d^{-1}	0.063	−0.997			
NEC	0.7769 mg L^{-1}	0.031	0.118	−0.139		
Tolerance concentration	3.458 mg L^{-1}	0.307	−0.582	0.578	−0.524	
Mean deviation	7.213 $\cdot 10^3$ cells mL^{-1}					

above. The same problem applies to the NOEC. The power model $EC_{50} = a(P_{ow})^b$ is usually fitted, but it is only based on empirical arguments.

Size of the Organism

The EC_{50} and NOEC not only relate in a more complex way to the solubility in fat—compared to the dynamic parameters—but also to the body size of the test organism, by exactly the same argument. The larger the test organism, the longer it takes for effects to build up. This means that the LC_{50} for daphnia and fish are difficult to compare. The comparison of the parameters of the DEB-based model is easy, because the NEC and the killing rate (or tolerance concentration) do not depend on body size. The elimination rate is inversely proportional to a volumetric length, on the assumption that the exchange rate between organism and environment are proportional to the surface area of the organism. It is the only parameter that depends on body size. Figure 9.6 illustrates that these simple considerations make sense.

Comparisons on the Basis of Parameters

The logit model has three parameters: the blank response, the EC_{50} (LC_{50}) and the slope. The NOEC is inconsistent with the logit model. Nonetheless, it is frequently presented with the logit parameters in practice. It can be viewed as a fourth parameter which is "estimated" in a rather odd way. (It can take a very limited number of values and its estimation procedure has regrettable properties.) An extrapolation factor is used (except in population growth tests) to extrapolate to ultimate effects. This factor counts as a fixed parameter. The problem with this 'parameter' is that it relates to toxic effects, rather than to responses in the blank, which implies that it should not be the same for all chemicals.

The DEB-based model has four parameters: a blank response, a NEC, a tolerance concentration (or killing rate), and an elimination rate. Respectively, for growth and reproduction tests only, it also has three or four parameters that are not estimated from the data. All these fixed parameters refer to details in the description of what happens in the blank. The blank response in the test does not provide the proper information for the fixed parameters. These parameters are

- scaled length at birth, i.e., the length at birth as a fraction of the maximum length in the blank, if the test were to continue in time. This parameter applies to the toxicity

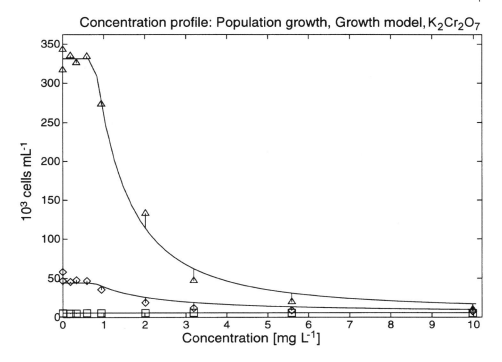

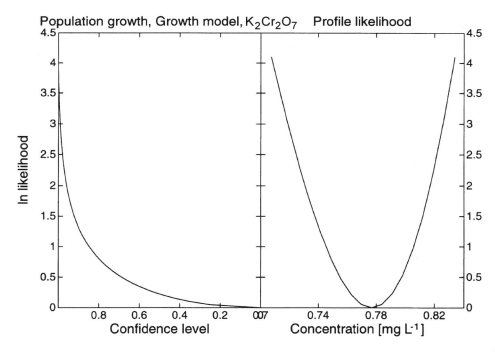

Figure 9.4a,b. Effects of $K_2Cr_2O_7$ on the population growth rate of *Skeletonema costatum*. Data are from the ISO ring test. Cost of growth has been selected as the target parameter. The profile ln likelihood function for the NEC, given in the graph on the bottom right, is close to its large sample shape: the parabola.

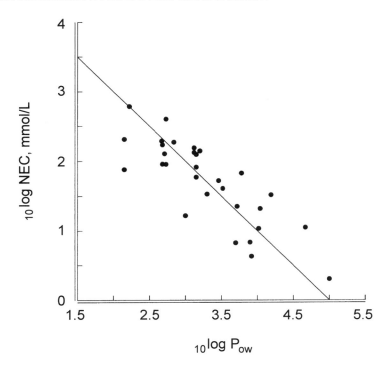

Figure 9.5a. The NEC of alkylbenzenes as a function of the octanol/water partition coefficient. The nitrobenzenes have been excluded. The slope of the line, i.e., −1, follows from simple theoretical considerations. The data are from the 4d toxicity tests on survival of the fathead minnow, as presented in Geiger et al. (1985; 1986; 1988; 1990). The partition coefficients were obtained from Richardson and Gangolli (1995) or calculated according to Rekker (1977).

test for reproduction. That for body growth uses a related parameter: the actual length at the start of the experiment. Although this fixed parameter can be treated as a free parameter, DEBtox treats it as a fixed one, because the size at the start of the experiment is frequently not measured in the case of early life stage tests.

- scaled length at puberty, i.e., the length at puberty as a fraction of the maximum length in the blank, if the test were to continue in time. This parameter only applies to reproduction. Puberty is defined as the moment at which allocation to reproduction starts, which is somewhat earlier than the moment of first reproduction.
- energy investment ratio is a dimensionless parameter with a rather complex interpretation (Kooijman, 1993). Realistic values for animals such as daphnia and fish are around 1. Numerical studies indicate that variations around this value have very little effect on the toxicity parameters.
- von Bertalanffy growth rate, with dimension $time^{-1}$. The value of this parameter is more important than the previous one, but it can be measured easily and is known for hundreds of species.

The fixed parameters depend on the species, or even the strain, the culture conditions, and details of the experimental protocol. If a particular laboratory has standardized its experimental protocol, these parameters need to be tuned only once, and not for each test. Part

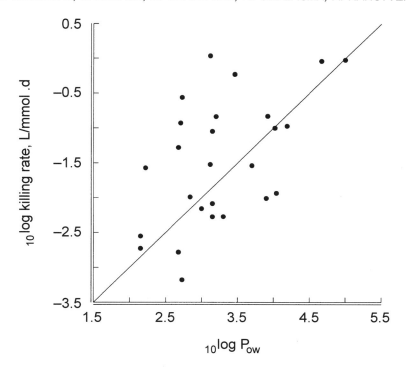

Figure 9.5b. The killing rate of alkylbenzenes as a function of the octanol/water partition coefficient. The nitrobenzenes have been excluded. The slope of the line, i.e., 1, follows from simple theoretical considerations. The data are from the 4-d toxicity tests on survival of the fathead minnow, as presented in Geiger et al. (1985; 1986; 1988; 1990). The partition coefficients were obtained from Richardson and Gangolli (1995) or calculated according to Rekker (1977).

of the differences in toxicity results among laboratories can be attributed to differences in fixed parameters. Good estimates can be given for the standard choices of species on the basis of existing ecophysiological data (Kooijman and Bedaux, 1996).

The elimination rate in the DEB-based model stands for the rate at which effects build up during exposure. Its conceptual role is more or less comparable to the extrapolation factor in the static approach, but differs in a statistical sense. It is an ordinary model parameter, for which point-estimates as well as interval-estimates are available. The slope parameter and the EC_{50} in the logistic model together play the same role as the tolerance concentration in the DEB-based model.

We can conclude that the DEB-based model has fewer parameters that are to be estimated from the results of the toxicity test than the logit model plus NOEC. On the basis of this measure for the complexity of a model, the simplest dynamic model is, therefore, simpler than the static one for all four toxicity tests.

Numerical Comparisons

In this section we present numerical comparisons between the DEB-based model and the logit model plus the NOEC. Since the logit model only uses the response at the end of the

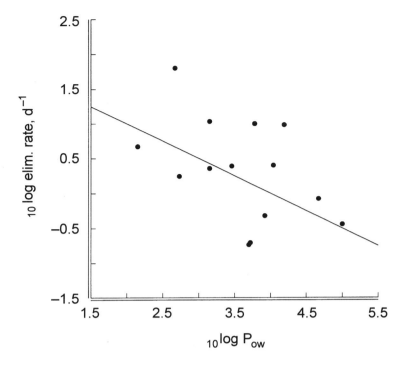

Figure 9.5c. The elimination rate of alkylbenzenes as a function of the octanol/water partition coefficient. The nitrobenzenes have been excluded. The slope of the line, i.e., −0.5, follows from simple theoretical considerations. The data are from the 4-d toxicity tests on survival of the fathead minnow, as presented in Geiger et al. (1985; 1986; 1988; 1990). The partition coefficients were obtained from Richardson and Gangolli (1995) or calculated according to Rekker (1977).

experiment (apart from the algal growth inhibition test), we restrict the comparisons to EC_x values for that exposure time. We compare the NOEC with the DEB-based EC_0 for that exposure time, and not with the NEC, because the NEC relates to long-term exposure. The aim of this section is to compare statistics that are familiar to users of the static approach, not to show the potential of the dynamic one.

All calculations for the DEB-based model have been done with the software package DEBtox, as provided in Kooijman and Bedaux (1996). All calculations are based on nominal concentrations, except for the survival analyses, which are based on measured concentrations.

We fitted models for different modes of action of the compounds (i.e., target parameters) and noted that the differences in goodness of fit were generally small, and resulted in very similar values for the NEC.

Survival Experiments

Figure 9.7 compares the $LC_{x.4d}$ values for fathead minnows on the basis of the logit and the DEB-based model for many different chemicals. The logit parameters have been esti-

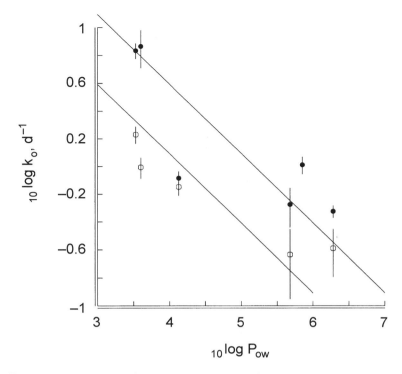

Figure 9.6. The elimination rate of alkylbenzenes as a function of the octanol/water partition coefficient for juvenile *Daphnia pulex* with a body length of 1 mm (•) and for adults of 3 mm (°). The lines correspond to the elimination rate = $392 / \sqrt{P_{ow}}$ d^{-1} for juveniles and $124 / \sqrt{P_{ow}}$ d^{-1} for adults. The ratio between these elimination rates, i.e., 3.17, corresponds well with the ratio of the body lengths, as expected from simple theoretical considerations. The data are from the 2-d toxicity tests on survival, as presented in Hawker and Connell (1986).

mated according to the maximum likelihood method, as described in Kooijman (1981). We can conclude that the LC_{50} and LC_{20} values are very similar for both models, but the LC_5 values for the DEB-based model tend to be higher than those of the logit model. This is to be expected, because $LC_{x \to 0}$ for x → 0 for the logit model, but $LC_{0.4d} \geq$ NEC for the DEB-based model. The fact that the LC_x values of the logit and DEB-based models correlate well is not surprising, because both models are fitted to the same data; very toxic chemicals result in low LC_x values with both methods.

Body Growth Experiments

We used the data from the final ring test, organized by the OECD in 1988, with 3,4-dichloroaniline (DCA), for the rainbow trout *Oncorhynchus mykiss*. Since the volumetric lengths (i.e., the cubic root of weights) at the end of the toxicity tests differed little from those at the start, an analysis in terms of EC_x values for the logit model is less appropriate: The EC_{50} would far exceed the highest tested concentration and the results would be most unreliable. We only present the results of the comparison of the NOEC with the $EC_{0.28d}$ in Figure 9.8.

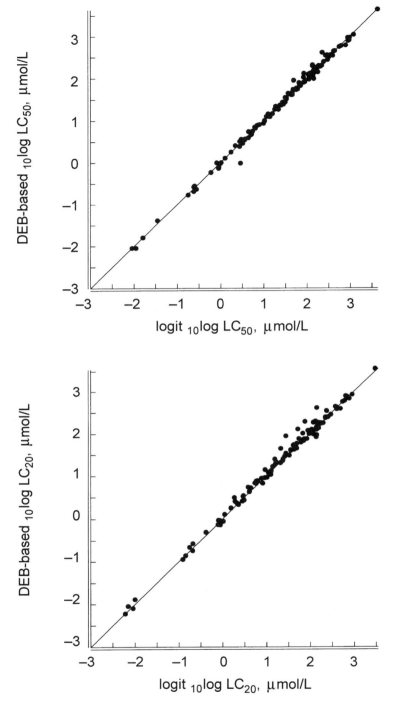

Figure 9.7a,b. $LC_{x.4d}$ comparisons for the logit and the DEB-based model for fathead minnows. The 121 data sets are from Brooke et al. (1984) and Geiger et al. (1985; 1986; 1988; 1990).

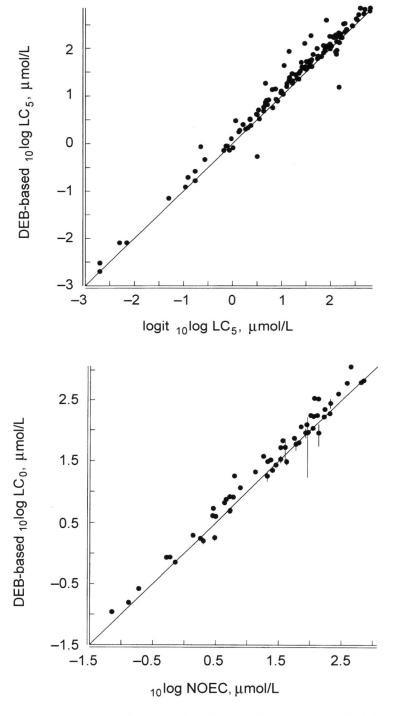

Figure 9.7c,d. $LC_{x.4d}$ comparisons for the logit and the DEB-based model for fathead minnows. The DEB-based EC_0 is plotted against the NOEC in the bottom graph. The bars indicate the estimated standard deviations. The 121 data sets are from Brooke et al. (1984) and Geiger et al. (1985; 1986; 1988; 1990).

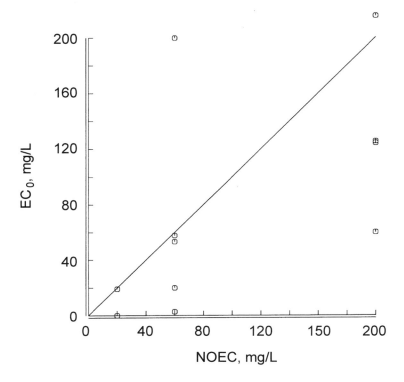

Figure 9.8. The $EC_{0.28d}$ for the DEB-based model for effects of DCA on the body growth of the rainbow trout is plotted against the NOEC on the basis of the Williams test for 11 toxicity tests of the OECD ring test 1988. Effects on growth via effects on maintenance costs turned out to be the best fitting DEB-based model.

The NOECs were identified on the basis of the test by Williams (Williams, 1972), using a significance level of 5%. There were 16 individuals for each concentration of DCA.

The DEB-based model for effects on maintenance fitted best, although the differences in goodness of fit with the models for effects on growth and assimilation were usually small. The fixed parameters have been set at: initial length = mean initial length, energy investment ratio = 1, von Bertalanffy growth rate = 0.01 d^{-1}.

Reproduction Experiments

We used the data from the final ring test, organized by the OECD in 1994/5, with 3,4-dichloroaniline (DCA), cadmium chloride, and phenol (Figures 9.9 and 9.10). The NOEC values for these data were taken from Anonymous (1995), as listed for the case where the reproduction of females that die during the test are excluded.

The logit model has been fitted with SYSTAT version 5.02 using nonlinear regression on the number of juveniles per adult for individuals that did not die in 21 days. The EC_x values and their standard deviations have been obtained via reparameterization of the logit model.

The data from DEBtox represent the cumulative total reproduction per living female, including all observation times. The reproduction of a female that died during the test has been included. No delay of reproduction could be observed for cadmium, and we selected the model for effects on the hazard of the ovum as the best fitting model for direct effects on reproduction. This model is rather similar to the other direct effect model, i.e., effect on costs per young. Where difference in goodness of fit for both models was relatively large (but still small absolutely), the hazard model fit best. Some delay of reproduction could be observed for DCA and phenol, and we selected the model for effects on the growth costs as the best fitting model for indirect effects on reproduction. Since no data on the size of the adults are available, we could not test the predicted effects on growth. The fixed parameters have been set to the default settings of DEBtox: i.e., scaled length at birth = 0.13, scaled length at puberty = 0.42, energy investment ratio = 1, von Bertalanffy growth rate = 0.1 d^{-1}.

The growth model (or hazard model, respectively) could be fitted to all data sets—without problems, using DEBtox. The NECs and tolerance concentrations for data sets—without effect of the toxicant—were large with huge standard errors, as could be expected. We did not include data sets into the comparisons where the $EC_{50.21d}$ was much larger than the highest tested concentration.

Figures 9.9 and 9.10 present the comparisons of the EC_x values and the NOECs for the chemicals DCA, cadmium, and phenol. The EC_x values are quite comparable for both models if $x \geq 5$. The standard deviations of the EC_x values of the logit model tend to be somewhat larger than for the DEB-based model. The EC_1 values of the logit model tend to be lower than that of the DEB-based one. This can be expected, because the $EC_{x \to 0}$ if $x \to 0$ for the logit model, while $EC_0 \geq NEC$ for the DEB-based model. We also see that the scatter in the NOECs is larger than the scatter in the EC_0 for two of the three chemicals.

Population Growth Experiments

We analyzed the data on the effects of 3,5-dichlorophenol (DCP) and potassium dichromate from the ISO ring test with the diatoms *Skeletonema costatum* and *Phaeodactylum tricornutum* (marine algal growth inhibition test, ISO 10253, Geneva, 1995).

The logit model was fitted to the data using nonlinear regression, as described in Kooijman et al. (1983), where the population growth rate decreases logistically with the concentration of test chemical. The standard deviations of the EC_x values were obtained from those of the EC_{50}, the slope, and the covariance of both parameters, on the basis of the formulae presented in Kooijman and Bedaux (1996).

The DEB-based model for effects on growth costs fitted best for dichromate, while that for effects on the hazard rate fitted best for DCP. Since the DEB-based model assumes that toxicant kinetics are fast with respect to population growth, we have that $EC_0 = NEC$.

Figure 9.11 compares the EC_x values and the NOECs. The two data sets for dichromate, where the EC_{50} and the EC_{20} are small for the logit model and large for the DEB-based one, have already been indicated by Hanstveit (1991) as outliers. For DCP we have the opposite situation: two data sets for which the EC_{50} is large for the logit model and small for the DEB-based one. We see that the logit model has relatively small standard deviations for the EC_{50}, but they rapidly become larger for the smaller effect levels to the extent that they become meaningless in quite a few data sets.

No proper NOEC could be identified in 4 of the 12 data sets for DCP and 6 of the 15 data sets for dichromate. These values have been set to zero in the graphs. The NEC was estimated to be zero in 1 data set for DCP and 1 for dichromate.

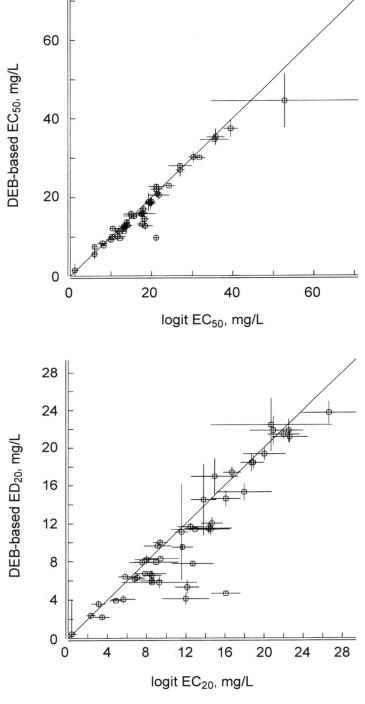

Figure 9.9a,b. $EC_{x.21d}$ comparisons for the logit and the DEB-based model for the Daphnia reproduction test on DCA. The DEB-based $EC_{0.21d}$ is plotted against the NOEC. The bars indicate the estimated standard deviations. The data sets are from the final ring test of the OECD 1994/5. Four NOECs could not be obtained, and have been set to zero in the graph.

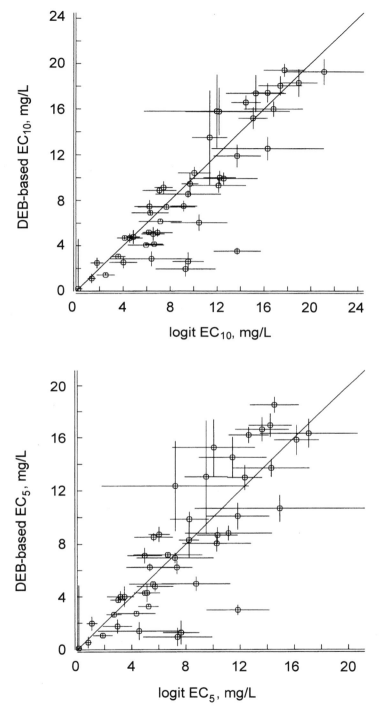

Figure 9.9c,d. $EC_{x.21d}$ comparisons for the logit and the DEB-based model for the Daphnia reproduction test on DCA. The DEB-based $EC_{0.21d}$ is plotted against the NOEC. The bars indicate the estimated standard deviations. The data sets are from the final ring test of the OECD 1994/5. Four NOECs could not be obtained, and have been set to zero in the graph.

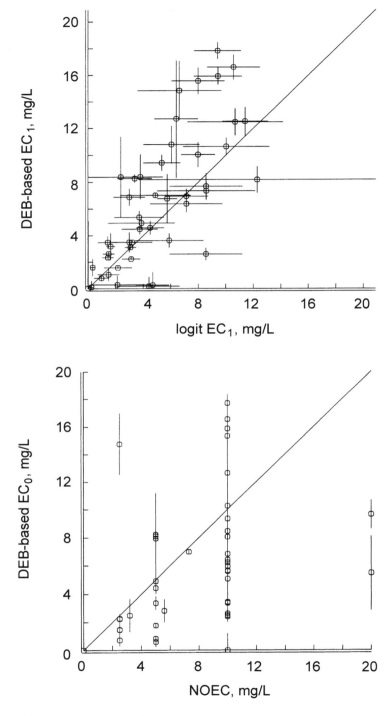

Figure 9.9e,f. $EC_{x.21d}$ comparisons for the logit and the DEB-based model for the Daphnia reproduction test on DCA. The DEB-based $EC_{0.21d}$ is plotted against the NOEC. The bars indicate the estimated standard deviations. The data sets are from the final ring test of the OECD 1994/5. Four NOECs could not be obtained, and have been set to zero in the graph.

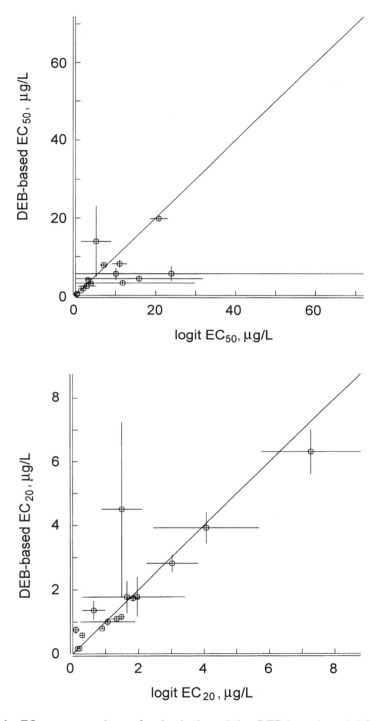

Figure 9.10a,b. $EC_{x.21d}$ comparisons for the logit and the DEB-based model for the Daphnia reproduction test on cadmium and phenol. The DEB-based $EC_{0.21d}$ is plotted against the NOEC. The bars indicate the estimated standard deviations. The data sets are from the final ring test of the OECD 1994/5.

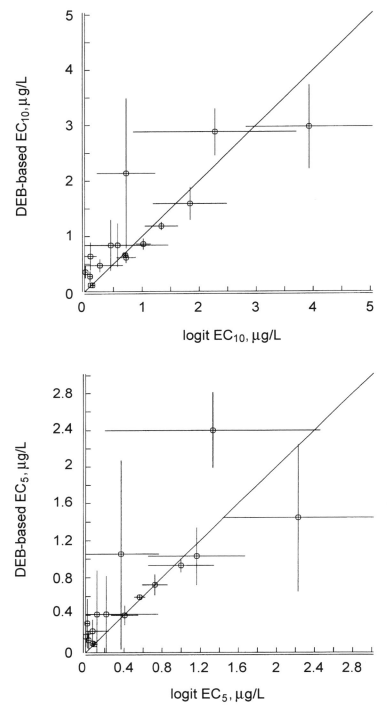

Figure 9.10c,d. $EC_{x.21d}$ comparisons for the logit and the DEB-based model for the Daphnia reproduction test on cadmium and phenol. The bars indicate the estimated standard deviations. The data sets are from the final ring test of the OECD 1994/5.

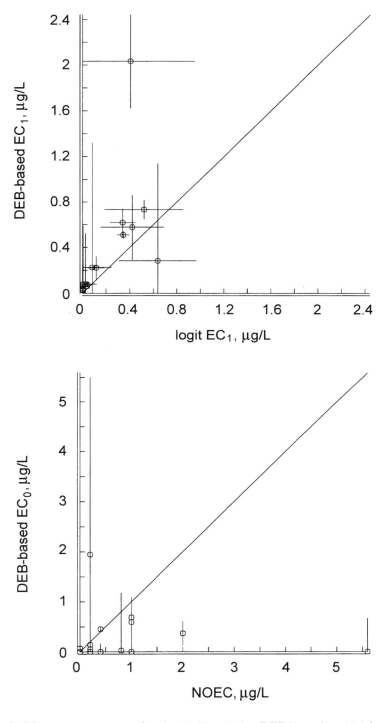

Figure 9.10e,f. $EC_{x.21d}$ comparisons for the logit and the DEB-based model for the Daphnia reproduction test on cadmium and phenol. The DEB-based $EC_{0.21d}$ is plotted against the NOEC. The bars indicate the estimated standard deviations. The data sets are from the final ring test of the OECD 1994/5.

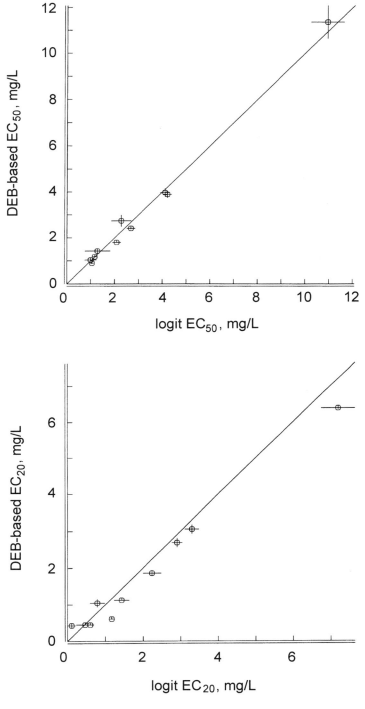

Figure 9.10g,h. $EC_{x.21d}$ comparisons for the logit and the DEB-based model for the Daphnia reproduction test on cadmium and phenol. The DEB-based $EC_{0.21d}$ is plotted against the NOEC. The bars indicate the estimated standard deviations. The data sets are from the final ring test of the OECD 1994/5.

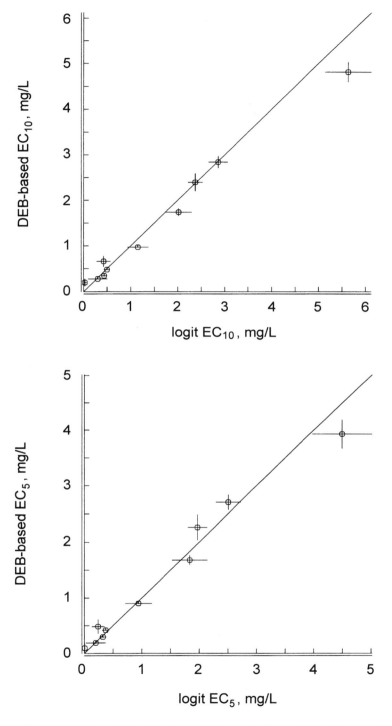

Figure 9.10i,j. $EC_{x.21d}$ comparisons for the logit and the DEB-based model for the Daphnia reproduction test on cadmium and phenol. The DEB-based $EC_{0.21d}$ is plotted against the NOEC. The bars indicate the estimated standard deviations. The data sets are from the final ring test of the OECD 1994/5.

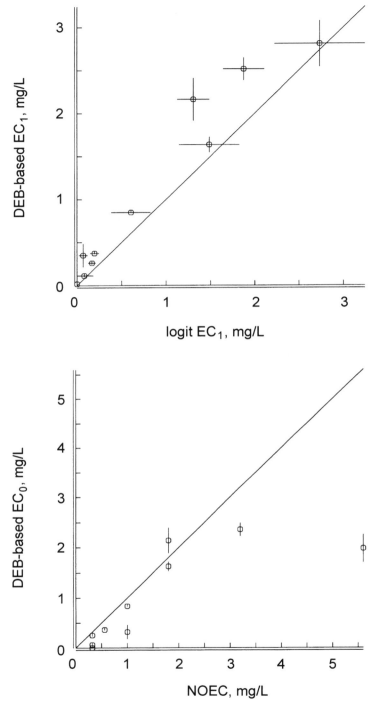

Figure 9.10k,l. $EC_{x.21d}$ comparisons for the logit and the DEB-based model for the Daphnia reproduction test on cadmium and phenol. The DEB-based $EC_{0.21d}$ is plotted against the NOEC. The bars indicate the estimated standard deviations. The data sets are from the final ring test of the OECD 1994/5.

Advantages of the Dynamic Approach

- The DEB-based model allows the estimation of the NEC, which has good statistical properties, including an interval estimate.
- The dynamic approach needs no extrapolation factor to arrive at long-term effects. The elimination rate plays this role.
- The comparison of results of different toxicity tests is easy on the basis of the DEB-based model, because (i) its parameters are independent of the exposure time, (ii) it gives simpler QSARs and (iii) it gives simpler relationships with the body size of the test animals.
- The static parameters (EC_x values) can be obtained from dynamic ones, but not vice versa, which illustrates that the dynamic approach extracts more information from the same toxicity data.
- The dynamic approach uses all available data, while the static approach uses only the data at the end of the exposure. It has fewer parameters to be estimated from the data than the static approach.
- Additional information, such as the elimination rate, derived from toxicokinetic data, can readily be used in the dynamic approach.
- The effects of repeated exposures and of pulse exposures can be evaluated readily, because of the dynamic properties of the DEB-based model.
- The toxicity of mixtures of compounds can readily be evaluated, including interactions between different compounds, due to the linearity of effect component of the DEB-based model.
- The effects of mutagenic compounds and ionogenic radiation can be quantified via effects on the target parameter "ageing acceleration" (Kooijman, 1993).
- Environmental risk assessment concerns effects of toxicants on "natural" populations, not on individuals in a laboratory. To evaluate the consequences of toxic effects on individuals for population dynamics, we have to know how survival and reproduction change during the lifetime of an organism. This requires a dynamic approach.
- The DEB-based model is mechanistic, based on biological ideas. It allows a series of models from simple (routine testing) to complex (research) by changing the kinetics component.

Disadvantages of the Dynamic Approach

- The dynamic approach requires rather advanced numerical techniques for statistical evaluations. This problem is solved by the software package DEBtox.
- The blank component in the dynamic approach is more complex than in the static approach and involves biological knowledge. This knowledge also must be used in the application of the toxicity results in environmental risk assessment. Fixed parameters for the blank response reflect this biological realism. Supplementary ecophysiological data are required for each species of test organism to determine the values. These data are available for the species that are frequently used, and appropriate values for the fixed parameters are known.
- The different modes of action of the various compounds relate to different target parameters, which hampers the comparison of the toxicity of such compounds to some extent. The proper identification of the mode of action is not always feasible with the present experimental design. Additional observations, such as body length at the end

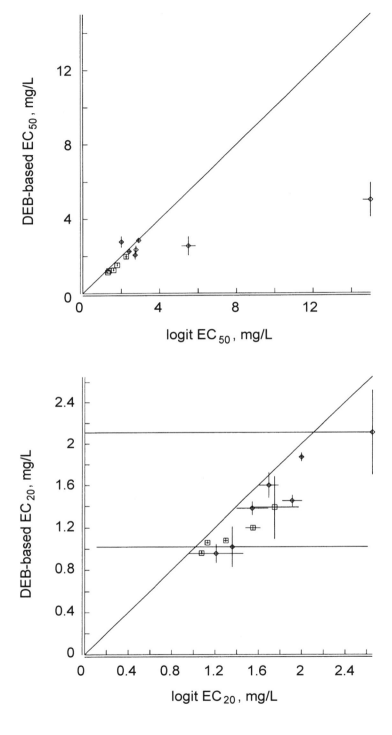

Figure 9.11a,b. EC_x comparisons for the logit and the DEB-based model for the algal growth inhibition test on dichromate and DCP. The bars indicate the estimated standard deviations. The data sets are from ISO ring tests.

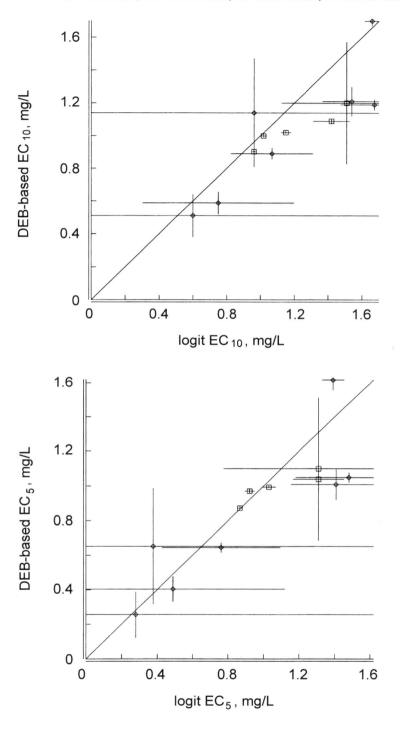

Figure 9.11c,d. EC_x comparisons for the logit and the DEB-based model for the algal growth inhibition test on dichromate and DCP. The bars indicate the estimated standard deviations. The data sets are from ISO ring tests.

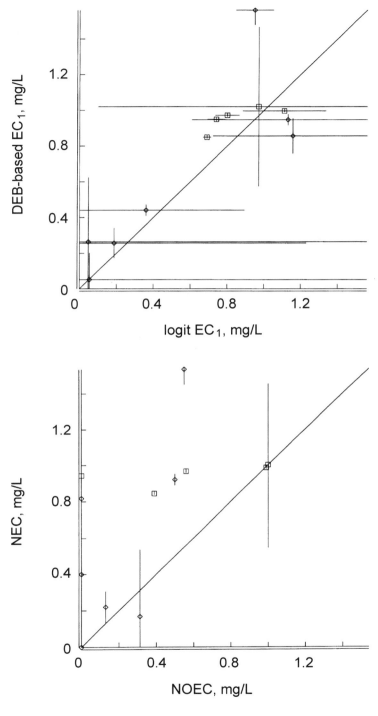

Figure 9.11e,f. EC_x comparisons for the logit and the DEB-based model for the algal growth inhibition test on dichromate and DCP. The DEB-based NEC is plotted against the NOEC in the bottom graph. The bars indicate the estimated standard deviations. The data sets are from ISO ring tests.

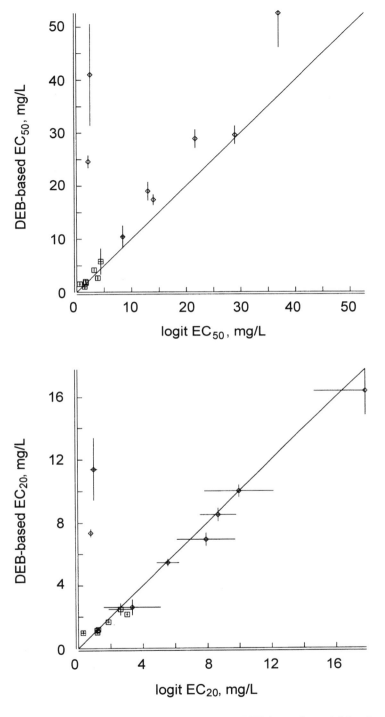

Figure 9.11g,h. EC_x comparisons for the logit and the DEB-based model for the algal growth inhibition test on dichromate and DCP. The bars indicate the estimated standard deviations. The data sets are from ISO ring tests.

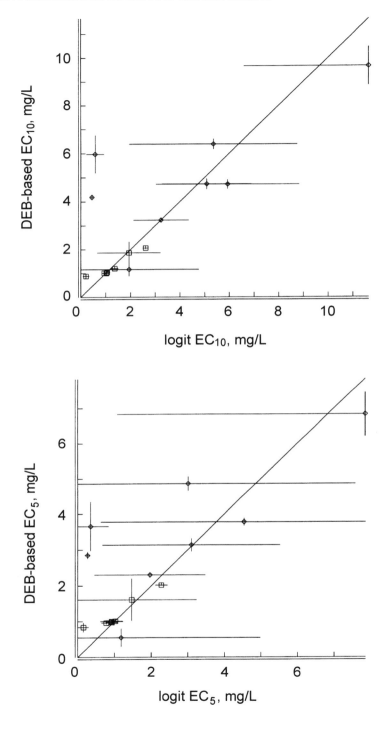

Figure 9.11i,j. EC_x comparisons for the logit and the DEB-based model for the algal growth inhibition test on dichromate and DCP. The bars indicate the estimated standard deviations. The data sets are from ISO ring tests.

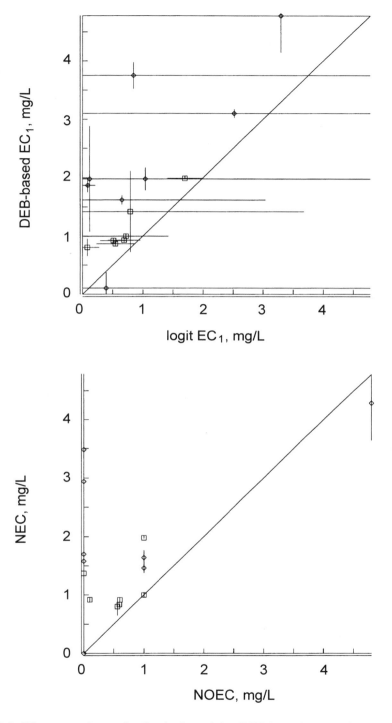

Figure 9.11k,l. EC_x comparisons for the logit and the DEB-based model for the algal growth inhibition test on dichromate and DCP. The DEB-based NEC is plotted against the NOEC in the bottom graph. The bars indicate the estimated standard deviations. The data sets are from ISO ring tests.

of the daphnia reproduction experiment and/or the measurement of the feeding rate, would help substantially. Numerical results indicate that the proper identification of the mode of action is not essential for a reliable estimation of the NEC.

- The dynamic approach does make assumptions about the effects—assumptions that can be wrong. One should realize, however, that every approach is based on assumptions. The use of a fixed extrapolation factor to transform the estimated toxicity into a long-term toxicity in the static approach, for instance, assumes that this factor is the same for all compounds. The assumptions are more explicit and more realistic in the dynamic approach.

REFERENCES

Anonymous. Draft Report of the Final Ring Test of the *Daphnia magna* Reproduction study. Draft OECD test guideline 202, 1995.

Brooke, L.T., D.L. Geiger, D.J. Call and C.E. Northcott. Acute toxicities of organic chemicals to fathead minnow (*Pimephales promelas*) Vol 1. Center for Lake Superior Environmental Studies, University of Wisconsin–Superior, USA, 1984.

Geiger, D.L., C.E. Northcott, D.J. Call, and L.T. Brooke. Acute toxicities of organic chemicals to fathead minnow (*Pimephales promelas*) Vol 2. Center for Lake Superior Environmental Studies. University of Wisconsin–Superior, USA, 1985.

Geiger, D.L., S.H. Poirier, L.T. Brooke, and D.J. Call. Acute toxicities of organic chemicals to fathead minnow (*Pimephales promelas*) Vol 3. Center for Lake Superior Environmental Studies. University of Wisconsin–Superior, USA, 1986.

Geiger, D.L., D.J. Call, and L.T. Brooke. Acute toxicities of organic chemicals to fathead minnow (*Pimephales promelas*) Vol 4. Center for Lake Superior Environmental Studies. University of Wisconsin–Superior, USA, 1988.

Geiger, D.L., L.T. Brooke, and D.J. Call. Acute toxicities of organic chemicals to fathead minnow (*Pimephales promelas*) Vol 5. Center for Lake Superior Environmental Studies. University of Wisconsin–Superior, USA, 1990.

Hanstveit, A.O. *The Results of an International Ring Test of Marine Algal Growth Inhibition Test According to ISO/DP 10253.* TNO Report 91/236, 1991.

Hawker, D.W. and D.W. Connell. Bioconcentration of lipophilic compounds by some aquatic organisms. *Ecotoxicol. Environ. Saf.* 11, pp. 184–197, 1986.

Kooijman, S.A.L.M. Parametric analyses of mortality rates in bioassays. *Water Res.* 15, pp. 107–119, 1981.

Kooijman, S.A.L.M. *Dynamic Energy Budgets in Biological Systems. Theory and Applications in Ecotoxicology.* Cambridge University Press, Cambridge, 1993, p. 350.

Kooijman, S.A.L.M. and J.J.M. Bedaux. *The Analysis of Aquatic Toxicity Data.* VU University Press, Amsterdam, 1996, p. 160 + floppy disk.

Kooijman, S.A.L.M., A.O. Hanstveit, and H. Oldersma. Parametric analyses of population growth in bioassays. *Water Res.* 17, pp. 727–738, 1983.

Rekker, R.F. The hydrophobic fragmental constant. Its derivation and application. A means of characterizing membrane systems. In *Pharmacochemistry Library.* Nauta, W.Th. and R.F. Rekker, Eds., Elsevier, Amsterdam, 1977.

Richardson, M.L. and S. Gangolli. *The Dictionary of Substances and Their Effects.* Publication of the Royal Society of Chemistry, Cambridge, 1995, p. 6844.

Williams, D.A. The comparison of several dose levels with a zero dose control. *Biometrics* 28, pp. 519–531, 1972.

10 | Application of Community Level Toxicity Testing to Environmental Risk Assessment

Robin A. Matthews, Geoffrey B. Matthews, and Wayne G. Landis

INTRODUCTION

This chapter reviews quantitative methods used in community-level toxicity testing, emphasizing methods that assist with understanding environmental risk. We use a broad definition of community-level toxicity testing, including laboratory multispecies toxicity tests and effluent testing in the field. We will discuss both large-scale studies that measure many physical, chemical, and biological parameters, as well as "reductionist" studies that measure only a small number of community indicators. Large-scale ecotoxicology studies are patterned after the descriptive, holistic philosophy of traditional ecological studies, where the effect must first be defined before it is tested for causal relationships (Munkittrick and McCarty, 1995). Large-scale studies have been the basis for environmental monitoring of effluents, spills, widespread pollution, and situations where the effects of a pollutant are not well-known. These studies produce complex, multivariate data sets filled with random, redundant, and noisy variables, an odd assortment of measurement units, and many other features that make them difficult to analyze.

Reductionist studies are more closely related to human health toxicology, where laboratory studies are used to link chemical exposure to a specific physiological response. The classic reductionist approach is to use results from single-species laboratory tests to set discharge standards, using the rationale that if we set discharge levels low enough to protect sensitive "indicator" species, we will also protect the less sensitive species in the community. The proponents and opponents of this rationale argue heatedly over whether single-species data are effective for determining ecological risk, but that debate is outside the scope of this chapter. Our focus will be on evaluating the use of biotic indices in ecotoxicology. In particular, we will review biotic indices that attempt to quantify community-level effects without the complexity and expense of a large-scale study.

In the next sections we describe three major tasks that are needed to apply community-level toxicity testing to environmental risk assessment: 1) creating an appropriate community-level data set; 2) extracting information from the data set; and 3) converting the information from the specific data set into a set of expectations or recommendations that address the question of risk. Each of these tasks is critical to the success of the environmental risk assessment, and each presents difficult challenges to scientists and regulators. Our emphasis will be on water quality monitoring and aquatic toxicity testing. This is partly due to our own expertise in these areas, but also reflects the 90-year history of using biotic indices to evaluate community-level responses to changes in water quality. We acknowledge the innovative, multivariate approaches that are used by our terrestrial colleagues, and point out that many of the conventional multivariate procedures now being used were developed to analyze terrestrial data.

CREATING A DATA SET

Creating an appropriate community-level data set involves selecting measurements that are likely to show a direct or indirect effect resulting from exposure to the contaminant. This requires some *a priori* knowledge (or a professional guess) about how different organisms may react to the pollutant and which ones are important. However, even large-scale studies are severely limited in the number of variables that can be measured compared to the number of possible responses in a community. We can make a fairly good guess about some pollutants, particularly if we have relevant information from single-species toxicity tests, historical biomonitoring studies, or models of single-species or community responses. For example, almost any variable that measures algal productivity or community structure should show a response to high levels of copper. But often we don't know precisely how the community will respond. Many of the most damaging environmental pollutants have had serious, *unanticipated* impacts at the community and ecosystem level. Therefore, community-level studies often use as many different types of measurements as possible, including measurements that detect structural changes, such as loss of species diversity, as well as functional changes, such as altered primary productivity rates (see Cairns et al., 1995, for a review of functional endpoints). As a result, the data set often contains a mixture of no-response, expected-response, and unexpected-response variables.

Defining the objectives of a study is the most important step toward creating an appropriate data set. This process cannot be overemphasized! Most people who work with multivariate data have experienced the frustration of trying to find answers to questions that were not posed during the experimental design, and to which the data are ill suited to providing answers. After defining the objectives, the next step is to design an appropriate sampling protocol that will provide relevant data. There are many obvious limitations at this step, including budget and time constraints, as well as sampling and analytical limitations. Inasmuch as the number and type of samples that can be collected will always be a tiny fraction of the actual community response, choosing when, where, and what to sample will have a major impact on the results. Furthermore, the natural environment is highly variable over both space and time. Knowing this dictates that we either: a) minimize the variance; b) quantify the variance; or c) reduce the impact of variance on the data analysis process. Variance can be minimized by selecting homogeneous sampling areas, measuring variables that don't change rapidly during the sampling period, or creating an artificial, less variable community. These all have the disadvantage of underrepresenting the range of responses possible in a community. Variance can be quantified by increasing the sample size (adding replicates, sites, sampling times, etc.), but this adds to the cost of the experiment, and still may not produce enough data to define the variance for rare species or uncommon events. Often the only alternative is to reduce the impact of variance on the data analysis process. This may be accomplished through the use of nonparametric procedures, data transformations, developing ways to handle outliers, and by careful examination of the data to identify noisy and random variables.

Once samples have been collected and analyzed, and the data have been entered into a data matrix and verified, there are a number of additional considerations. Many of the parameters will have large variances and outliers. There may be missing values and censored data (above or below the detection limits). The data set may include a combination of continuous and discrete data, including categorical, binary, and count data. Measurement units may be different, even for similar types of data, such as counts. There may be random, redundant, or dependent measurements. All of these features are common in multivariate, community-level data sets, and require data analysis decisions that affect the data analysis process. Each choice should be given careful consideration, and decisions should be carefully documented.

One final comment is that environmental data are often entered into a simple spreadsheet, with emphasis placed on the ease of data entry, rather than on the ultimate need to analyze and archive the data. This may be fine for simple data sets, but it causes problems for the community-level data set. Most multivariate statistical procedures require that the data be in a single file, in ASCII format, with only numerical entries in the file. In spreadsheets, however, important documentation (e.g., dates, sites, comments, units) is usually mixed with the data, and the data are often entered into several spreadsheets. Before any multivariate statistics can be generated, the data need to be exported into a single file, and inconsistencies in "key" variables (site names, dates, treatments, etc.) must be resolved. Comments and entries such as "bdl" or ">" must be replaced with appropriate numerical codes, and the data must be reverified. This tedious, inefficient, error-prone process can be avoided if the data entry step is thoughtfully designed.

EXTRACTING INFORMATION

Extracting information from the data set involves using appropriate analytical tools to help find both expected and unexpected patterns. It is usually good to start with simple graphical displays and descriptive statistics, even for large, multivariate data sets. For a good introduction to graphical procedures, including suggestions for multivariate plotting, see Cleveland (1985) and Helsel and Hirsch (1992). However, a graphical approach is rarely sufficient by itself, in part because of the large number of plots that can be generated (for example, with 50 variables there are over 1000 bivariate plots), and because many of the patterns in the data will be nonlinear and multivariate. Univariate pattern analysis tools are also helpful, particularly for detecting simple linear patterns such as correlations between paired variables, and for confirming the statistical significance of these patterns (see Newman, 1995, for a summary of methods applicable to ecotoxicology). Many researchers stop at this point—rather than proceeding with multivariate tests—in the belief that the data won't fit the assumptions for parametric, multivariate tests (e.g., normal distributions and homogeneity of variances) or because they are unfamiliar with multivariate procedures. However, the reason for conducting a community-level test is to find complex patterns that do not show up in single-species testing. These patterns are multivariate by their very nature. Because of this, if we are interested in looking for complex patterns, we need to become familiar with, and rely upon, multivariate statistical tools.

Multivariate statistical tools can be used to discover a wide variety of regularities in data. In ecotoxicology we are usually concerned with the same kind of question as in hypothesis testing: are two groups of samples significantly different? (However, see Chapter 12 by Dixon). In ecotoxicology our data generally have a large number of variables. The number of variables measured is often proportional to, or even much larger than, the number of samples taken. This is sometimes referred to as the "curse of dimensionality." The number of samples is grossly insufficient to describe the possibilities among the variables. For example, if each sample has 10 variables, and we decide to smooth it by binning the data into histogram cells, taking only two cells per variable will result in over a thousand histogram cells. We will need to have at least a thousand samples to generate a reliable histogram. If there are 20 variables, the number of cells jumps to over a million, and this is with the coarsest possible histogram. Few ecotoxicological studies are rich enough to contemplate collecting a million samples while, at the same time, counting 20 different species or other variables, for each sample would be regarded as a modest effort. All intuitions gained in two-, three-, or four-dimensional space are thoroughly out of place when dealing with high dimensions. In this multivariate situation, therefore, we must be very cautious to avoid overfitting the data with our model, for the data underdetermine the analysis. Almost any model framework will be able to fit the data reasonably well. The problem of finding patterns that are spurious (type I errors) is much bigger than in the univariate case.

There are essentially four techniques for finding significant differences in multivariate data: indices, metrics (similarity indices), classifiers (supervised learning, or discriminants), and clustering (unsupervised learning). Only the first one, indices, is widely known and used within ecotoxicology, even though the other three are generally better, i.e., they are more robust and more informative.

Indices

Indices reduce a multivariate data set to a univariate one. Each sample in the data set, which may consist of measurements from dozens of variables, is reduced to a single number using a formula, given *a priori*. To determine if two groups of samples are different, we need only determine whether the two index values are different. The multivariate comparison has been reduced to a univariate comparison, and all the standard univariate methods can be used. The advantages of indices lie in their simplicity. Their appeal to human intuition is obvious: "On a scale of one to ten, how would you rate this community?" Indices facilitate comparisons between communities, and even give a certain amount of information about a single community, as having a large or small value, without comparing that value to any other site. The disadvantages of indices also stem from their simplicity, and are just as apparent. Talking about a system that ranks "high" on an index is like talking about a "large" animal, without specifying whether the context is Rodentia or Cetacea. These indices are often used to classify systems into categories such as "healthy" and "unhealthy," without much regard for the limitations and assumptions that are hidden within the simplicity.

Species Richness, Diversity, and Equitability Indices

Some of the most popular measures of community change include indices that define biological diversity as a function of species richness and evenness. Although most ecologists profess to be skeptical about using diversity indices to assess community responses to pollutants, diversity indices are still widely used as measures of community "health."

Species diversity is a function of the number of species present (species richness or species abundance) and the evenness with which the individuals are distributed among these species (Hurlbert, 1971). The simplest measure of species richness is S, the number of species in a sample. However, the total number of species in a sample tends to increase, up to a point, with the sample size (N). Species richness indices that are a function of S and N have been used since the early 1900s (e.g., Gleason, 1922; Margalef, 1958; Menhinick, 1964):

$$\text{Gleason's index} = \frac{S}{\ln N} \tag{1}$$

$$\text{Margalef's index} = \frac{S-1}{\ln N} \tag{2}$$

$$\text{Menhinick's index} = \frac{S}{\sqrt{N}} \tag{3}$$

Odum et al. (1960) proposed standardizing the collection process, so that S represented the number of species in a sample of 1000 individuals. This type of standardization is recommended in the U.S. Environmental Protection Agency's Rapid Bioassessment Protocols, which suggest a 100-organism subsample for benthic invertebrate counts (Plafkin et al., 1989). Barbour and Gerritsen (1996) defend the fixed-count approach for determining species richness, arguing that in most bioassessments, rare taxa (which are unlikely to be collected in small samples) are not essential for determining the condition of the site. A contrasting argument is presented by Courtemanch (1996), who cautions that fixed-count methods devalue the information provided by rare species, and should not be used except for preliminary screening to detect possible disturbances.

Patrick et al. (1954) used Preston's $\log_2$-normal distribution (Preston, 1948) to evaluate changes in species richness, under the hypothesis that disturbed communities would display a dramatic decrease in the number of rare species, an increase in the number of common and extremely common species, and that these changes could be predicted and verified using a fitted $\log_2$-normal distribution. The $\log_2$-normal distribution is estimated using Preston's equation:

$$S(R) = S_0 e^{-a^2 R^2} \tag{4}$$

where S(R) is the number of species in the Rth octave, S_0 is the number of species in the modal octave, R is the octave number, and a is the inverse measure of the width of the distribution. The theoretical total number of species is estimated as the area under the $\log_2$-normal curve:

$$\hat{S} = 1.77\frac{S_0}{a} \tag{5}$$

Hurlbert's rarefaction index (1971) is another variation of estimating species richness from an assumed distribution. In this equation, the expected number of species, $E(S_n)$ is estimated for a random sample of n_i individuals drawn from a population of N individuals:

$$E(S_n) = \sum_{i=1}^{S}\left\{1 - \left[\frac{\left[\begin{matrix} N - n_i \\ n \end{matrix}\right]}{\left[\begin{matrix} N \\ n \end{matrix}\right]}\right]\right\} \tag{6}$$

Vinson and Hawkins (1996) used this equation to show the effects of various subsample sizes on species richness. Their results indicated that S increased rapidly with samples of up to 200 individuals, and much more slowly in subsamples of 200–1000 individuals. As to whether a 200 individual, fixed-count approach results in an unacceptable loss of information, Vinson and Hawkins (1996) concluded that it really depends on the question asked. If rare taxa are critical to the assessment, such as might be the case for certain indicator species, the fixed-count is not likely to be the best approach. However, in many instances, such as when cost-effectiveness is critical and the degree of disturbance is measurable within the common taxa, the fixed-count method offers many advantages.

One of the first diversity indices to be used was Simpson's index, often referred to as Simpson's λ:

$$\hat{\lambda} = \sum_{i=1}^{S}\frac{n_i(n_i - 1)}{n(n - 1)} \tag{7}$$

where n is the number of individuals in the sample and n_i is the number of individuals of species i in the sample. Simpson's λ is a measure of the probability that two individuals chosen at random from a population will be from the same group (Simpson, 1949). The index is inversely related to diversity, increasing to a maximum of 1.0 when all individuals belong to the same group. Simpson's λ is influenced more strongly by abundant species than rare species.

The most commonly used diversity index, Shannon's H' (Shannon and Weaver, 1949) is based on information theory, and is a measure of the amount of information that can be transmitted as a single decision between equally probably alternatives.

$$H' = \sum_{i=1}^{S}p_i \ln p_i = -\sum_{i=1}^{S}\frac{n_i}{n} \ln \frac{n_i}{n} \tag{8}$$

Shannon's H' continues to be used extensively to measure changes in community structure and as an indicator of community disturbance despite many criticisms (e.g., Washington, 1984; Guhl, 1987; Ludwig and Reynolds, 1988; Pontasch et al., 1989; Whitton and Kelly, 1995).

Diversity indices are notoriously affected by differences in sample size and species richness. In response to this, "evenness" measures have been developed to allow comparisons between different samples and to assess the degree to which the diversity index approaches its maximum value for that particular sample. Pielou's J is a ratio of H' to the maximum value that H' can attain when all species in the sample are represented by an equal number of individuals (Pielou, 1975).

$$\text{Pielou's } J = \frac{H'}{\ln S} \tag{9}$$

Sheldon (1969) proposed an equitability index based on $e^{H'}$, which is the minimum number of equally common species that could yield the observed H' diversity. The resulting index represents the relative proportion of common species in the sample.

$$\text{Sheldon's index} = \frac{e^{H'}}{S} \tag{10}$$

One problem with Sheldon's index is that as $e^{H'}$ approaches 1.0, so does S. As a result, samples with a very low species richness will have an index value approaching 1.0. Heip (1974) suggested a variation of Sheldon's index that still represents the proportion of common species in the sample, but approaches zero as H' decreases and 1.0 as H' increases.

$$\text{Heip's index} = \frac{e^{H'} - 1}{S - 1} \tag{11}$$

Hill's index (Hill, 1973), and the modification proposed by Alatalo (1981), provide measures of the relative proportions of common and very common species in the sample.

$$\text{Hill's index} = \frac{\dfrac{1}{\lambda}}{e^{H'}} \tag{12}$$

$$\text{modified Hill's index} = \frac{\dfrac{1}{\lambda} - 1}{e^{H'} - 1} \tag{13}$$

Although diversity and evenness indices appear to provide a simple way to assess the effects of pollution on a community, many researchers question their use in biological monitoring. Guhl (1987) concluded that "indices developed on a purely theoretical-mathematical or cybernetic basis can neither characterize a body of water completely nor can they serve for the comparison of two of them." Similarly, Washington (1984), in an evaluation of 18 diversity indices, 19 biotic indices, and five similarity indices, concluded that none of the diversity indices except for Simpson's λ, Hurlbert's "encounter" index, indices based on the theory of "runs" (Cairns and Dickson, 1971; Keefe and Bergersen, 1977), and possibly McIntosh's ecological distance index (McIntosh, 1967), were suitable for evaluating the effects of pollution stress on community structure.

$$\text{Hurlbert's index} = \left(\frac{n}{n-1}\right)\left(1 - \sum_{i=1}^{S}\left(\frac{ni}{n}\right)^2\right) \tag{14}$$

$$\text{Cairns' SCI} = \left(\frac{\sum \frac{\#\text{ runs}}{\#\text{ specimens}}}{\#\text{ runs needed to be significant}}\right) \times (\#\text{ of taxa}) \tag{15}$$

where # of runs is the number of adjacent, dissimilar taxa in a sample, with the individual organisms lined up in rows.

$$\text{Keefe's TU} = 1 - \left(\frac{n}{n-1}\right)\left[\sum_{i=1}^{S}\left(\frac{n_i}{n}\right)^2 - \frac{1}{n}\right] \tag{16}$$

$$\text{McIntosh's index} = \frac{n - \sqrt{\sum_{i=1}^{S} n_i^2}}{n - \sqrt{n}} \tag{17}$$

A further problem with diversity indices, and other indices based on taxonomy, is the effect of keying organisms to different taxonomic levels. This problem takes on several forms. First, not all organisms are readily identifiable to the same level. For example, most people, with a minimum of training, can separate stoneflies, mayflies, and caddisflies to genus, and sometimes species. By contrast, a high degree of training is required to identify chironomids beyond family. However, Chironimidae is an astonishingly diverse family, with many species likely to be present at both polluted and unpolluted sites. When Chironimidae are counted as a single taxon, the result is an apparent loss of species diversity, usually at polluted sites. Second, the amount of effort put into taxonomic identification varies among projects, so that taxa that are identified to species in one study may be identified only to order in another. Osborne et al. (1980) evaluated the effects of differential identification by comparing the results for H′ to the Hierarchical Diversity Index (HDI), where H′ (F), H′$_F$(G), and H′$_{FG}$(S) are, respectively, the family, genus, and species components of the total diversity. They concluded that identification of taxa to the family level was sufficient, at least at their sites, to detect intersite diversity differences. They also evaluated a trophic-based Hierarchical Diversity Index (HTDI) in which the species diversity was weighted by general trophic level, functional trophic level, and lowest taxonomic unit, $H'(T1)$, $H'_{T1}(T2)$, and $H'_{T1T2}(T3)$, respectively. Their results were inconclusive, in part due to the problem of assigning trophic categories for most taxa.

$$HDI = H'(F) + H'_F(G) + H'_{FG}(S) \tag{18}$$

$$HTDI = H'(T_1) + H'_{T1}(T_2) + H'_{T1T2}(T_3) \tag{19}$$

Whittaker (1972) described three levels of diversity: α- (within-habitat), β- (between habitat), and γ- (large-scale, landscape) diversity. Titmus (1981) suggested using β-diversity to evaluate the extent of faunal change between two sites. The β-diversity is calculated by subtracting α-diversity, which is a weighted sum of the diversities at both sites, from γ-diversity, which is the H' for the combined taxa from both sites:

$$H'_\beta = H'_\gamma - H'_\alpha \tag{20}$$

$$H'_\gamma = \sum_{i=1}^{S} p_i \ln p_i \tag{21}$$

$$H'_\alpha = -\sum_j p_{\bullet j} \sum_i p_{i|j} \ln p_{i|j} \tag{22}$$

where $p_{i|j}$ is the proportion of species i at station j and $p_{\bullet j}$ is the proportion of the total fauna at station j. Since the β- and γ-diversity measures are, in practice, independent, they measure different aspects of the community composition, namely, structural and taxonomic changes (Titmus, 1981).

Biotic Indices

Unlike diversity indices, which were developed to measure uncertainty and coopted by ecologists to describe changes in community structure, biotic indices were developed directly to measure specific community changes. Because of this, we avoid the question as to whether biotic indices are appropriate for measuring community changes. However, we are still left with questions about the validity of our assumptions, difficulties in interpreting the index, and the possibility that our index obscures (or doesn't measure) important changes in the community.

One of the first biotic indices to be used extensively was the saprobity index, which defines a polysaprobic zone of gross organic pollution; an α-mesosaprobic zone where some oxidation takes place and some species are present; a β-mesosaprobic zone where decomposition approaches mineralization and many species occur; and an oligosaprobic zone where oxygen concentrations are high and species that are intolerant of low oxygen conditions are present. The saprobity index was introduced by Kolkwitz and Marsson (1908; 1909), and has been described more recently by Sládecek (1973) and Guhl (1987):

$$\text{saprobity index} = \frac{\sum_{i=1}^{n} g_i \times s_i \times h_i}{\sum_{i=1}^{n} g_i \times h_i} \tag{23}$$

where h_i is the frequency of species i, g_i is the weighting factor of species i, and s_i is the saprobity of species i. A major disadvantage of the saprobity index is that it requires detailed taxonomic identifications as well as arbitrary judgments (weighting factors and saprobity value for each species). Nevertheless, the saprobity index is still being used, especially in European biomonitoring studies, where well-established species lists are available for the local flora and fauna. One additional criticism is that different lists are available for the same region so that index values differ for the same organisms. However, Guhl (1987) compared the saprobity index for 48 samples, using different lists, and found the maximum difference to be 0.2 units. This difference was less than the variability inherent in sample collection.

Many of the early biotic indices used ratios of pollution-tolerant organisms to total invertebrate abundance. One of the simplest indices was the number of oligochaetes per m^2, proposed by Wright and Tidd (1933). Similar indices were proposed by Goodnight and Whitley (1960), King and Ball (1964), and Raffaelli and Mason (1981).

$$\text{Goodnight and Whitley's index} = \frac{\text{Tubificid number}}{\text{Number of benthic organisms}} \tag{24}$$

$$\text{King and Ball's index} = \log\left(\frac{\text{Insect weight}}{\text{Tubificid weight}}\right) \tag{25}$$

$$\text{Raffaelli and Mason's index} = \left(\frac{\text{Nematode number}}{\text{Copepod number}}\right) \times 100 \tag{26}$$

Guhl (1987) criticized these types of indices on the basis that not all tubificid and oligochaete species live in polluted waters, and there were no constraints on where samples were collected. Clean lake bottom samples containing mostly oligochaetes and chironomids could receive a lower rating than a dirty stream containing a variety of pollution-tolerant organisms. A broader criticism is that there is no single species that has proven to be reliable for all sites and all types of environmental disturbance. As with the fixed-count vs. full-sample species richness argument, the solution is to use appropriate methods for the question being asked. Indicator species should be selected from species that would normally be found at the site and that are known to respond to the type of disturbance that is at question.

Whittaker (1965) proposed a dominance index that measures relative importance (y) of any given species (n) to the total sample (N):

$$\text{Whittaker's dominance index} = \sum_{i=1}^{n}\left(\frac{y}{N}\right)^{2} \tag{27}$$

Whittaker's index can be based on any type of measurement (numerical abundance, biomass, productivity, percent cover, etc.). However, this index has the same problems as the others: we rarely know enough about the factors affecting species dominance in an unpolluted community to use the index to assess the effects of pollution.

Patrick (1949) suggested using seven histograms of species abundance in the following categories: I. bluegreen algae, *Stigeoclonium*, *Spirogyra*, *Tribonema*, bdelloid rotifers, *Cephalodellamegalo cephala*, and *Proales decipiens*; II. oligochaetes, leeches, and pulmonate snails; III. protozoa; IV. diatoms, red algae, and most green algae; V. all rotifers not included in histogram I, plus clams, prosobranch snails, and tricladid worms; VI. all insects and crustacea; VII. all fishes. Patrick's histograms never gained popularity because they were considered to be too labor-intensive to be useful (Washington, 1984). However, these histograms are quite similar in concept to some of the currently used multivariate indices that rely on changes in broad taxonomic groups rather than a few indicator organisms. Although generating the histograms required extensive sampling, the taxonomic requirements were relatively simple.

Beck's index (1955) was one of the first to be called a "biotic" index. It was based on defining clean water indicator species (class I) and species tolerant of moderate amounts of organic pollution (class II). A grossly polluted stream would have an index of zero, while a clean stream would have an index greater than 10.

$$\text{Beck's biotic index} = 2(S \times \text{class I}) + (S \times \text{class II}) \tag{28}$$

Heister's modification of Beck's biotic index used all taxa, not just Beck's class I and class II indicator taxa (Heister, 1972). Heister added classes III–V for organisms that were tolerant of gross levels of organic pollution, independent of dissolved oxygen, and unclassifiable. The class III–V organisms were assigned a value of zero, so the resulting index value was the same as for Beck's index.

Beak et al. (1959) introduced a "bivariate control chart" based on the ratio of the density of species found near a waste outfall to its density in concentric rings at different distances from the outfall. This approach was similar to some of the earlier biotic indices based on pollution-tolerant species, but could be used for any taxon, allowing it to be adapted to the particular site. Beak's bivariate control chart index was expanded into a "river index" (Beak, 1965). Beak's river index assigned pollution status scores ranging from 0–6 based on a variety of changes that commonly occur in aquatic communities subjected to organic and other types of pollution. Sites were subjectively assigned index scores based on categorical descriptions and population surveys done at the site. For example, an index score of 6 would be given to unpolluted sites that were characterized by the presence of sensitive macroinvertebrates and a "normal" fish community, while a score of zero would be assigned when macroinvertebrates and fish were absent.

In the 1970s a variety of biotic indices based on benthic macroinvertebrate species richness were introduced, including the Trent index, Graham's biotic index, and Chandler's biotic score. All of these biotic indices worked best in the regions where they were developed because they were based on the presence or abundance of specific taxa. One of the simplest of these was Graham's biotic index, which was an adaptation of the Trent index (see Ballock et al., 1976). Sites were assigned index values ranging from 1 (clean) to 6 (polluted) based on the following taxonomic distinctions: an index score of 1 = more than 10 species of stoneflies and nonbaetid mayflies; 2 = stoneflies and nonbaetid mayflies present (fewer than 10 species) or more than 10 species of caddisflies and shrimp; 3 = caddisflies and shrimp present (fewer than 10 species) or more than 10 species of *Baetis*, shrimp, *Asellus*, snails, or leeches; 4 = fewer than 10 species of *Baetis*, shrimps, *Asellus*, snails, or leeches; 5 = invertebrates restricted to Tubifex, *Nais*, midge larva, or bloodworms; 6 = no macroinvertebrates.

Chandler's CBS index (Chandler, 1970) used a much more complex point scale that took into account both species richness and abundance. Point scores ranging from 0–100 were assigned to each taxon at the site, with the highest scores given to clean-water taxa that were also abundant at the site. The CBS index scores were the sum of the separate point scores, and ranged from zero (severe pollution) to 300 for moderately polluted areas, and from 300 to over 3000 for mildly polluted to unpolluted areas. Because the upper CBS index value varied based on site-specific features, the CBS$_{average}$ was proposed by Ballock et al. (1976) to adjust the scale to 0–100.

$$\text{CBS}_{average} = \frac{\sum\limits_{i=1}^{G} \text{weighted scores}}{\text{total number of groups (G)}} \tag{29}$$

The Chutter and Hilsenhoff indices (Chutter, 1972; Hilsenhoff, 1977) assigned pollution tolerance quality values (Q_i) to macroinvertebrate species in much the same way as the saprobity index:

$$\text{Chutter / Hilsenhoff index} = \sum\limits_{i=1}^{S} \frac{n_i Q_i}{n} \tag{30}$$

The Chutter index used quality values ranging from 0–10; sites with index values of 0–2 were considered to be clean, while values from 7–10 were considered to be polluted. The Hilsenhoff index, which is still widely used in North America, uses quality values ranging from 0–5. Index values of less than 1.75 are considered to be clean, and values greater than 3.75 are considered to be polluted. Local quality value lists are usually available, although the quality values may be based on anecdotal evidence because pollution tolerance data are lacking for many macroinvertebrate species.

Although many biotic indices are based on macroinvertebrate taxonomic data, Whitton and Kelly (1995) discussed the values of using algae and other plant data in biomonitoring. Descy (1979) described a diatom index that is similar to the Chutter and Hilsenhoff indices, except that it is based on the pollution sensitivity of diatoms:

$$\text{Descy's diatom index } = \frac{\sum_{j=1}^{n} a_j v_j i_j}{\sum_{j=1}^{n} a_j v_j} \tag{31}$$

where a_j is the proportion of species j; v_j is the indicator value (1–3); and i_j is the pollution sensitivity (1–5). The index ranges from 1–5, with values greater than 4.5 being characteristic of very clean sites and values in the 1–2 range characterizing polluted sites.

The Index of Biological Integrity (IBI) was developed by Karr (1981) using fish taxa as the major biotic component in the index. The IBI has been incorporated into the U.S. Environmental Protection Agency's Rapid Bioassessment Protocols (Table 10.1), and is being used as a biomonitoring tool in many parts of the United States (e.g., Barbour et al., 1996; Fore et al., 1996). The IBI is a flexible index that provides some compensation for regional and site-specific variations in important habitat features such as substrate, canopy cover, and riparian vegetation type. The IBI is based on assigning a rank of 5, 3, or 1 to a group of variables, or "metrics," that are predetermined to be correlated with increasing levels of pollution.[1] The choice of "metrics" and criteria used to determine the rank for each "metric" are flexible and are determined based on initial sampling to establish the best "metrics" and scoring criteria for the region. Because of this, the IBI is a data-intensive index, even though its field application after developing the "metrics" and scoring criteria is relatively simple. The IBI, like many other biotic indices, is actually a simple classifier, as described below in our section on classifiers, and although it appears to be gaining popularity, we question whether it is truly an improvement over other types of classifiers.

The Pollution-Induced Community Tolerance (PICT) (Blanck et al., 1988) is a distinctly different approach to evaluating the community-level effects of a toxicant. The hypothesis behind PICT is that long-term exposure to toxicants will result in the selection of tolerant species. However, rather than measuring changes in species richness or abundance, the PICT measures changes in functional parameters—such as photosynthesis—by comparing the ratio of the functional measurement from tolerant and intolerant communities exposed to high concentrations of the toxicant. Blanck et al. (1988) reported finding a pronounced arsenate tolerance in marine periphyton that were cultured in the presence of low concentrations of arsenate for three weeks. Molander et al. (1990; 1992), also found PICT evident in marine periphyton communities exposed to tri-n-butyl tin, diuron (3-(3,4-dichlorophenyl)-1,1-dimethylurea), and 4,5,6-trichloroguaiacol. In theory, the PICT could be used successfully to measure a wide variety of sublethal effects. In practice, however, it will be difficult to quantify the variability associated with reference-level functional responses, and to separate the variability from changes that are caused by the pollutant.

---

[1] The IBI "metric" is a biological measurement that changes in some predictable way with increased human influence (Barbour et al., 1996). It is not related to mathematical metrics, which are multivariate distance measures. Inasmuch as the mathematical use of the term predates the IBI, we will distinguish IBI "metrics" with quotations.

Table 10.1. IBI Metric Scoring for Fish Communities.

Metric	Scoring Criteria		
	5	3	1
1. Number of native fish species	>67%	33–67%	<33%
2. Number of darter or benthic species	>67%	33–67%	<33%
3. Number of sunfish or pool species	>67%	33–67%	<33%
4. Number of sucker or long-lived species	>67%	33–67%	<33%
5. Number of intolerant species	>67%	33–67%	<33%
6. Proportion of green sunfish or tolerant individuals	<10%	10–25%	>25%
7. Proportion of omnivorous individuals	< 20%	20–45%	>45%
8. Proportion of insectivores	>45%	20–45%	<20%
9. Proportion of top carnivores	>5%	1–5%	<1%
10. Total number of individuals	>67%	33–67%	<33%
11. Proportion of hybrids or exotics	0%	0–1%	>1%
12. Proportion with disease/anomalies	<1%	>1–5%	>5%

From Plafkin, et al., *Rapid Bioassessment Protocol V*, U.S. EPA, Washington, DC, 1989. Reproduced with permission.

Metrics

Metrics do not attempt to put a single sample on a linear scale, but rather give the distance (or similarity) between two samples. A metric cannot be used with a single sample, and thus at least two samples must be compared, giving some context to how one sample compares with the other. Metrics can be abused—like indices—in the service of simplicity. For instance, we can pick one sample as a "reference" and then use "distance from the reference" as an index for all other samples. However, the artificiality of this technique is much more apparent than with an index. Properly used, distances are relative only to other distances. Two samples are significantly different from each other when the between-sample distances are significantly larger than the within-sample distances. In fact, the ratio of the average within-sample distance to the average between-sample distance is an excellent statistic, can be used with any metric, and can easily be used with a randomization test to find statistically significant distances (Good, 1982; Noreen, 1989; Pontasch et al., 1989). Multivariate analysis of variance (MANOVA) could be regarded as another example of this kind of technique, generating a statistic from the relative positions of all samples in the multivariate space, but it is less intuitive and less general than the simple ratio statistic. MANOVA is also related to classifier techniques, discussed below.

An index can, of course, be converted into a metric by simply subtracting the index values for two systems. Indices, then, can be regarded as a restricted subclass of the class of metrics. But this approach severely limits the kinds of metrics that can be used, and has no real advantages.

Similarity Indices

Similarity indices, and their counterpart, dissimilarity indices or distance measures, are the basis of gradient analysis and multivariate clustering and ordination. They can also be used for general comparisons between two or more sites. In this section, we will describe some of the commonly used similarity and dissimilarity indices. For a complete discussion, refer to one of the many texts available that deal extensively with multivariate analyses (e.g., Sneath and Sokal, 1973; Pielou, 1985; Ludwig and Reynolds, 1988; Everitt, 1993; Tabachnick and Fidell, 1996).

Similarity indices have been used for environmental assessments, particularly in conjunction with (or comparison to) diversity and biotic indices (e.g., Wolda, 1981; Washington, 1984; Guhl, 1987; Pontasch et al., 1989). Similarity indices computed from taxonomic data were found to be affected by sample size, diversity, and the problems associated with measuring species richness (Wolda, 1981). Washington (1984) stated that similarity indices were also limited by the need to have a clean water reference site for comparison; however, we have used similarity to look at differences among groups of polluted sites, as well as between unpolluted and polluted sites (Matthews et al., 1995).

Similarity indices are available for continuous data and discrete data, including binary (e.g., presence/absence), categorical (e.g., common, uncommon, rare), or count data. The binary indices are calculated for pairs of sites or samples, and compare the number of species (or other variables) that are present at both sites (a) to the number of species present at only one of the sites (b or c). A few of the binary indices also include the number of species missing from both sites (d), but unless there is a reasonable way to estimate the total number of species that should be present, d is not meaningful.

Five simple, binary similarity indices are described by Everitt (1993), including Jaccard's index (Jaccard, 1908), which is one of the simplest and most frequently used similarity indices.

$$\text{Jaccard's index} = \frac{a}{a + b + c} \qquad (32)$$

Everett's similarity indices:

$$\frac{a + d}{a + b + c + d} \qquad (33)$$

$$\frac{2(a + d)}{2(a + d) + b + c} \qquad (34)$$

$$\frac{2a}{2a + b + c} \qquad (35)$$

$$\frac{a}{a + a(b + c)} \qquad (36)$$

The indices range from zero (no species in common) to 1.0 (all species present at both sites), and may be converted to percent similarity by multiplying by 100.

Wolda (1981) reviewed five other binary indices that were described by Dice (1945), Sorensen (1948), Morisita (1959), Mountford (1962), and Baroni-Urbani and Buser (1976), concluding that only Morisita's index (a discrete, but not binary index) was more or less independent of the effects of sample size.

$$\text{Baroni - Urbani and Buser's index} = \frac{\sqrt{cd} + c}{\sqrt{cd} + a + b - c} \tag{37}$$

$$\text{Dice's index} = \frac{c}{\min(a, b)} \tag{38}$$

$$\text{Mountford's index} = \frac{2c}{2ab - (a + b)c} \tag{39}$$

$$\text{Sorensen's index} = \frac{2c}{a + b} \tag{40}$$

$$\text{Morisita's index} = \frac{2\sum n_{1i} n_{2i}}{(\lambda_1 + \lambda_2) N_1 N_2}, \quad \lambda_j = \frac{\sum n_{ji}(n_{ji} - 1)}{N_j(N_j - 1)} \tag{41}$$

where n_{ji} is the number of individuals of species i in sample j and N_j is the number of individuals in sample j.

Categorical data that have more than two possible outcomes for a variable can be converted into binary form (each outcome is assigned a separate, yes-no binary code), but this results in many negative matches. Everitt (1993) presented an alternative approach for a simple, categorical similarity index, where s_{ijk} is assigned a score of zero or 1, depending on whether two samples have the same categorical value for variable k.

$$s_{ij} = \frac{\sum_{k=1}^{p} s_{ijk}}{p} \tag{42}$$

Gower (1971) published a similarity index that could be used for data sets containing a mixture of continuous and discrete data (including categorical or binary data):

$$s_{ij} = \frac{\sum_{k=1}^{p} w_{ijk} s_{ijk}}{\sum_{k=1}^{p} w_{ijk}} \quad \text{quantitative data } s_{ijk} = 1 - \frac{\left| x_{ik} - x_{jk} \right|}{R_k} \tag{43}$$

In this index, s_{ijk} is still the measure of similarity between the ith and j^{th} individuals for the kth variable. The value for w_{ijk} is usually 1.0, but can be assigned a value of zero for comparisons that are considered to be invalid when the variable k is unknown for one or both individuals, or to exclude negative matches for binary variables. For categorical data (and binary data), s_{ijk} is assigned the value of 1.0 when the two individuals have the same value, and zero when they do not. For continuous data, s_{ijk} is calculated as indicated, where x_{ij} and x_{jk} are the individual values for variable k, and R_k is the range of variable k. The ability to incorporate mixed data in a similarity index is a valuable feature!

The Bray-Curtis index (1957) is one of the best known similarity indices:

$$\text{Bray - Curtis index} = \left(\frac{2W}{a + b} \right)$$

$$\text{where } W = \sum_{i=1}^{S}\left[\min(x_{ij}, x_{ik})\right], \quad a = \sum_{i=1}^{S} x_{ik}, \quad b = \sum_{i=1}^{S} x_{ik} \tag{44}$$

The Bray-Curtis index ranges from zero (no species in common) to 1.0 (all species found at both sites and in identical abundances). The quantitative data used in the Bray-Curtis index (and others) are often transformed to minimize the influence of scale.

Pinkham and Pearson (1976) argued that for biological assessments it is more appropriate to compare species abundances for two sites simultaneously. They define x_{ia} and x_{ib} as the number of individuals in the ith taxon for sites a and b, and measure the average $x_{ia} \cap x_{ib}$.

$$\text{Pinkham and Pearson' s index} = \frac{1}{K} \sum_{i=1}^{s} \frac{\min(x_{ia}, x_{ib})}{\max(x_{ia}, x_{ib})} \tag{45}$$

Distance Metrics

A standard method for calculating similarity is to subtract a distance measure, or metric, from its maximum value, usually 1.0. Distance metrics form the foundation for many classification and clustering algorithms, as discussed below. The following commonly used distances are described by Ludwig and Reynolds (1988), Pontasch et al. (1989), and Everitt (1993):

$$\text{Euclidean distance} = \sqrt{\sum_{i=1}^{S}(x_{ij} - x_{ik})^2} \tag{46}$$

$$\text{squared Euclidean distance} = \sum_{i=1}^{S}(x_{ij} - x_{ik})^2 \tag{47}$$

$$\text{mean Euclidean distance} = \sqrt{\frac{\sum_{i=1}^{S}(x_{ij} - x_{ik})^2}{S}} \tag{48}$$

$$\text{Canberra distance} = \frac{\sum_{i=1}^{S}\left[\frac{|x_{ij} - x_{ik}|}{(x_{ij} + x_{ik})}\right]}{S} \tag{49}$$

$$\text{absolute distance (City Block)} = \sum_{i=1}^{S}|x_{ij} - x_{ik}| \tag{50}$$

$$\text{mean absolute distance} = \frac{\sum_{i=1}^{S}|x_{ij} - x_{ik}|}{S} \tag{51}$$

$$\text{relative Euclidean distance} = \sqrt{\sum_{i=1}^{S}\left[\left(\frac{x_{ij}}{\sum_{i=1}^{S}x_{ij}}\right) - \left(\frac{x_{ik}}{\sum_{i=1}^{S}x_{ik}}\right)\right]} \tag{52}$$

$$\text{relative absolute distance} = \sum_{i=1}^{S}\left|\left(\frac{x_{ij}}{\sum_{i=1}^{S}x_{ij}}\right) - \left(\frac{x_{ik}}{\sum_{i=1}^{S}x_{ik}}\right)\right| \tag{53}$$

$$\text{chord distane} = \sqrt{2(1 - \text{ccos}_{jk})} \qquad \text{ccos}_{jk} = \frac{\sum_{i=1}^{S}(x_{ij}x_{ik})}{\sqrt{\sum_{i=1}^{S}x_{ij}^2 \sum_{i=1}^{S}x_{ik}^2}} \tag{54}$$

$$\text{geodesic distance} = \text{arccos}(\text{ccos}_{jk}) \tag{55}$$

All of the distance metrics are sensitive to outliers, scale, transformations, etc. Some, like Euclidean distance, are sensitive to magnitudes. A sample with large numbers (see **A** below) will dominate the distance metric, often to the point where patterns such as those in **B** are obscured. Others, such as the relative distances and the chord distances, are not sensitive to magnitude. Samples like **A** will only be different from samples with different relative proportions for each variable, such as in **C**.

A	100	100	100	0	0
B	10	10	10	0	0
C	0	0	10	10	10

There are also some measures that attempt to correct for covariance as well as relative variance, such as Mahlanobis distance (D^2; computations summarized by Norusis, 1988):

$$D^2_{ab} = (n - g)\sum_{i=1}^{p} \sum_{j=1}^{p} w^*_{ij}(\overline{X}_{ia} - \overline{X}_{ib})(\overline{X}_{ja} - \overline{X}_{jb}) \tag{56}$$

where p is the number of variables; g is the number of groups; n is the sample size; $\overline{X}_{ia}$, $\overline{X}_{ib}$, $\overline{X}_{ja}$, and $\overline{X}_{jb}$ are means of variables i or j for groups a or b; and w_{ij}^* is the element from the inverse of the within-group covariance matrix.

Classifiers

Classifiers are computer systems that attempt to induce a classification rule that discriminates among groups of points in a set of data. Classifier systems, therefore, involve a two-stage process. In the first (training) stage the system looks at the data and induces a formula or rule for distinguishing which samples belong in which group. In the second stage, this rule can be used to classify new, previously unseen, samples into groups. Classifier systems are often used in diagnostic situations. For instance, a classifier system might be trained based on hospital records, and then used to aid diagnosis of new patients. Their advantage in ecotoxicology is that they go beyond testing for the existence of a difference, to the point of trying to discover the nature of that difference. This can be invaluable in subsequent risk assessments, where human judgment is needed. If the human expert can be given a succinct description of the nature of the difference among groups, as well as its existence, the risks can be more realistically assessed. The type of classifier can be selected with this in mind.

Because ecotoxicological data are usually underdetermined, we have to be careful in getting an honest estimate of a classifier's performance, and we also have to be careful not to find differences where none truly exist. Cross validation, where only a portion of the data is used for training and the remainder reserved as "unseen" cases for testing, can be used to determine the reliability of the classifier. If the reliability of the induced classifier, as measured by cross validation, is better than that obtained by chance, then there is a statistically significant difference between the groups.

There are four major kinds of classifiers: discriminant functions, decision-tree and rule-based induction, nearest neighbor classifiers, and neural nets. To illustrate the different kinds of classifiers, we follow the discussion by R.J. Henry, pp. 6–16 in Michie et al. (1994) and use the well-known iris data set, which contains 150 samples, with 50 samples from each of three species of iris: *I. setosa*, *I. versicolor*, and *I. virginica*. Four variables were measured from each sample: sepal length, petal length, sepal width, and petal width. In order to graph the data, we will use only petal length and petal width.

Fisher's linear discriminant functions, developed in the 1930s, are among the oldest and most widely used classifiers. A linear discriminant is a line dividing one group from another as best as possible. In higher-dimensional space, the line becomes a hyperplane, but the idea is the same. This is illustrated in Figure 10.1. One of the lines discriminates between *I. versicolor* and *I. virginica*, with the rules:

a. If petal width < 3.272 − 0.3254 × petal length then the sample is *I. versicolor*.
b. If petal width > 3.272 − 0.3254 × petal length then the sample is *I. virginica*.

Another line is needed to distinguish *I. setosa* from *I. versicolor* and *I. virginica*. Even with the best possible linear discriminant, six of the samples would be misclassified. The major disadvantage of linear discriminants is that with a large number of dimensions the classification rule is difficult to interpret. Further, with more than two classes, more than one discriminant is needed.

Decision-tree and rule-based induction attempt to discriminate groups using logical—rather than arithmetic—combinations of the variables. For the iris data the following set of rules will discriminate better than the linear function:

a. If petal length < 2.65 then the sample is *I. setosa*
b. If petal length > 4.95 then the sample is *I. virginica*
c. If 2.65 < petal length < 4.95 then
   —if petal width < 1.65 then the sample is *I. versicolor*
   —if petal width > 1.65 then the sample is *I. virginica*

As illustrated in Figure 10.2, this classifier has only three misclassifications. Logical rules like this tend to be easier for humans to interpret and understand. Graphically, they imply a partitioning that is orthogonal to the axes. The ease of interpretation of rule-based classifiers holds great promise for their subsequent use in risk assessments; however, currently there has been little use of this technique in ecotoxicology.

Perhaps the most direct way to use a data set in the building of a classifier is the k-nearest-neighbor method. Rather than building a set of rules, the data set itself is used to classify new samples. For example, the new data point in Figure 10.3 would be classified as *I. virginica*, because, out of its nine nearest neighbors (within the circle), six are *I. virginica*. The circle appears to be an ellipse, because the horizontal and vertical axes are in different scales. However, the k-nearest-neighbor method relies on a distance metric, and proper scaling for such metrics is an inherent problem. Other problems stem from having to decide which points to include as being "typical" or "representative" of their class, and which points to exclude. For ecotoxicological purposes their primary defect is that they do not provide a guide to the interpretation of the groups.

Neural nets can also be used to build discriminant functions. We do not plot the results here because neural nets can produce arbitrarily complex, nonlinear functions in order to discriminate among groups. Choosing a level of complexity for the net itself is a major problem in the design of neural nets. Further, neural nets provide no guidance as to why they work. Just like k-nearest-neighbor systems, they can perform well, if tuned properly, but do not give the scientist insight into the differences they discriminate among.

Only the tree and rule-based inductive techniques show real promise as new tools to assist with the environmental risk analysis process. Michie et al. (1994) reviewed all of these classification techniques in an extensive project (STATLOG) that compared the per-

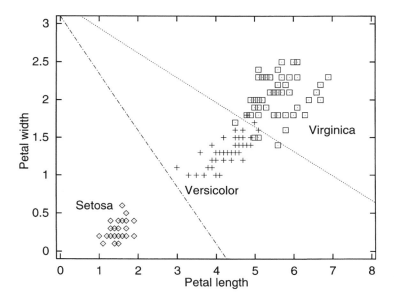

Figure 10.1. Iris data classified by linear discriminants.

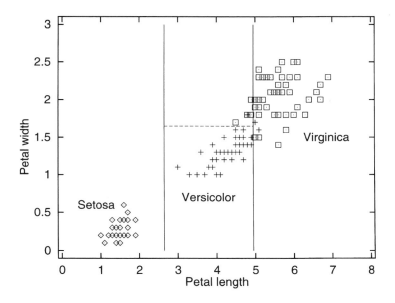

Figure 10.2. Iris data classified by rule-based system.

formance of many different algorithms on many different kinds of data sets. A more elementary book, suitable as an introduction to these ideas, is by Weiss and Kulikowski (1991). Probably the two most important books on the subject are by Breiman et al. (1984), which describes CART, and Quinlan (1993), which describes C4.5. These two inductive computer programs, and dozens of variants on them, are widely used, readily available for

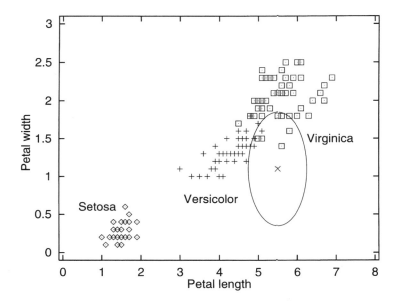

Figure 10.3. Iris data used to classify a new point by 9-nearest-neighbor method.

use on a wide variety of computers, and eminently suitable for use in ecotoxicological risk assessment.

Simple classifiers such as the Index of Biological Integrity (IBI) bear a strong resemblance to biotic indices, as discussed above. The IBI uses a two-phase process. First, extensive data are gathered from a range of polluted and unpolluted sites. These data are used to identify a subset of approximately 10–12 measurements ("metrics") that best delineate the polluted and unpolluted sites. These "metrics" are then used to evaluate new sites by ranking the site on each "metric" and summing the ranks. Essentially, the IBI is a simple, "hand-generated" classifier system—in fact, a linear discriminant. One of the strengths of the IBI is that it addresses many of the problems inherent in biotic indices, such as the need to use regional data for evaluating community responses and the need to use ranked or proportional measurements. The major problem with the IBI is that it fails to take into account the extensive literature demonstrating that for complex systems machine-generated classifiers are often more accurate, more robust, and easier to understand than hand-generated systems. As mentioned above, for instance, rule-based systems are universally acknowledged to be more useful and intuitive than linear functions such as the IBI's summation. In addition, the IBI and other indices are usually built on predetermined judgments about which classes of species or parameters must be used in the discriminant, precluding the use of other parameters that might, for any particular type of community, be more accurate and more intuitive.

Clustering

Clustering algorithms attempt to divide samples into two or more natural groups—the clusters—such that samples within a single cluster are very similar to each other, and samples from two different clusters are less similar. Determining what is "similar" varies among the clustering techniques (metrics are often used), as does the choice of how the samples are divided up into the clusters.

Clustering differs most significantly from the previous techniques in that it is "blind." Just as blind testing involves examining the data without knowing what the treatment was, a clustering analysis will examine all samples in the data set without knowing which group the samples come from. Samples are pooled and group tags are ignored. This results in two distinct advantages. First, the strength of association between group and cluster forms a natural statistic for a significance test for differences in groups. Second, a blind analysis can reveal patterns and regularities in the data that the scientist was not looking for. All techniques that rely on the *a priori* grouping of samples into "treatment" and "control," for example, can only find patterns (if they exist) that are related to treatment and control. If there is a confounding factor, such as geographical location or time of day, regularities associated with these other factors will not be found without performing a test for each one. A single clustering, on the other hand, blind to treatment, location, and time, can be used to investigate possible effects of all three.

There are a large number of clustering algorithms described in the statistical literature, including agglomerative, divisive, and k-means. All rely on a metric to determine "similarity" of points within a cluster. Our discussion of metrics, above, is relevant here. These techniques are well described in Jain and Dubes (1988) and Everitt (1993), and available with most commercial statistical packages (SAS, SPSS, etc.).

Of more interest are the recent clustering algorithms from the machine learning literature, such as Fisher's COBWEB (1987), Matthews and Hearne's RIFFLE (1991), and Zembowicz and Zytkow's 49er (Fayyad et al., 1996). All of these search for equivalence classes in the data, i.e. partitions of the points such that many (if not all) of the variables will (approximately) discriminate among the clusters. In the following hypothetical example containing three variables (political party, hair color, and sex) the contingency table demonstrates that political party and sex are approximately concordant, while political party and hair color are not:

	Male	Female	Blonde	Brunette
Democrat	8	76	43	41
Republican	94	2	50	46

For these samples, a division along party lines or sex will be more "natural" than a division by hair color, because it will capture more differences in the samples. More sophisticated patterns can also be searched for using contingency tables, such as functions, multifunctions, and general relations. The clustering algorithms mentioned here will search among many variables for these kinds of equivalencies. They overcome the curse of dimensionality in this manner by ignoring many variables, such as hair color, that do not form equivalencies with other variables. Statistical metrics might be able to weight or scale variables, but they are not able to ignore them completely. The advantage of these techniques over statistical clustering algorithms for risk assessment is that they provide, again, some guidance as to the nature of the clusters. If the clustering is strong and shows a statistically significant association with the groups, then the variables that are descriptors for the clusters can tell the scientist something about the nature of the effect. Thus, these techniques have a similar advantage to the tree and rule-based classifiers. In fact, they can be used in conjunction with those methods as a blind check on the variables used to generate the rules.

Any clustering can be used with a test of the significance of the association between groups and clusters to obtain a hypothesis test. In addition, the strength of association (i.e., significance level or p-value) is not the only information that can be gleaned from a clustering approach. For example, suppose that for three treatment groups (low, medium, and high doses) the data fall into three clusters, and a significant association is found. If the contingency table for group versus cluster looks like the following, we can observe that each dose was distinct from the others:

	Cluster 1	Cluster 2	Cluster 3
Low dose	49	1	0
Medium dose	2	47	1
High dose	0	2	48

If, on the other hand, the contingency table looks like the one below, we can conclude that only the high dose made any real difference:

	Cluster 1	Cluster 2	Cluster 3
Low dose	24	26	0
Medium dose	22	28	0
High dose	0	0	50

More complex relationships, and further hypotheses, can also be gleaned from examining the contingency tables produced by these algorithms.

CONTRIBUTING TO THE ENVIRONMENTAL RISK ASSESSMENT

The curse of dimensionality has been an annoyance to ecotoxicological risk assessment throughout its history. Its implications create two major problems: multivariate data are extremely difficult to analyze formally, and monitoring all the possible responses in an ecosystem is economically unfeasible. The overwhelming complexity of nature is only part of the problem. Even when we can identify changes in the variables, we often can't predict how the rest of the community will respond to those changes. One reason for this is that environmental toxicology, as a science, is only a few decades old. Most of the research that has been done in the field has dealt with examining the effects of specific effluents or chemicals. Toxicologists have not developed an extensive theoretical foundation, as others have for chemistry and physics, to help explain how communities respond to pollutants.

Currently, most of our regulatory policies are based on demonstrating that a pollutant will have no measurable (or statistically significant) effect at the population or community level. This is an elusive goal, easily affected by choices in measured variables, statistical tools, and professional interpretation of the results. Furthermore, it is contrary to what we do know about communities: all actions, including the introduction of a contaminant into the community or the alteration of the physical structure of the community, produce some sort of a response. In reality, we aren't trying to regulate against all possible changes, only those that cause harm or disrupt community "integrity." Here we find the true goal: defining the kinds of changes that are acceptable, and those that are not. This goal is not just a statistical exercise. It requires judgments and interpretations from natural and social scientists, economists, environmental engineers, policymakers, and the public.

There are guidance documents for conducting ecological risk assessments (e.g., EPA, 1992; 1993). To be of any value, however, the risk assessment must be based on an accurate understanding of the community-level response to a pollutant. Therefore, the most important tasks in the statistical analysis are to distinguish the real community responses from the spurious ones, and to increase our understanding of the nature of the response (i.e., the relationship between the measured variable and the community response). Multivariate procedures should be evaluated based on how well they meet these two goals.

We propose the following strategies. First, consider the amount of information available concerning the potential community-level effects of a pollutant. For situations where the effects of a pollutant are relatively unknown, the initial data collection should be as extensive as possible. Because of the likelihood of unexpected responses at the community level, the sampling effort should include variables that measure general community functions as well as those that are expected to show a specific response. Approaches that are designed to save time and money, such as the fixed-count method for measuring species richness, most biotic indices, and the IBI, should be used only in situations where there are reasonable expectations about how a community will respond to the pollutant. These methods are not designed to detect new patterns, but rather to confirm the expected.

Second, the multivariate data set should be examined in collaboration with a data analysis expert to reduce the data to a reliable set of rules describing the community response. The importance of this step cannot be overemphasized, and the required level of effort will be at least as great as that for sampling. Multivariate analyses require a data analysis expert to be used effectively. There is little chance of success in their application if they are used as "canned programs." Similarly, only the environmental scientist has the expertise to interpret the ecological significance of the patterns in the data. Fayyad et al. (1996) describe state-of-the-art methods for collaborative data analysis in their book, *Advances in Knowledge Discovery and Data Mining.* Included in this book is a case study using inductive logic programming to assist with the biological classification of river water quality (Dzeroski, 1996).

Finally, the information gained at the community level needs to be incorporated into both the risk assessment and risk management processes. The data collection and analysis goal was to learn more about how the community responded to a pollutant, thus helping to characterize environmental risk. The risk management goal, however, will be to reduce, eliminate, mitigate, or accept the environmental impacts of the pollutant. Knowledge about how the community responds to the pollutant can be used to set better discharge limits and define monitoring programs. The reality is that the world is extraordinarily complex and responds to pollutants in myriad ways. Estimating the risk of a pollutant and setting safe discharge standards is a daunting task that will require our best professional judgment, followed by constant vigilance. The cycle of data collection, finding patterns, interpreting them, refining the analysis, and collecting more data will be repeated many times over before a good picture of the true risks can emerge.

REFERENCES

Alatalo, R.V. Problems in the measurement of evenness in ecology. *Oikos* 37, pp. 199–204, 1981.

Ballock, D., C.E. Dames, and F.H. Jones. Biological assessment of water quality in three British rivers the North Esk (Scotland), the Ivel (England) and the Taf (Wales). *Water Pollut. Control* 75, pp. 92–114, 1976.

Barbour, M.T. and J. Gerritsen. Subsampling of benthic samples: a defense of the fixed-count method. *J. North Am. Benthol. Soc.* 15, pp. 386–391, 1996.

Barbour, M.T., J. Gerritsen, G.E. Griffith, R. Frydenborg, E. McCarron, J.S. White, and M.L. Bastian. A framework for biological criteria for Florida streams using benthic macroinvertebrates. *J. North Am. Benthol. Soc.* 15, pp. 185–211, 1996.

Baroni-Urbani, C. and M.W. Buser. Similarity of binary data. *Syst. Zool.* 25, pp. 251–259, 1976.

Beak, T.W. A biotic index of polluted streams and its relationship to fisheries. *Adv. Water Pollut.* 1, pp. 191–210, 1965.

Beak, T.W., C. De Courval, and N.E. Cooke. Pollution monitoring and prevention by use of bivariate control charts. *Sewage Ind. Wastes* 31, pp. 1383–1394, 1959.

Beck, W. M. Suggested method for reporting biotic data. *Sewage Ind. Wastes* 27, pp. 1193–1197, 1955.

Blanck, H., S.Å. Wängberh, and S. Molander. Pollution-induced community tolerance—a new ecotoxicological tool, in *Functional Testing of Aquatic Biota for Estimating Hazards of Chemicals*, ASTM STP 988, Cairns, J., Jr. and J.R. Pratt, Eds., American Society for Testing and Materials, Philadelphia, PA, 1988.

Bray, J.R. and J.T. Curtis. An ordination of the upland forest communities of Southern Wisconsin. *Ecol. Monogr.* 27, pp. 325–349, 1957.

Breiman, L, J.H. Friedman, R.A. Olshen, and C.J. Stone. *Classification and Regression Trees.* Wadsworth, Inc., Monterey, CA, 1984, p. 358.

Cairns, J., Jr. and K.L. Dickson. A simple method for the biological assessment of the effects of waste discharges on aquatic bottom-dwelling organisms. *J. Water Pollut. Control Fed.* 43, pp. 755–772, 1971.

Cairns, J., Jr., B.R. Neiderlehner, and E.P. Smith. Ecosystem effects: Functional end points, in *Fundamentals of Aquatic Toxicology: Effects, Environmental Fate, and Risk Assessment*, 2nd ed., Rand, G.M., Ed., Taylor & Francis, Washington, DC, 1995.

Chandler, J.R. A biological approach to water quality management. *Water Pollut. Control* 69, pp. 415–421, 1970.

Chutter, F.M. An empirical biotic index of the quality of water in South African streams and rivers. *Water Res.* 6, pp. 19–30, 1972.

Cleveland, W.S. *The Elements of Graphing Data.* Wadsworth, Inc., Monterey, CA, 1985, p. 323.

Courtemanch, D.L. Commentary on the subsampling procedure used for rapid bioassessments. *J. North Am. Benthol. Soc.* 15, pp. 381–385, 1996.

Descy, J.P. A new approach to water quality estimation using diatoms. *Nona Hedargia* 64, pp. 304–323, 1979.

Dice, L.R. Measures of the amount of ecological association between species. *Ecology* 26, pp. 297–302, 1945.

Dzeroski, S. Inductive logic programming and knowledge discovery in databases, in *Advances in Knowledge Discovery and Data Mining*, Fayyad, U.M., G. Piatetsky-Shapiro, P. Smyth, and R. Uthurusamy, Eds., AAAI Press, Menlo Park, CA, 1996.

Everitt, B.S. *Cluster Analysis.* John Wiley & Sons Inc., New York, 1993, p. 136.

EPA. *Framework for Ecological Risk Assessment*, EPA/630/R-92/001, Risk Assessment Forum, U.S. Environmental Protection Agency, Washington, DC, 1992.

EPA. *A Review of Ecological Assessment Case Studies from a Risk Assessment Perspective*, EPA/630/R-92/005, Risk Assessment Forum, U.S. Environmental Protection Agency, Washington, DC, 1993.

Fayyad, U.M., G. Piatetsky-Shapiro, P. Smyth, and R. Uthurusamy. *Advances in Knowledge Discovery and Data Mining*, AAAI Press, Menlo Park, CA, 1996, p. 611.

Fisher, D.H. Knowledge acquisition via incremental conceptual clustering. *Machine Learning* 2, pp. 139–172, 1987.

Fore, L.S., J.R. Karr, and R.W. Wisseman. Assessing invertebrate responses to human activities: Evaluating alternative approaches. *J. North Am. Benth. Soc.* 15, pp. 212–231, 1996.

Gleason, H.A. On the relation between species and area. *Ecology* 3, pp. 158, 1922.

Good, I.J. An index of separateness of clusters and a permutation test for its significance. *J. Statist. Comp. Simul.* 5, pp. 81–84, 1982.

Goodnight, C.J. and L.S. Whitley. Oligochaetes as indicators of pollution, in *Proceedings of the 15th Industrial Waste Conference*. Purdue University, pp. 139–142, 1960.

Gower, J.C. A general coefficient of similarity and some of its properties. *Biometrics* 27, pp. 857–872, 1971.

Guhl, W. 1987. Aquatic ecosystem characterization by biotic indices. *Int. Rev. Gesamten Hydrobiol.* 72, pp. 431–455, 1987.

Heip, C. A new index measuring evenness. *J. Mar. Biol. Assoc.* 54, pp. 555–557, 1974.

Heister, R.D. The biotic index as a measure of organic pollution in streams. *Am. Biol. Teach.* February, pp. 79–83, 1972.

Helsel, D.R. and R.M. Hirsch. *Statistical Methods in Water Resources*. Elsevier Science Publishing Company, Inc., New York, 1992, p. 522.

Hill, M.O. Diversity and evenness: a unifying notation and its consequences. *Ecology* 54, pp. 427–432, 1973.

Hilsenhoff, W.L. *Use of Arthropods to Evaluate Water Quality of Streams*. Technical Bulletin No. 100, U.S. Department of Nature Research, 1977.

Hurlbert, S.H. The nonconcept of species diversity: a critique and alternative parameters. *Ecology* 52, pp. 577–586, 1971.

Jaccard, P. Nouvelles recherches sur la distribution florate. *Bull Soc. Vaud. Sci. Nat.* XLIV, 163, pp. 223–269, 1908.

Jain, A.K. and R.C. Dubes. *Algorithms for Clustering Data*. Prentice Hall, Englewood Cliffs, NJ, 1988, p. 320.

Karr, J.R. Assessment of biotic integrity using fish communities. *Fisheries* 6, pp. 21–27, 1981.

Keefe, T.J. and E.P. Bergersen. A simple diversity index based on the theory of runs. *Water Res.* 11, pp. 689–691, 1977.

King, D.L. and R.C. Ball. A quantitative biological measure of stream pollution. *J. Water Pollut. Control Fed.* 36, pp. 650–653, 1964.

Kolkwitz, R. and M. Marsson. Okologie der pflanzlichen Saprobien. *Ber. Dtsch. Bot. Ges.* 26a, pp. 505–519, 1908.

Kolkwitz, R. and M. Marsson. Okologie der tierischen Saprobien. *Int. Rev. Gesamten Hydrobiol. Hydrogr.* 2, pp. 126–152, 1909.

Ludwig, J.A. and J.F. Reynolds. *Statistical Ecology. A Primer on Methods and Computing*. John Wiley & Sons Inc., New York, 1988, p. 337.

Margalef, R. Information theory in ecology. *Gen. Syst.* 3, pp. 36–71, 1958.

Matthews, G.B. and J. Hearne. Clustering without a metric. *IEEE Trans. Pattern Anal. Machine Intelligence* 13, pp. 175–184, 1991.

Matthews, G., R. Matthews and W. Landis. *Exxon Valdez Data Analysis, NOAA Final Report.* Submitted by the Institute of Environmental Toxicology and Chemistry to the National Oceanic and Atmospheric Administration, Seattle, WA, August 30, 1995.

McIntosh, R.P. An index of diversity and the relations of certain concepts to diversity. *Ecology* 48, pp. 392–404,1967.

Menhinick, E.P. A comparison of some species-individuals diversity indices applies to samples of field insects. *Ecology* 45, pp. 859–861, 1964.

Michie, D., D.J. Spiegelhalter, and C.C. Taylor. *Machine Learning, Neural and Statistical Classification.* Ellis Horwood Ltd., Hemel Hempstead, England, 1994, p. 289.

Molander, S., H. Blanck, and M. Söderström. Toxicity assessment by pollution-induced community tolerance (PICT), and identification of metabolites in periphyton communities after exposure to 4,5,6-trichloroguaiacol. *Aquat. Toxicol.* 18, pp. 115–136, 1990.

Molander, S., B. Dahl, H. Blanck, J. Jonsson, and M. Söderström. Combined effects of tri-n-butyl tin (TBT) and diuron on marine periphyton communities detected as pollution-induced community tolerance. *Arch. Environ. Contam. Toxicol.* 22, pp. 419–427, 1992.

Morisita, M. Measuring of interspecific association and similarity between communities. *Mem. Fac. Sci. Kyushu Univ. Ser. E Bio.* 3, pp. 65–80, 1959.

Mountford, M.D. An index of similarity and its application to classificatory problems, in *Progress in Soil Zoology*, Murphy, P.W., Ed., Butterworths, London, 1962.

Munkittrick, K.R. and L.S. McCarty. An integrated approach to aquatic ecosystem health: top-down, bottom-up, or middle-out? *J. Aquat. Ecosyst. Health* 4, pp. 77–90, 1995.

Newman, M.C. *Quantitative Methods in Aquatic Ecotoxicology.* Lewis Publishers, Boca Raton, FL, 1995, p. 426.

Noreen, E.W. *Computer Intensive Methods for Testing Hypotheses.* Wiley-Interscience, New York, 1989, p. 229.

Norusis, M.J. *SPSS-X Advanced Statistics Guide*, 2nd Ed. SPSS Inc., Chicago, IL, 1988, p. 527.

Odum, H.T., J.E. Cantlon, and L.S. Kornicker. An organizational hierarchy postulate for the interpretation of species-individuals distribution, species entropy and ecosystem evolution and the meaning of a species-variety index. *Ecology* 41, pp. 395–399, 1960.

Osborne, L.L., R.W. Davies, and K.J. Linton. Use of hierarchical diversity indices in lotic community analysis. *J. Appl. Ecol.* 17, pp. 567–580, 1980.

Patrick, R. A proposed biological measure of stream conditions, based on a survey of the Conestoga Basin, Lancaster County, Pennsylvania. *Proc. Acad. Nat. Sci. Phila.* 101, pp. 277–341, 1949.

Patrick, R.M. H. Hohn, and J.H. Wallace. A new method for determining the pattern of the diatom flora. *Notulae Naturae* 259, pp. 1–12, 1954.

Pielou, E.C. *Ecological Diversity.* Wiley Publishing Co., New York, 1975, p. 165.

Pielou, E.C. *The Interpretation of Ecological Data.* Wiley Publishing Co., New York, 1985, p. 263.

Pinkham, C.F. and J.G. Pearson. Applications of new coefficient of similarity to pollution surveys. *J. Water Pollut. Control Fed.* 48, pp. 717–723, 1976.

Plafkin, J.L., M.T. Barbour, K.D. Porter, S.K. Gross, and R.M. Hughes. *Rapid Bioassessment Protocols for Use in Streams and Rivers. Benthic Macroinvertebrates and Fish*, EPA/444/4-89-001, U.S. Environmental Protection Agency, Assessment and Watershed Protection Division, Washington, DC, 1989.

Pontasch, K.W., E.P. Smith, and J. Cairns, Jr. Diversity indices, community comparison indices, and canonical discriminant analysis: interpreting the results of multispecies toxicity tests. *Water Res.* 23, pp. 1229–1238, 1989.

Preston, R.W. The commonness and the rarity of species. *Ecology* 29, pp. 254–283, 1948.

Quinlan, J.R. *C4.5: Programs for Machine Learning*. Morgan Kaufmann Publ., San Mateo, CA, 1993, p. 302.

Raffaelli, D.G. and C.F. Mason. Pollution monitoring with meiofauna, using the ratio of nematodes to copepods. *Mar. Pollut. Bull.* 12, pp. 158–163, 1981.

Shannon, C.E. and W. Weaver. *The Mathematical Theory of Communication*. The University of Illinois Press, Urbana, IL, 1949, p. 117.

Sheldon, A.L. Equitability indices: dependence on the species count. *Ecology* 50, pp. 466–467, 1969.

Simpson, E.H. Measurement of diversity. *Nature* 163, pp. 4148, 1949.

Sládecek, V. System of water quality from the biological point of view. *Arch. Hydrobiol. Beiheft* 17, pp. 1–218, 1973.

Sneath, P.H.A. and R.R. Sokal. *Numerical Taxonomy*. W.H. Freeman, San Francisco, CA, 1973, p. 573.

Sorensen, T.A. A method of establishing groups of equal amplitude in plant sociology based on similarity of species content, and its application to analyses of the vegetation on Danish commons. *K. Dan. Vidensk. Selsk. Biol. Skr.* 5, pp. 1–34, 1948.

Tabachnick, B.G. and L.S. Fidell. *Using Multivariate Statistics*, Third Edition. Harper Collins Publishers Inc., New York, 1996, p. 746.

Titmus, G. The use of Whittaker's components of diversity in summarizing river survey data. *Water Res.* 15, pp. 863–865, 1981.

Vinson, M.R. and C.P. Hawkins. Effects of sampling area and subsampling procedure on comparisons of taxa richness among streams. *J. North Am. Benthol. Soc.* 15, pp. 392–399, 1996.

Washington, H.G. Diversity, biotic and similarity indices. *Water Res.* 18, pp. 653–694, 1984.

Weiss, S.M. and C.A. Kulikowski. *Computer Systems That Learn*. Morgan Kaufmann Publ., San Mateo, CA, 1991, p. 223.

Whittaker, R.H. Dominance and diversity in land plant communities. *Science* 147, pp. 250–260, 1965.

Whittaker, R.H. Evolution and measurement of species diversity. *Taxon* 21, pp. 213–251, 1972.

Whitton, B.A. and M.G. Kelly. Use of algae and other plants for monitoring rivers. *Aust. J. Ecol.* 20, pp. 45–56, 1995.

Wolda, H. Similarity indices, sample size and diversity. *Oecologia* 50, pp. 296–302, 1981.

Wright, S. and W.M. Tidd. Summary of limnological investigations in western Lake Erie in 1929 and 1930. *Trans. Am. Fish. Soc.* pp. 271–285, 1933.

11 ||| Landscape Perspectives on Ecological Risk Assessment

Carl Richards and Lucinda B. Johnson

INTRODUCTION

The National Environmental Policy Act of 1969 (NEPA) requires an Environmental Impact Statement (EIS) for major federal actions significantly affecting the quality of the human environment. Initially the emphasis was on point-source emissions, and there was little or no regard for cumulative impacts over time and space or the regional effects of emissions. From this perspective, we have evolved to the realization that consideration of scale is critical to all aspects of the ecological risk assessment process. Ecological risk assessment (ERA) is a flexible process for predicting the likelihood of adverse ecological effects resulting from environmental problems ranging from global change, acid precipitation, widespread use of toxic chemicals, to habitat loss (EPA, 1996). Ecological risk assessment begins with a definition of the problem, followed by analysis, and ends with risk characterization. The realistic goals and assessment endpoints are constrained first by the boundaries of the problem, then by the spatial and temporal character of the ecosystem properties used as assessment endpoints.

In order for the ERA process to be useful for addressing a diverse set of environmental problems, both stressors and endpoints must be measured at appropriate spatial and temporal scales, and stressor effects must be separated from natural system variability. Even when assessing the potential impact of a small-scale spill on an ecosystem, the temporal characteristics of the stressor (e.g., rate of spill, season), spatial variability of the receiving system (e.g., size of stream, position in watershed), and the temporal dynamics of ecosystem processes (e.g., annual variation) must be considered. Tools and concepts arising from the emerging discipline of landscape ecology can be used to assess the influence of spatial variability on ecosystem structure and function as it responds to various stressors (O'Neill et al., 1988; Hunsaker et al., 1992; Johnson and Gage, 1997).

In this chapter we will discuss the importance of landscapes to ecological risk assessment at all spatial scales, review some fundamental concepts of landscape ecology as they relate to understanding ecosystem structure and function, examine how landscape approaches influence some of the fundamental steps of ecological risk assessment, and explain how

Scales of processes in aquatic systems

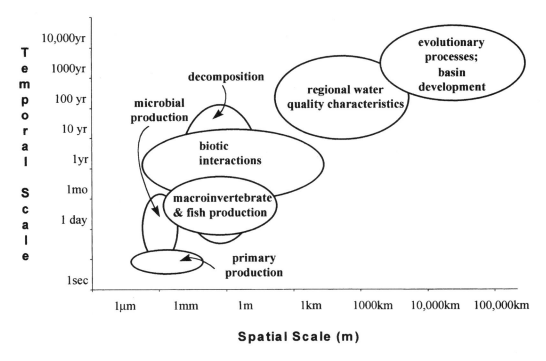

Figure 11.1. Spatial and temporal scales associated with representative ecosystem processes in aquatic systems.

some new technologies, tools, and approaches can be used to include landscape perspectives in the ecological risk assessment process.

LANDSCAPE PERSPECTIVES

Scale, in ecological terms, refers to the distance or time frame over which some phenomenon occurs. Ecological processes are ordered into nested levels defined by their temporal and spatial scales (Figure 11.1). This is known as hierarchy theory or the hierarchy framework (Allen and Starr, 1982; O'Neill et al., 1986). The boundaries at each level are based on similarities in process rates (O'Neill, 1989). At lower levels of organization process rates are rapid, while those at higher levels are slower. Relationships between elements in the set are asymmetrical, in that lower levels are constrained by higher levels. The hierarchy framework suggests that ecological phenomena must be observed at appropriate spatial and temporal scales that are defined by their size and context. Observations at small scales lead to an understanding of mechanisms or dynamics, while those at larger scales give context or significance (Allen and Hoekstra, 1992). Measurements made at spatial and temporal scales that are smaller than appropriate will appear as "noise," while those made at larger spatial and temporal scales may appear constant and thus may not capture the phenomenon being studied.

The concepts embedded in hierarchy framework require that ecosystems be examined in the context of landscapes. A landscape is loosely defined as a mosaic of ecosystems having a similar geomorphology and disturbance regime. Thus, landscapes are separated by distinct boundaries (ecotones), generally reflected by changes in geology, vegetation, and land use. Just as an ecosystem can range in size from the minuscule to the gargantuan, landscape sizes also can range over orders of magnitude. Landscapes can be characterized on the basis of their structure, function, and response to change. Landscape structure describes the number, types, and spatial configuration (including the shape, connectivity, and arrangement) of ecosystem patches. Landscape function embodies the interactions among ecosystems with respect to the flow of energy, materials, and organisms. Temporal change is a fundamental characteristic of landscapes, and the trajectory and magnitude of change over time reflect some of the underlying mechanisms regulating landscape structure and function. For example, fire alters vegetation patterns, resetting ecosystem structure and function to that characteristic of early successional stages. The intensity and frequency of fire are important determinants of community structure and productivity in many ecosystems, and result in characteristic patterns on the landscape. Similarly, stressors can influence an ecosystem's component processes, as well as its physical structure. A stressor's impact on a landscape will depend on the type, magnitude, stressor frequency, and feedbacks between adjoining ecosystems.

The first step in the risk assessment process involves establishing assessment goals and defining the nature and boundaries of the problem (Suter and Barnthouse, 1993). The boundaries of the problem must be defined by the temporal and spatial characteristics of both the stressor and endpoints. For example, stream ecosystem processes and properties (e.g., primary production, decomposition, and community structure) are controlled by local factors, such as the physical structure of the stream channel and physicochemical characteristics of water, and external factors, such as riparian zone vegetative composition and width. Local factors, in turn, are regulated by higher-order factors, including climate, land use, and geology. In addition, populations and communities are directly influenced by feeding behavior, reproductive tactics, dispersal mechanisms of individuals, by the biotic interactions among species, and landscape patterns, which regulate the flow of energy and organisms between habitat patches.

STRESSORS

Stressors are the perturbations that disrupt ecosystem structure and function and result in the unacceptable conditions that are addressed by the ERA process. Stressors can either be the result of a natural disturbance, or they may result directly or indirectly from anthropogenic activities. As with ecological processes, perturbations can also be arrayed on temporal and spatial scales that characterize the nature of their impacts on ecological systems (Figure 11.2). Spills and point-source discharges are found at the smallest spatial and temporal scales. Typically, these perturbations have ecological effects that are of short duration and occur over limited areas, although some compounds (e.g., radioactive materials, persistent toxic substances) may have prolonged temporal influences. Of potentially greater temporal and spatial dimensions are land use conversions and other alterations to physical and chemical components of habitats. These ecological stressors are persistent and are often promulgated over large areas. Examples of such stressors are large-scale agricultural and

Effects of Environmental Stressors

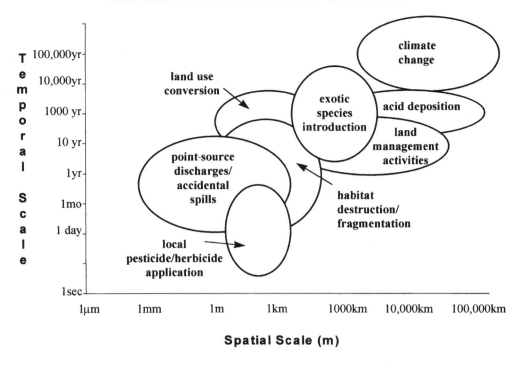

Figure 11.2. Spatial and temporal scales associated with categories of environmental stressors.

forest management practices, wetland drainage, stream channelization, and urbanization. Landscape fragmentation, which alters the flow of materials among ecosystem components and the interaction of organisms among landscape patches, also occurs at large temporal and spatial scales. Once these activities have occurred, recovery of ecosystem processes that are regulated by the physical characteristics of a system may take several decades. Nonpoint-source pollutants, often the result of land conversions, affect systems at this spatial scale as well. However, recovery from stress associated with nonpoint pollutants such as nutrients may be faster than stress associated with land use management practices that alter physical habitat. Introduced species can quickly colonize large geographic areas and are rarely, if ever, extirpated from ecosystems once they are established. Therefore, these stressors also have relatively large spatial and temporal scales. At the largest spatial and temporal scale, climate change results from multiple physical and chemical processes interacting over global scales. Effects persist over long periods of time due to the size and complexity of the processes involved and the area affected.

The deleterious effect of some stressors may not be apparent until the stressors are viewed at an appropriate spatial or temporal scale. Such stressors may result from the cumulative impact of repeated disturbances at the same location over time, or at distinct but proximal locations at the same time. A disturbance that recurs at intervals shorter than the recovery time exacerbates the effects of the original disturbance and may lengthen the recovery process (Yount and Niemi, 1990). The loss of a single wetland may not have a discernable effect on flood frequency and magnitude, or storage capacity in a watershed; however, loss of many wetlands in the lower reaches of the watershed may have a large impact on flood events (Anonymous, 1994). Such temporally and spatially disjunct disturbances may also carry indirect effects, which are difficult to identify and quantify.

Both nonanthropogenic and anthropogenic stressors can increase the susceptibility and magnitude of a system's response. The presence of another stressor is an important factor determining system susceptibility. For example, vegetation that is under stress due to nutrient or water limitation is more susceptible to disease and insect infestation (e.g., Payne, 1980). At regional scales, fragmentation of previously intact ecosystems (e.g., Amazonia) by roads can increase susceptibility to other disturbances, such as erosion, slash and burn agriculture, mining, and unsustainable hunting of valuable game species. When stressors occur in combination, resulting effects on ecological systems are difficult to understand and quantify, since effects can be additive or multiplicative. Finally, landscape pattern can potentially have a large impact on the effect of a stressor by influencing its mode of action and magnitude (see Graham et al., 1991; Hunsaker et al., 1992). For example, small forest stands have a high edge-to-interior ratio and are more susceptible to disturbances such as blowdown than larger stands (Franklin and Forman, 1987). Thus, many aspects of the ERA process may be influenced by the structure of the landscape being examined.

ENDPOINTS

Endpoints are the components of ecosystems and communities used to quantify and monitor effects of stressors. The selection of endpoints appropriate to the environmental stressor at hand is a complex process (Suter, 1990), and the choice and appropriate use of these measures is fundamental to the logic and success of the ERA process. Since some assessment endpoints are conceptual (e.g., nutrient cycling) or are extremely complex (e.g., animal or plant community dynamics), measurement endpoints are often used as surrogates to denote the condition of an ecosystem component. Societal and biological relevance, unambiguous operational definition, accessibility to prediction and measurement, and susceptibility to the hazardous agent must be considered when selecting endpoints (Suter and Barnthouse, 1993). Two additional factors must be considered in the selection process: (1) assessment endpoints must be matched in scale to the assessment goal and the stressor, and (2) natural sources of variation in assessment endpoints must be understood so they may be separated from effects of stressors.

As with ecological processes and stressors, endpoints have inherent spatial and temporal scales due to the system components they measure (Figure 11.3). Application of endpoints of one scale to address questions at another confounds the interpretation of effects. Responses at one scale may be registered as noise if the stressor operates at larger or smaller (spatial or temporal) scales than the assessment endpoint. This does not imply that assessment endpoints at one scale cannot be used to address a problem at a smaller or larger spatial scale, but rather, that the nested hierarchy of influences on the assessment endpoint in question must be considered explicitly to avoid either missing the effect altogether or misinterpreting the signal as noise in the system.

Biological and ecological endpoints are inherently variable because they respond to a variety of biotic and abiotic factors, in addition to the stressors of interest in an ERA. Therefore, variability in ecological endpoints must be partitioned among environmental

Scales of Endpoints

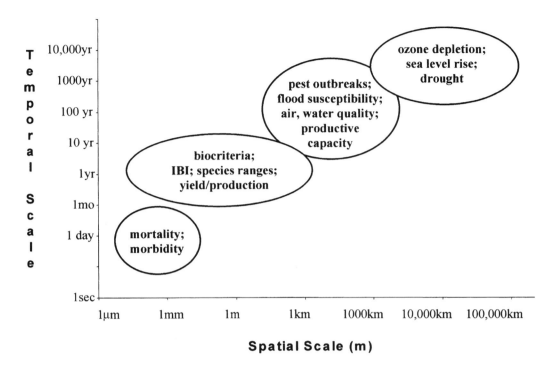

Figure 11.3. Spatial and temporal scales associated with typical endpoints used in environmental assessments.

stressors and other natural sources of variation. For example, natural cycles of population density must be separated from potential stressors affecting population density versus effects of xenobiotic substances. The ability to predict the uncertainty associated with an environmental stressor requires an understanding of all sources of endpoint variability.

Ecosystem Endpoints

All endpoints are influenced by landscape characteristics to varying degrees. Endpoints that measure ecosystem processes (e.g., primary productivity, nutrient uptake, decomposition rates) are constrained by regional climatic and geologic conditions as well as site-specific physicochemical conditions. Exposure to a stressor can induce a measurable effect, such as mortality, increased morbidity, or reduced reproductive potential, which is subsequently quantified in the assessment process. Alternatively, the exposure can induce a change (increase or decrease) in the variability of that endpoint as a result of exposure. However, before variation in ecosystem endpoints can be used for impact assessment, it must be calibrated with respect to regional patterns. For example, considerable differences might exist in the primary production of lakes resulting from differences in geology within watersheds. Once again, such differences must be differentiated from anthropogenic stressors.

Physical and Chemical Endpoints

Physical and chemical endpoints are frequently employed as surrogates of ecosystem parameters or for monitoring known pollutants. Chemical parameters are the most commonly used endpoints in this category. A variety of factors operating at different spatial scales may moderate the observed concentrations of chemicals. For example, the concentrations of atmospherically derived chemicals (e.g., NO_x, SO_2, some PCBs) follow airsheds that may be orders of magnitude larger than an area of concern in an environmental assessment. Similarly, concentrations of chemicals in streams and lakes are often a function of activities distant from a sampling point. Undisturbed lakes on Isle Royale (a remote island in Lake Superior), for example, have measurable concentrations of PCBs (Czuczwa et al., 1984) that could only have originated from distant atmospheric sources. The constant hydraulic movement of materials through watersheds and into streams suggests that spatially explicit analyses are important for these systems. In a study conducted within a large 15,000 km^2 area of the midwestern United States, Johnson et al. (1997) found that land use and land cover characteristics within stream buffers had a strong influence on some water chemistry parameters, but not others. For example, total phosphorous was more closely related to ecotone conditions (Figure 11.4a). Nitrogen, on the other hand, was as, or more, strongly related to whole-watershed conditions than to local conditions, suggesting that overland and groundwater flow from distant parts of the watershed had significant effects on local conditions. Other evidence for importance of spatial position of landscape elements comes from the work of Johnston et al. (1990) who showed that the spatial position of wetlands influenced stream water quality. Wetlands in stream headwaters were less effective at removing some constituents, such as suspended solids, than wetlands located lower in the watershed. These studies indicate that different portions of a watershed may have different influences on chemical constituents, and that cumulative land use characteristics of watersheds can be used to describe some aspects of stream condition.

Since the physical dimensions of habitat set the stage for ecosystem, community, and population dynamics, measures of physical status are often used as endpoints in assessment programs. These endpoints are particularly insightful when stressors are known to alter the physical character of a site. Simple measures that quantify habitat components often can provide estimates of disturbance. For example, in streams, morphological structure of stream channels has profound effects on ecosystem and community parameters, and forested and agricultural land uses are known to influence these parameters (Richards and Host, 1994; Richards et al., 1996). Consequently, physical endpoints such as depth, substrate composition, and channel morphology are often employed to monitor stream systems (Osborne et al., 1991; Ohio EPA, 1987).

Processes regulating physical endpoints also have spatial and temporal dependencies. For example, physical disturbance may result from local activities such as soil disturbance at a construction site, but it may also be influenced by large-scale activities—such as watershed land use changes—that influence local conditions. In a study of over 40 watersheds in the Midwest, Richards et al. (1996) found that the amounts of fine sediment abundance in substrates were influenced more by surficial geology and land use in stream buffers than in the whole watershed (Figure 11.4b). On the other hand, bank full depth, which has a large influence on stream biota, was more strongly influenced by whole watersheds due to its relationship with hydrologic characteristics of watersheds rather than ecotones. Consequently,

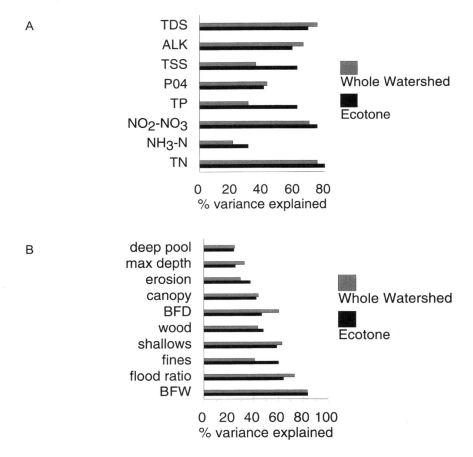

Figure 11.4. Comparison of total variation explained and (a) stream water chemistry variables, and (b) physical habitat variables by landscape data in whole watersheds and ecotones (100 meter buffer) adjacent to stream courses. Landscape data included land use, geology, and landform variables. Data are the results of redundancy analysis (see Richards et al., 1996, Johnson et al., 1997). (BFD = bank full depth; wood = coarse woody debris abundance; fines = amount of fine substrates; BFW = bank full width.)

land use changes that influence the movement of water in landscapes (e.g., agricultural development, wetland draining, urbanization) have strong influences on stream health through alteration of regional hydrology.

Population Endpoints

The most frequently used endpoints in almost any type of environmental assessments are measures of organism abundance or population dynamics (e.g., growth, fecundity, mortality). Populations are a natural level of organization in biological systems and consequently are central to many monitoring programs. In addition, information about animal and plant populations is often relatively easy to communicate to scientific and lay audiences, and populations are the original source of concern in the public eye (e.g., spotted owls and game species). Populations are influenced by landscape characteristics in a variety of ways. One of the most obvious relates to the overall behavior or range of a species (Pulliam, 1994). Migratory species that spend large portions of their lives in different ecosystems or biomes (e.g., Pacific salmon and neotropical migrant birds) are exposed to very different sets of environmental stresses for extended periods of their life cycles. Stresses may differ at each location, potentially having a significant effect on productivity or survival. Similarly, movement of animals within their ranges can also influence exposure to stressors by varying the type and number of stressors encountered. All species have definable ranges varying in size from large (e.g., gray wolf, neotropical migratory birds, and lake trout) to more restricted ranges (e.g., many insects and plants). Movements among landscape patches during foraging or migration influence total exposure. For example, organisms occurring in contiguous patches may be more vulnerable to disease vectors having low dispersal abilities than those occurring in disjoint patches. Wide-ranging species may encounter a larger number of stressors, and thus may adjust their use of the landscape (e.g., by avoiding areas with roads or unsuitable land cover patch types) to minimize exposure. Certain species, such as the gray wolf, are known to favor regions with low road density and low land cover patch boundary complexity, and evidence suggests that landscape fragmentation restricts the dispersal of this species in the Upper Great Lakes states (Mladenoff et al., 1995). Homogeneous patches, such as plantations or agricultural fields, are more vulnerable to disease and pest outbreaks than patches containing multiple species and age classes. Thus, knowledge of the distribution of a stressor across a landscape, coupled with knowledge of an organism's use of the landscape during various phases of its life cycle, can lead to more precise estimates of exposure to potential stressors.

Community Endpoints

Community or assemblage endpoints are also widely used in environmental assessments and have clear spatial and temporal influences. Since the information in community endpoints is ultimately derived from individual populations, landscape characteristics that influence individual populations are also relevant to community measures. However, additional concepts are germane to some community measures. Derived community measures (e.g., species richness, diversity, evenness) are particularly responsive to landscape characteristics and the scale of processes controlling populations. For example, species richness generally increases with the size of an area sampled, due to parallel increases in habitat heterogeneity (MacArthur and Wilson, 1967). The size of the sampling area strongly influences richness and diversity values, since many species may be restricted to certain subhabitats within a sampling area. Community structure observed at small spatial scales can often be controlled by features at larger spatial scales. Richards et al. (1997) demonstrated that invertebrate community composition in stream reaches was partly a function of the relationship between species life history traits and landscape characteristics such as dominant surficial geology (Figure 11.5). As a result, several important aspects of community structure bore little or no relation to local conditions. These points underscore the importance of understanding the functional grain size of the landscape, particularly with respect to features that control habitat quantity, quality, and distribution.

There is a growing recognition of the role of landscape position, as well as composition, in regulating plant and animal population dynamics. The proximity and degree of con-

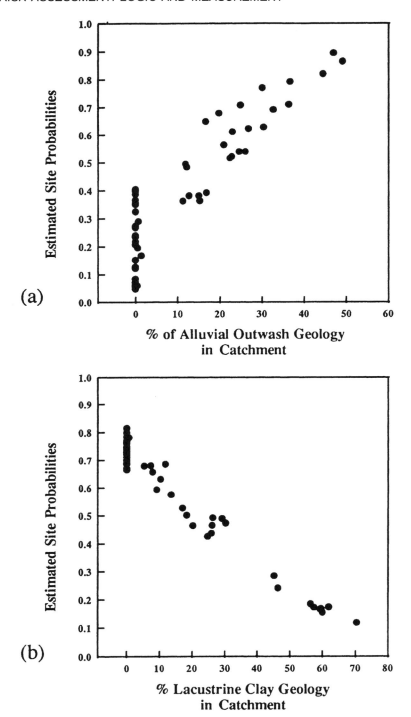

Figure 11.5. The influence of catchment-scale properties on the relative abundance of macroinvertebrate (a) obligate erosional taxa and (b) merovoltine taxa in streams (from Richards et al., 1997, with permission from Blackwell Science, Ltd.). Probabilities were estimated with logistic regression models.

nectivity of landscape patches that serve as sources or sinks for animal populations can strongly influence population dynamics at another location. Movements and dispersal among habitat patches influence extinction, population genetics, and total abundance of organisms (see Hanski and Gilpin, 1991). Similarly, community parameters are also influenced by landscape pattern, including patch shape, diversity, and connectivity. The theoretical and empirical basis for the influence of habitat or landscape unit proximity on community structure is well documented in the literature of island biogeography (MacArthur and Wilson, 1967). The potential influence of the position of landscape patches on risk assessment is large. For example, the effects of a localized disturbance on invertebrate or fish communities may be minimized when favorable habitat that can support a constant supply of immigrants is located upstream. Streams typically recover more quickly from environmental stresses when upstream habitat remains intact (Niemi et al., 1990; Yount and Niemi, 1990).

Regional Endpoints

Regional endpoints are used specifically to track the influences of large-scale environmental stressors or to report on the condition of natural resources in regional areas (Suter, 1990; Hunsaker et al., 1990). Acid rain, cumulative loading of nutrients and other pollutants in rivers, and ozone depletion are a few of these types of stressors. With a few exceptions, the endpoints used in such assessments are the same as those described previously. Many endpoints (e.g., production, abundance, species abundance) can be used for regions merely by altering the boundaries considered. Other endpoints—such as some landscape indices (O'Neill et al., 1988)—can only be applied to large areas. Obviously, the implications of knowing the spatial scales relevant to the underlying ecological processes within a region are paramount to understanding the behavior of endpoints in response to large-scale stressors.

APPLICATIONS OF LANDSCAPE ECOLOGY TO RISK ASSESSMENT

An important step in environmental risk assessment involves identifying and quantifying the sources of uncertainties associated with each stage of the assessment process. Only in this manner can a meaningful projection of the magnitude of impact be determined. In ecological risk assessment, substantial error may be associated with the databases commonly used for these projections. Since ecological phenomena involve numerous biotic and physical parameters, and each is associated with a unique sampling method and innate sampling bias, errors are rapidly propagated when making projections. In most ecological systems, multiple stressors act on endpoints, thus interactive or synergistic effects may mask the expression of individual effects. In addition, the incorporation of phenomena with long temporal and large spatial scales provides additional sources of uncertainty not encountered with relatively short-term assessments of smaller ecological systems. Clearly, there is a need to develop techniques for differentiating and quantifying sources of uncertainty.

An environmental risk assessment cannot be conducted successfully without quantifying the major sources of variation in endpoints and landscape structure. Quantifying underlying variation allows the risk assessor to construct a valid conceptual model of the system and identify those system components of the landscape that must be partitioned to constrain risk projections to specific stressors. The techniques for decomposing ecological variance at large scales are in their infancy. Most utilize available databases and statistical techniques in novel ways to address specific questions. One approach for decomposing variance in regional landscapes of the Midwest is described in Richards et al. (1996) and Johnson et al. (1997). Landscape components that influence stream water quality, habitat, and biota were identified utilizing readily available digital landscape data and conventional stream survey techniques. Using GIS technology, statistical questions concerning the influence of different landscape components on endpoint variance were posed. In addition to the positional aspects of landscapes previously described (Figure 11.4), landscape components were classified into those influenced by land management (e.g., land use and land cover characteristics, land cover patch density; hereafter referred to as "land use" factors) and those fixed upon the landscape but which may influence the behavior of endpoints (e.g., geology, land form, topography; hereafter referred to as "geo/struc" factors). Approximately equal portions of variance in physical variables important to stream biota were attributed to "land use" and "geology" (Figure 11.6). However, physical variables differed with respect to the relative influence of these landscape categories (Figure 11.7). Bank full width (BFW) was primarily explained by "geo/struc" and was not greatly influenced by "land use." Woody debris, on the other hand, was more influenced by "land use." Other factors, such as flood ratio—a measure of the intensity of floods—were equally influenced by "land use" and "geo/struc." Additional analyses revealed that watershed size, extent of row crop agriculture, and contrasts in dominant surficial geology within watersheds were the most important landscape elements influencing stream habitats and biota (Richards et al., 1996).

Common water chemistry parameters were also significantly influenced by both "land use" and "geology" features (Johnson et al., 1997; Figure 11.6). "Land use" and "geo/struc" accounted for similar proportions of the variance during summer. The relative strength of those relationships, however, changed seasonally. At baseflow conditions in autumn, stream water chemistry was influenced less by "land use" than "geo/struc," whereas in the summer "land use" had the greatest effect (Figure 11.5), demonstrating the importance of temporal variation on processes regulating landscape/stream water chemistry relationships. As with in-stream physical factors, there were distinct differences among individual water quality parameters in terms of the relative influence of "land use" versus "geo/struc." Nitrate-nitrogen (NO_3-N) was strongly influenced by "land use" and minimally influenced by "geology" (Figure 11.8). Further analyses showed that the dominant landscape factor explaining variation in summer water chemistry was the proportion of row crop agriculture in the watershed. In contrast, total alkalinity and total dissolved solids responded to geologic factors on the landscape. In-stream phosphorus concentrations were not strongly related to landscape factors.

The relevance of the above studies to risk management in the midwestern United States is that the fundamental scales at which major ecosystem components vary were identified. Local subunits of geology—in this case remnants of glacial movements—accounted for a large portion of the variation in some stream water chemistry parameters, habitats, and biota variables. Land use, especially row crop agriculture, accounted for most of the variation in water chemistry variables. Without knowledge of regional patterns, endpoint behavior reflecting land management changes at smaller spatial scales might be misinterpreted. For example, a commonly used biological metric such as EPT taxa (Ephemerotpera,

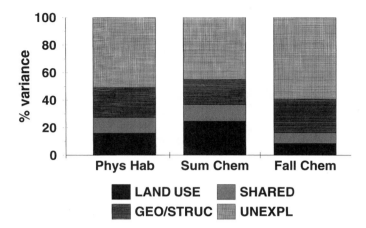

Figure 11.6. Decomposition of variance associated with physical habitat and stream water chemistry variables (from Richards et al., 1996; Johnson et al., 1997). Numbers represent total variance described for 10 physical habitat variables at base flow conditions and 8 water chemistry variables in summer and fall. Variance categories are land use, geology and landscape structure, shared variation between land use and geology/structure variables, and unexplained variance. Data are the results of redundancy analysis.

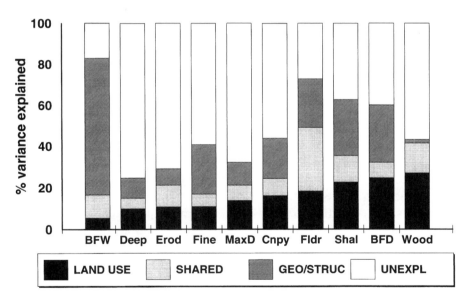

Figure 11.7. Decomposition of variance associated with physical habitat variables important to stream biota in Midwestern streams (from Richards et al., 1996, with permission from NRC Research Press). Percent variance refers to the total variance explained by landscape variables. Categories are land use, geology and landscape structure, shared variation between land use and geology/structure variables, and unexplained variance. Data are the results of redundancy analysis. (BFW = bank full width; deep = % deep water habitat; Erod = % eroding banks; Fine = % fine substrate; MaxD = maximum depth; Cnpy = % open canopy; Fldr = flood ratio; Shal = % shallow water habitat; BFD = bank full depth, Wood = abundance of coarse woody debris.)

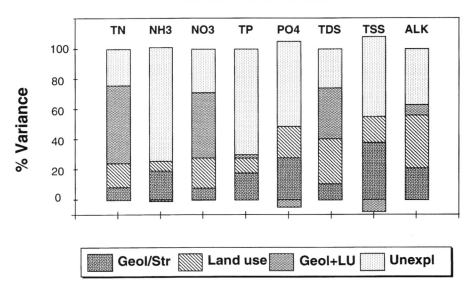

Figure 11.8. Decomposition of variance associated with stream chemistry variables in Midwestern streams (from Johnson et al., 1997). Percent variance refers to the total variance explained by landscape variables. Categories are land use, geology and structure, shared variation between geology and structure, and unexplained variation. Data are the results of redundancy analysis.

Plecoptera, Trichoptera), is often used to assess land use effects on stream communities. However, in some circumstances (e.g., watersheds dominated by erodible soils) these taxa may be influenced more by geologic characteristics of watersheds than by land use. Since these taxa often require hard substrate, a stream with few EPT taxa may reflect geologic characteristics of the watershed, rather than land use (Richards et al., 1997). When land use influences on stream parameters are strong, the appropriate scale of influence must be recognized. For example, the value of altering nitrogen-based fertilizer application or tillage practices at the field scale must be weighed against the total amount of row crop agriculture in a watershed, since the influence of local management practices on total nitrogen and nitrate concentrations is masked by the cumulative effects of watershed-wide land management activities.

TOOLS FOR REGIONAL ECOLOGICAL RISK ASSESSMENT

While methods for site-based ecological risk assessment are relatively well established (Westman, 1985), those for ecological risk assessment at larger scales are in the early stages of development. The greatest shortcomings of regional analyses lie in the lack of long-term data sets for large geographic regions and the lack of appropriate models for predicting ecosystem endpoints at regional scales. Spatial data are necessary to quantify environmental variables, describe stressors, and assess endpoints. Newer tools, such as geographic information systems (GIS) and image processing, have greatly expanded the availability of spatially explicit maps linked to databases that can be used to assess the status and trends of

Table 11.1. Some Potential Applications of GIS Technology for Regional Risk Assessment.

Application	Examples
Generate secondary data from primary data	slope and aspect from elevation points; contour lines from point data
Quantify associations between ecological endpoints and stressors	biotic index, population density, species distributions, or water quality criteria, with respect to wetland loss, riparian land conversion, or regional use of herbicide or pesticide
Quantify landscape patterns and spatial relationships	shape, connectivity, juxtaposition, or diversity of landscape patches
Quantify temporal patterns	greenness, ice/snow cover, wetland area
Quantify temporal change	land use conversion (e.g., wetland loss; agricultural land conversion)
Link spatial data with models	hydrologic, ecosystem, climate change

landscape components (Table 11.1; Johnson and Gage, 1997). These technologies have permitted ecologists to examine the effects of spatial heterogeneity on ecological processes over much larger areas than was previously possible. GIS technology is not limited to regional applications, and can be used effectively for ecological risk assessment at any spatial scale. The widespread availability of standardized databases from federal and state government agencies (Table 11.2), satellite imagery, and inexpensive computing and visualization technologies now puts these tools in the hands of most researchers, government agencies, and policymakers. When linked to mechanistic and stochastic models, new data can be generated to assess ecological conditions (e.g., air or water quality) or to predict the effects of stressors on communities or ecosystems. In addition to recent enhancements in computing and mapping technologies, new and improved statistical methods have contributed to the expansion of the field of landscape ecology (reviewed by Johnson and Gage, 1997).

CONCLUSIONS

The integration of ecological principles into risk assessment is a rapidly evolving process, yet further development of protocols that meet rigorous standards is still needed. Conceptual models have been developed for assessing ecological risk at landscape scales. For example, Graham et al. (1991) presented a methodology for examining ozone-triggered land cover changes that influenced water quality in lakes. Their approach included stochastic simulations of land cover change that were driven by spatial variation in air quality and variations in landscape pattern. The unique aspect of this conceptual approach is that it utilized spatially explicit landscape databases to predict variation in endpoints or observed effects on a regional scale. Lowrance and Vellidis (1995) presented a detailed conceptual model for examining the potential influences of a variety of stressors to three ecosystems within bottomland hardwood forests. They presented a rationale for utilizing a wide array of assessment and measurement endpoints and stressed the importance of linking field and watershed-scale processes. They also suggested potential sources of data for these analyses.

Table 11.2. Spatial Data Available for Most Regions of the United States. Additional Data May Be Available from State Agencies and Non-Governmental Organizations.

Data Theme	Data Source	Internet Address	Scale	Metadata	Program Name
Bathymetry	NOAA	www.glerl.noaa.gov/data/bathy.html	variable	NO	Great Lakes Environmental Lab
Bedrock geology	USGS; individual states	wwwflag.wr.usgs.gov/cgi-bin/tiger_form_alchi_box.pl	1:2,500,000	YES	
Boundaries	USGS	edcwww.cr.usgs.gov/nsdi/gendlg.html	1:100,000	YES	Digital Line Graph
Ecoregions	U.S. EPA	nsdi.usgs.gov/ndsi/wais/water/ecoregion.html	1:7,500,000	YES	
Elevation	USGS	nsdi.usgs.gov/nsdi/gendem.htm	1:24,000	YES	Digital Elevation Model
Forest Cover	USDA, Forest Service	www.epa.gov/docs/grd/forest_inventory	1:100,000 1:250,000 ~1:250,000	YES	
Hydrography	USGS	edcwww.cr.usgs.gov/nsdi/gendlg.html	1:24,000	YES	Digital Line Graph
Hydrology	USGS	nsdi.usgs.gov/nsdi/wais/water/hcdn.html	1:100,000	YES	Stream Flow Data
Hydrologic units	USGS; individual states	www.cr.usgs.gov/doc/edchrome/datasets/edcdata.html	1:2,000,000	YES	
Land use/land cover	USGS; individual states	edcwww.cr.usgs.gov/nsdi/gendlg.htm	1:100,000 1:250,000	YES	LUDA
NPDES Sites	U.S. EPA				
Orthophoto quads	USGS	nsdi.usgs.gov/nsdi/products/doq.html	1:12,000 1:24,000	YES	
Population/Census	U.S. Department of Interior	www.census.gov/ftp/pub/www/tiger		YES	TIGER

Quaternary Geology	individual states	nsdi_usgs.gov/nsdi/products/quaternary_geol.html	1:1,000,000	YES	
Soils	USDA; individual states	www.ncg.nrcs.usda.gov/nsdi/nsdi_node.html	1:12,000–63,360 1:250,000	YES	STATSGO; SURGO NATSGO
Transportation Network	USGS; individual states	edcwww.cr.usgs.gov/nsdi/gendlg.htm	1:24,000 1:100,000	YES	Digital Line Graph
Toxics Release Inventory	U.S. EPA	earth1.epa.gov/tri_cover93		YES	Toxics Release Inventory
Water Quality	U.S. EPA	nsdi.usgs.gov/nsdi/products/nawqa.html		YES	STORET, WATSTORE
Wetlands	U.S. Department of Interior, Fish and Wildlife Service	www.nwi.fws.gov/data.html	1:24,000 +	YES	National Wetland Inventory

A fully developed risk assessment that utilized some of the widely available landscape data was described by Solomon et al. (1996). They employed standard approaches for assessing effects of exposure to atrazine on aquatic organisms. In addition they also characterized the relative risk of atrazine exposure in different regions of the United States by combining some widely available spatial databases on atrazine usage and rainfall. This approach demonstrated the power of simple spatial data analysis techniques to address complex problems. The extrapolation of traditional laboratory toxicology studies to larger spatial scales will continue to be a subject requiring considerable attention.

The key problems limiting integration of landscape ecology with ecological risk assessment include extrapolation across scales, quantification of uncertainty, and validation of predictive tools (National Research Council, 1993). In addition, we lack the long-term databases that allow us to understand ecosystem dynamics in response to stressors occurring over long time periods. Full integration of these concepts to risk assessment will require considerable research and effort from the scientists, resource managers, and policymakers involved in the risk assessment process. Although many research needs remain, ongoing federally funded programs show promise for resolving some of these issues. The dialogue between toxicologists and ecologists is fostering an increased understanding of the mechanisms regulating ecosystem processes over landscape scales. The challenge remains to broaden this concept to encompass a wider array of stressors, to understand and quantify the effects of multiple stressors, and to discriminate between natural variability and that induced by human activities.

ACKNOWLEDGMENTS

We gratefully acknowledge the comments and suggestions of Cynthia Hagley and Jeffrey Schuldt, as well as those of three anonymous reviewers, on earlier versions of this manuscript. Portions of this work were funded by U.S. EPA grant numbers CR-818373-01-0 and CR-822043-01-0. This is Publication 203 of the Center for Water and Environment, Natural Resources Research Institute, and Contribution 43 of the Natural Resources Geographic Information System laboratory.

REFERENCES

Allen, T.F.H. and T.B. Starr. *Hierarchy.* University of Chicago Press, Chicago, 1982, p. 310.

Allen, T.F.H. and T.W. Hoekstra. *Toward a Unified Ecology.* Columbia University Press, New York, 1992, p. 384.

Anonymous. *Sharing the challenge: Floodplain management into the 21st Century.* Report of the Interagency Floodplain Management Review Committee to the Administration Floodplain Management Task Force, Washington, DC, 1994.

EPA (Environmental Protection Agency). Proposed guidelines for ecological risk assessment, FRL-5605-9. *Federal Register,* 61, pp. 47552–47631, 1996.

Czuczwa, R.A., J.M. McVeety, and B.D. Hites. Polychlorinated dibenzo-p-dioxins and dibenzofurans in sediments from Siskiwit Lake, Isle Royale. *Science* 226, pp. 568–569, 1984.

Franklin, J.F. and R.T.T. Forman. Creating landscape patterns by forest cutting: ecological consequences and principles. *Landscape Ecol.* 1, pp. 5–18, 1987.

Graham, R.L., C.T. Hunsaker, R.V. O'Neill, and B.L. Jackson. Ecological risk assessment at the regional scale. *Ecol. App.* 1, pp. 196–206, 1991.

Hanski, I. and M.E. Gilpin. Metapopulation dynamics: brief history and conceptual domain, in *Metapopulation Dynamics,* Gilpin, M.E. and I. Hanski, Eds., Academic Press, London, 1991.

Hunsaker, C.T., R.L. Graham, G.W. Suter, II, R.V. O'Neill, L.W. Barnthouse, and R.H. Gardner. Assessing ecological risk on a regional scale. *Environ. Manage.* 14, pp. 325–332, 1990.

Hunsaker, C.T., D.A. Levine, S.P. Timmins, B.L. Jackson, and R.V. O'Neill. Landscape characterization for assessing regional water quality, in *Ecological Indicators*, Vol. 2, McKenzie, D.H., D.E. Hyatt, and V.J. McDonald, Eds., Elsevier Applied Science, London, 1992.

Johnson, L.B. and S.H. Gage. Landscape approaches to the analysis of aquatic ecosystems. *Freshwater Biol.* 37, pp. 113–132, 1997.

Johnson, L.B., C. Richards, G.E. Host, and J.W. Arthur. Landscape influences on water chemistry in Midwestern stream ecosystems. *Freshwater Biol.* 37, pp. 193–208, 1997.

Johnston, C. A., N. E. Detenbeck, and G. J. Niemi. The cumulative effect of wetlands on stream water quality and quantity: a landscape approach. *Biogeochemistry* 10, pp. 105–141, 1990.

Lowrance, R., and G. Vellidis. A conceptual model for assessing ecological risk to water quality function of bottomland hardwood forests. *Environ. Manage.* 19, pp. 239–258, 1995.

MacArthur, R.H. and E.O. Wilson. *The Theory of Island Biogeography*. Princeton University Press, Princeton, NJ, 1967, p. 203.

Mladenoff, D.J., T.A. Sickley, R.G. Haight, and A.P. Wydevens. A regional landscape analysis and prediction of favorable gray wolf habitat in the northern Great Lakes Region. *Conserv. Biol.* 9, pp. 279–294, 1995.

National Research Council. *Issues in Risk Assessment*. National Academy Press, Washington, DC, 1993, p. 356.

Niemi, G.J., P. DeVore, N. Detenbeck, D. Taylor, A. Lima, J. Pastor, J.D. Yount, and R.J. Naiman. Overview of case studies on recovery of aquatic systems from disturbance. *Environ. Manage.* 14, pp. 571–588, 1990.

Ohio EPA. *Biological Criteria for the Protection of Aquatic Life: Volume II. User's Manual for Biological Assessment of Ohio Surface Waters*. Ohio Environmental Protection Agency, Columbus, OH, 1987.

O'Neill, R.V. Perspectives in hierarchy and scale, in *Perspectives in Ecological Theory*, Roughgarden, J., R.M. May, and S.A. Levin, Eds., Princeton University Press, Princeton, NJ, 1989.

O'Neill, R.V., D.L. DeAngelis, J.B. Waide, and T.F.H. Allen. *A Hierarchical Concept of Ecosystems*. Princeton University Press, Princeton, NJ, 1986, p. 253.

O'Neill, R.V., J.R. Krummel, R.H. Gardner, G. Sugihara, B. Jackson, D.L. DeAngelis, B.T. Milne, M.G. Turner, B. Zygmunt, S.W. Christensen, V.H. Dale, and R.L. Graham. Indices of landscape pattern. *Landscape Ecol.* 1, pp. 153–162, 1988.

Osborne, L.L., B. Dickson, M. Ebbers, R. Ford, J. Lyons, D. Kline, E. Rankin, D. Ross, R. Sauer, P. Seelback, C. Speas, T. Stefanavage, J. Waite, and S. Walker. Stream habitat assessment programs in states of the AFS North Central Division. *Fisheries* 16, pp. 28-3, 1991.

Payne, T.L. Life history and habits, in *The Southern Pine Beetle*, Thatcher, R.C., J.L. Searcy, J.E. Coster, and G.D. Hertel, Eds., Forest Service Technical Bulletin 1631, USDA, Washington, DC, 1980.

Pulliam, H.R. Incorporating concepts from population and behavioral ecology into models of exposure to toxins and risk assessment, in *Wildlife Toxicology and Population Modeling*, Kendall, R.J. and T.E. Lacher, Eds., Lewis Publishers, Boca Raton, FL, 1994.

Richards, C. and G. Host. Examining land use influences on stream habitats and macroinvertebrates: a GIS approach. *Water Resour. Bull.* 30, pp. 729–738, 1994.

Richards, C., L.B. Johnson, and G.E. Host. Landscape-scale influences on stream habitats and biota. *Can. J. Fish. Aquat. Sci.* 53 (Suppl. 1), pp. 295–311, 1996.

Richards, C., R.J. Haro, L.B. Johnson, and G.E. Host. Catchment and reach-scale properties as indicators of macroinvertebrate species traits. *Freshwater Biol.* 37, pp. 219–230, 1997.

Solomon, K.R, D.B. Baker, R.P. Richards, K.R. Dixon, S.J. Klaine, T.W. La Point, R.J. Kendall, C.P. Weisskopf, J.M. Giddings, J.P. Giesy, L.W. Hall, and W.M. Williams. Ecological risk assessment of atrazine in North American surface waters. *Environ. Toxicol. Chem.* 15, pp. 31–76, 1996.

Suter, G.W., II. Endpoints for regional ecological risk assessment. *Environ. Manage.* 14, pp. 9–23, 1990.

Suter, G., II and L. Barnthouse. Assessment concepts, in *Ecological Risk Assessment*, Suter, G.W. II, Ed., Lewis Publishers, Boca Raton, FL, 1993.

Westman, W.E. *Ecology, Impact Assessment and Environmental Planning*. John Wiley & Sons, New York, 1985, p. 532.

Yount, J.D. and G. J. Niemi. Recovery of lotic communities and ecosystems from disturbance—a narrative review of case studies. *Environ. Manage.* 14, pp. 547–570, 1990.

12 ||| Assessing Effect and No Effect with Equivalence Tests

Philip M. Dixon

INTRODUCTION

Testing for No Effect

Risk assessment can involve quantitative comparisons between groups, e.g., between an affected area and an appropriate background, or between an exposed group and an appropriate control. No Observed Effect Concentrations (NOEC), used in the quotient method of risk assessment, are derived from experimental comparisons of controls and groups exposed to a toxicant (Newman, 1995). Evaluation of effluents in the U.S. National Pollutant Discharge Elimination System (NPDES) Permits Program also includes an acute toxicity test which compares mortality in control waters to mortality in waters containing a specified concentration of effluent (Weber, 1993). The comparison might also be done as one of the final stages in a probabilistic risk assessment, in which case the quantity being compared is the estimated risk.

If formal statistical methods are used in these comparisons, usually they are tests of the null hypothesis of no difference. The details of the test (e.g., nonparametric test, t-test, or something more complicated) will depend on the form of the data and the set of assumptions that are made, but in all common cases the statistical test evaluates a null hypothesis (H_0) of no difference. For example, an analysis of an acute toxicity experiment might test the null hypothesis that there is no difference in mean toxicity between control and effluent-treated water. Because ecological and environmental data include many sources of random variation that are not eliminated by good experimental design, exposure to effluent may appear to increase mortality even if there is no true effect. The statistical null hypothesis of no true difference is rejected only if there is strong evidence (summarized by the α-level or p-value) of an effect. These tests are familiar to anyone who has had an introductory course in statistics and are reasonable in many clinical trials and agricultural experiments, where there is concern that random variation might be mistaken for true improvements (Fisher, 1960).

The classical null hypothesis (no difference between two treatments) is inappropriate if the intent is to prove that some treatment is "safe" or has no effect (Bross, 1985; Millard, 1987; Dixon and Garrett, 1994). If a treatment has no effect, then one expects a statistical test to accept the null hypothesis, but accepting the null hypothesis (of no difference) does not prove that the treatment has no effect. The causal "arrow" cannot be reversed. The null hypothesis may be accepted because the true difference is close to zero, because the number of replicates is too small, or because the random variation among experimental units is too large. Tests of the classical null hypothesis are biased in favor of concluding "no effect," even if there were an effect (Hoekstra and van Ewijk, 1993). Large-scale ecological and environmental experiments typically have few replicates because of the cost and practical difficulties of replicating at appropriate spatial scales (Weber, 1993). Finally, ecological quantities often include considerable measurement error, spatial variability, and temporal variability (Eberhardt, 1978; Osenberg et al., 1994). The result of large random variation and low replication is a low statistical power (low probability of rejecting the null hypothesis) even if the treatment has a moderate effect.

It is also possible to have an experiment that is too precise (Dixon and Garrett, 1994). If the sample size is large, an experiment or observational study is likely to reject the null hypothesis even if the treatment has a small, biologically insignificant effect. Statistical significance has been confused with biological significance. More generally, the classical null hypothesis, that the true difference is exactly zero, is almost certainly false. The true effect of a treatment may be very close to zero, but it is unlikely to be exactly zero. One solution to both problems—the interpretation of nonsignificant results, and the confusion of biological and statistical significance—is to define a more biologically relevant null hypothesis (McBride et al., 1993; McDonald and Erickson, 1994; Erickson and McDonald, 1995).

The U.S. Environmental Protection Agency has formalized the determination of the null hypothesis in its Data Quality Objectives (DQO) Process (EPA, 1993; 1994). Consider a situation where only one direction of change is of interest—e.g., does the mean contaminant concentration exceed an action level? The true mean concentration is unknown; the decision whether or not to remediate is based on the estimated mean concentration, so the wrong decision may be made. If the true concentration is above the action level, but the estimated concentration is low, then the site will not be cleaned up, resulting in some cost measured in dollars or human health. An incorrect decision in the other direction (cleanup when the true concentration is below the action level) also results in an extra cost. The DQO process defines the appropriate null hypothesis as the true state of nature associated with the more severe decision error (EPA, 1994). Once the appropriate null hypothesis is chosen, well known statistical methods can be used to test the one-sided hypotheses.

The Nonequivalence Hypothesis

When both directions of change are relevant, a more biologically relevant hypothesis is the nonequivalence hypothesis—that the true difference, δ, lies outside some equivalence region (d_l, d_u).

$$H_0: \delta \leq d_l \text{ or } \delta \geq d_u$$

The alternative to this is that the difference is so small that the groups are "equivalent," i.e., the true difference lies within the equivalence region.

$$H_a: d_l < \delta < d_u$$

The upper and lower bounds of the equivalence region, d_u and d_l, are chosen *a priori* to reflect those differences that are biologically or ecologically equivalent to zero. They may be symmetrical, i.e., $d_l = -d_u$, but they need not be. The nonequivalence hypothesis avoids problems caused by the interpretation of nonsignificant results because it reverses the usual null and alternative hypotheses. If an experiment provides good evidence that the true difference is close to zero (small error variance, large sample size), the hypothesis that the difference is large will be rejected.

In some situations, the bounds of the equivalence region can be specified as absolute numbers. In an evaluation of a remediated site, equivalence might be specified as a primary productivity, p_r, that is within 100 g/m^2/yr of that on background sites, p_b. This pair of hypotheses is

$$H_0: p_r \leq p_b - 100 \text{ or } p_r \geq p_b + 100$$
$$H_a: p_b - 100 < p_r < p_b + 100$$

These will be referred to as shift alternatives, after the form of the alternative hypothesis.

Often it is not possible to specify absolute bounds, especially if one is analyzing responses with different units. In such cases, relative bounds might be appropriate. In the evaluation of generic drugs, the U.S. Food and Drug Administration requires that characteristics of the generic drug (y_g) be within 20% of the name-brand drug (y_n) (Chow and Liu, 1992). This pair of hypotheses is

$$H_0: y_g \leq 0.8 \, y_n \text{ or } y_g \geq 1.2 \, y_n$$
$$H_a: 0.8 \, y_n < y_g < 1.2 \, y_n$$

Ratio alternatives may be more widely applicable than shift alternatives, but calculation of the relevant statistics is generally more complicated.

In this chapter, I will describe some statistical techniques for testing the shift and ratio nonequivalence hypotheses. Parametric tests (assuming normality), nonparametric tests and confidence intervals will be described. The calculations and interpretation of the results will be illustrated with an environmental example. Finally, I will briefly describe some extensions of the basic methods to trend studies and data with below detection limit values.

Application

The use and interpretation of equivalence tests will be illustrated with data on nitrate concentration in Upper Three Runs Creek, Aiken County, South Carolina, between 1967 and 1993. During the measurement period (in October 1989) the station was moved approximately 1.6 km downstream (Bennett et al., 1991). The new catchment basin included at least one creek not previously in the catchment. Did the station move change the recorded nitrate concentrations? Is the mean nitrate concentration in the creek equivalent

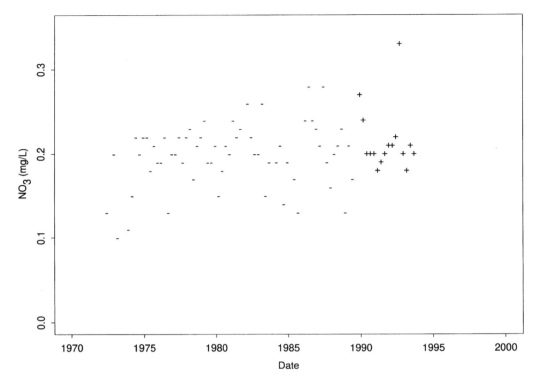

Figure 12.1. Time series plot of nitrate concentration measured at U.S.G.S. benchmark station on Upper Three Runs Creek, Aiken, South Carolina. Symbols mark observations made before (–) and after (+) the station was moved in October 1989.

before and after the move? Samples taken before October 1973 were analyzed with different analytical techniques; these data are omitted from this analysis. The comparison is between 59 observations taken between October 1973 and September 1989 and 16 observations taken between October 1989 and September 1993.

Inspection of the data indicates an apparent increasing trend and no visually obvious shift in October 1989 (Figure 12.1). Statistical tests of the trend, using either linear regression against date or Kendall's τ_b, indicate no statistically significant trend. Hence, the trend is assumed to be zero. Analysis of covariance could be used to eliminate linear or polynomial trends from the comparison between the two periods. Results from such an analysis are similar to the simpler unadjusted results presented here. There is a small change in mean nitrate concentration between the two periods (Table 12.1), but it is not statistically significant ($t = 1.39$, 73 d.f., $p = 0.17$). This result may come from either a small change or a large random variance, and is not sufficient evidence of no change in the true mean. We need to use an equivalence test to decide if the true change is "equivalent to zero."

EQUIVALENCE TESTS

The statistical literature on equivalence tests is large, partly because they raise some interesting theoretical statistical issues and partly because they are economically important,

Table 12.1. Summary of Nitrate Data Before and After the Benchmark Station was Moved in October 1989. These Numbers Include More Digits than are Appropriate for Their Precision So That Calculations in the Text Can be Verified.

	N	Median	Mean	s.d.	s.e.
Before	59	0.20	0.2008	0.0358	0.00466
After	16	0.20	0.2150	0.0378	0.00944
Difference		0	0.0142		0.0102

especially in the pharmaceutical industry (Chow and Liu, 1992). Much of the statistical development has been motivated by the regulatory requirement that a newly developed generic drug must be shown to be equivalent to the name-brand drug. In this context, the tests are often called bioequivalence tests. Although some details of the use of equivalence tests in drug development, e.g., the analysis of crossover designs, are not too useful in environmental applications, the principles are generally applicable. Recently, equivalence tests have been applied to environmental questions (Hoekstra and van Ewijk, 1993; Dixon and Garrett, 1994; McDonald and Erickson, 1994; Stunkard, 1994; Erickson and McDonald, 1995).

Many different equivalence tests have been proposed; all except the most recent are reviewed by Chow and Liu (1992). Those that are most useful for environmental applications include hypothesis tests and confidence intervals for two-group experimental designs (experimental or sampling designs where independent replicate observations are taken from two populations). McDonald and Erickson (1994) illustrate a test for data from a paired design, where each observation from the "treatment" group is paired with an observation from the "background" or control group.

All equivalence tests force the user to specify some region of equivalence before the data are analyzed (McDonald and Erickson, 1994). While this may seem to be a major difficulty, it really is not. Such decisions should be made by the user of a traditional hypothesis test of no difference. Traditional statistical tests just appear easier because the difficult ecological/ environmental decision can be postponed until after the statistical test is performed (McDonald and Erickson, 1994).

The details of the appropriate statistical test depend on how the equivalence region is specified. If the region is specified in absolute terms—e.g., the mean nitrate concentration after the benchmark station moved is within 0.03 mg/L of the earlier mean—then the appropriate statistical test is a test of shift alternatives. Other environmental situations where absolute equivalence regions might be useful include risk comparisons (e.g., excess risk less than 10^{-6}) and the evaluation of amphibian decline (e.g., annual population growth rate bounded by ± 0.06 (Dixon, 1997a). If the equivalence region is specified relative to some background or reference level, e.g., biomass on a remediated site is within 20% of biomass on the reference site, then the appropriate test is a test of ratio alternatives.

The tests described in this chapter all make one crucial assumption about the data. They assume that the observations are independent within groups. Environmental data are often spatially or temporally correlated, especially if the data do not come from a designed experiment. Although the nitrate data were collected over time, examination of the residuals suggests that the temporal correlation is close to zero. Hence, the data will be analyzed as if observations were independent.

METHODS FOR SHIFT ALTERNATIVES

Parametric Confidence Intervals

One of the earliest methods to evaluate equivalence is to compute a $100(1-2\alpha)\%$ confidence interval for the mean difference (Westlake, 1981; Schuirmann, 1987). If this confidence interval lies completely inside the equivalence region (d_l, d_u), then one concludes that the difference between the two groups is equivalent to zero. To calculate a confidence interval, compute the means of the two groups, $\overline{X}_A$ and $\overline{X}_B$, the within-group variances, $s_A{}^2$ and $s_B{}^2$, the pooled estimate of the variance, s^2, and the estimated standard deviation of the difference in the means, s_d. Each of these quantities is computed just as for a standard t-test of no difference.

$$s^2 = [(n_A - 1)s_A^2 + (n_B - 1)s_B^2]/(n_A + n_B - 2) \tag{1}$$

$$s_d = \sqrt{s^2(1/n_A + 1/n_B)} \tag{2}$$

Under the common assumptions of normally distributed observations with a common variance, a $100(1 - \alpha)\%$ confidence interval for the true difference is given by

$$(\overline{X}_A - \overline{X}_B + t_{\alpha/2,\, n_A+n_B-2}\, s_d,\, \overline{X}_A - \overline{X}_B + t_{1-\alpha/2,\, n_A+n_B-2}\, s_d)$$

where t_{α,n_A+n_B-2} is the $100\alpha\%$ percentile of a t-distribution with $n_A + n_B - 2$ degrees of freedom. Although it is common to compute a 95% confidence interval for the difference, for equivalence applications it will be more common to compute a 90% confidence interval. The reasons for the different confidence level will be clearer after discussion of hypothesis tests for nonequivalence.

The confidence interval approach can be generalized to unequal variances by using Welch's approximate t-statistic (Miller, 1986). The standard deviation of the difference is calculated slightly differently

$$s_d^* = \sqrt{s_A^2 / n_A + s_B^2 / n_B} \tag{3}$$

The confidence interval is given by

$$(\overline{X}_A - \overline{X}_B + t_{\alpha/2,\, df}\, s_d^*,\, \overline{X}_A - \overline{X}_B + t_{1-\alpha/2,\, df}\, s_d^*)$$

but the degrees of freedom for the t-statistic are calculated using Satterthwaite's approximation (details in Miller, 1986).

To evaluate equivalence in nitrate concentration, one must first specify the equivalence region. The size of this region may be based on biological and chemical information or it might be based on the magnitude of temporal or individual variability (Adler and Weinberg, 1978). For illustration, I will arbitrarily specify (−0.03 mg/L, 0.03 mg/L) as the boundaries of the equivalence region. The estimate of the pooled variance, s^2, is 0.00131, so the estimated standard deviation of the difference in means, assuming equal variances, is

$$s_d = \sqrt{0.00131}\sqrt{1/59 + 1/16} = 0.0102$$

I am interested in a 90% confidence interval and the variance estimate has $(n_A + n_B - 2)$ = 73 degrees of freedom, so the 5% and 95% percentiles are −1.666 and 1.666. The 90% confidence interval for the mean difference is $(0.0142 − 1.666 \times 0.0102, 0.0142 + 1.666 \times 0.0102) = (−0.0028, 0.0312)$. This interval includes one boundary of the equivalence region (0.03), so there is insufficient evidence of equivalence. If the definition of equivalence were broader, e.g., (−0.035, 0.035), the 90% confidence interval would fall entirely within the broader equivalence region. If (−0.035, 0.035) were defined *a priori* as equivalent to no change, there would be statistically significant equivalence.

The variance estimates for the before and after data are very similar (Table 12.3), so there is little evidence to reject the assumption of equal variances. However, if the assumption of equal variances was not reasonable, the Welch confidence interval could be computed. For the nitrate data, the standard deviation of the difference, s_d^*, is $\sqrt{0.0358^2/59 + 0.0378^2/16} = 0.0105$. The approximate degrees of freedom, using a Satterthwaite approximation, are 22.84, so the 5% and 95% percentiles of the appropriate t-distribution are −1.714 and 1.714. Hence, the 90% confidence interval, assuming unequal variances, is (−0.0039, 0.0322). This also includes one boundary of the equivalence region, so there is insufficient evidence of equivalence. In this example, the choice of equal variance or unequal variance confidence interval has little effect on the ecological conclusion.

Parametric Tests

If the data were normally distributed, or could be transformed to a normal distribution, then a parametric test of the nonequivalence hypothesis would be appropriate. The currently recommended method is the two one-sided tests approach (Schuirmann, 1987). Many other tests have been proposed (Anderson and Hauck, 1983; Metzler, 1974; Munk, 1993), but one strength of the two one-sided tests method is that it can be generalized to many circumstances. The two one-sided tests approach rejects the nonequivalence hypothesis at a specified α level

$$H_0: \delta \leq d_l \text{ or } \delta \geq d_u$$

only if both of the two subhypotheses are rejected at level α.

$$H_{0a}: \delta \leq d_l$$
$$H_{0b}: \delta \leq d_u$$

If the data were assumed to be normally distributed, each of the two subhypotheses would be tested using a one-sided t-test or unequal-variance t-test. Means, variances, and the pooled standard deviation (Equations 1 and 2) are calculated just as for a confidence interval or hypothesis test of no difference. To test H_{0a}, compute the t-statistic

$$T_a = \frac{\overline{X}_A - \overline{X}_B - d_l}{s_d}$$

and reject H_{0a} if T_a is larger than $t_{1-\alpha, df}$, where $t_{1-\alpha, df}$ is the $100(1-\alpha)\%$ critical value of a t-distribution with df degrees of freedom. To test the other subhypothesis, compute

$$T_b = \frac{\overline{X}_A - \overline{X}_B - d_u}{s_d}$$

and reject H_{0b} if T_b is smaller than $t_{\alpha, df}$. If the data were not assumed to have equal variances, each subhypothesis could be tested by an unequal variance t-test (Dannenberg et al., 1994).

Notice the similarities between these two t-statistics, the confidence interval for the mean difference, and the "usual" t-statistic, used to test the usual hypothesis of no difference,

$$T = \frac{\overline{X}_A - \overline{X}_B}{s_d}$$

However, the critical values are not the same as the "usual" critical values for a two-sided α-level test. An α-level test of nonequivalence requires that each one-sided test use α-level critical values. Just as the "usual" test of no difference is related to a particular confidence interval, the two one-sided tests method of evaluating equivalence is related to a confidence interval. The 5% α-level equivalence test is related to the 90% confidence interval—not a 95% confidence interval—because equivalence requires that two 5% level one-sided tests be significant.

For the nitrate example, the two t-statistics are

$$T_a = \frac{0.2150 - 0.2008 - (-0.03)}{0.0102} = 4.33$$

$$T_b = \frac{0.2150 - 0.2008 - (0.03)}{0.0102} = -1.55$$

The first subhypothesis, H_{0a}, is rejected because $4.33 > 1.665$. The second subhypothesis, H_{0b}, is not rejected because -1.55 is not < -1.665. Hence, the overall hypothesis of nonequivalence, H_0, is not rejected. There is insufficient evidence to conclude that the mean nitrate concentration after the station was moved is within ± 0.03 mg/L of the previous mean concentration. As expected by the relationship between the hypothesis tests and confidence intervals, this conclusion is identical to the confidence interval conclusion.

Influence of Experimental Precision

Experimental precision, s_d, has a big effect on the performance of an equivalence test. Its effects can be illustrated by considering three experiments that have the same equivalence region $(-1, 1)$ but different standard errors of the difference in means, s_d, and determining when the nonequivalence hypothesis is rejected. The first experiment provides a precise estimate of the difference in means, $\overline{d}$, perhaps because there were a lot of replicates, the measurement error was small, or the heterogeneity among individuals was small. The s.d. is 0.01, so the confidence interval for the difference is narrow (approximately $\overline{d} \pm 0.02$), so the interval will fall entirely within the equivalence region for many observed differences (from -0.98 to 0.98). The second experiment has a larger s_d (0.45) and wider confidence interval $(\overline{d} \pm 0.9)$. The interval falls entirely within the rejection region only when the observed difference is between -0.1 and 0.1. In the third experiment, $s_d = 0.75$. The confidence interval $(\overline{d} \pm 1.5)$ never lies entirely within the equivalence region, so the hypothesis of non-equivalence is never rejected.

Bayesian Approaches

The approaches in the previous section are grounded in the frequentist interpretation of statistics (Dixon and Ellison, 1996). This assumes that there is a fixed, but unknown, difference between the two statistical populations (nitrate concentration before the move, nitrate concentration after the move). The p-value and confidence interval are random quantities because they are computed from random observations. In the Bayesian interpretation, the difference between the two populations is random and the observed data are treated as fixed information. This approach requires an initial specification (prior distribution) of the difference between the two populations. In the absence of specific information, one often uses a noninformative prior distribution, i.e., that any difference is equally likely. The observed data are used to update the prior distribution into a posterior distribution, which describes the random distribution of the difference given the data that were actually observed. This posterior is used to calculate upper and lower bounds of a 90% highest posterior density interval, constructed so that the random true difference lies between those fixed bounds with 90% probability. The posterior distribution can also be used to directly calculate the probability that the true difference lies within the equivalence region. The results of a Bayesian analysis appear similar to familiar p-values and confidence intervals, but their interpretation is quite different (Dixon and Ellison, 1996).

Bayesian techniques for equivalence have been developed (Rodda and Davis, 1980; Mandallaz and Mau, 1981). One practically relevant and analytically tractable set of assumptions is that both means (μ_A and μ_B) and the variance (σ^2) have noninformative priors and the (possibly transformed) observations have normal distributions with different group means but equal within-group variances. Under these assumptions, the random variable

$$T_d = \frac{(\mu_A - \mu_B) - (\overline{X}_A - \overline{X}_B)}{s_d}$$

has a t-distribution with $n_A + n_B - 2$ degrees of freedom (Rodda and Davis, 1980). In the Bayesian framework, $\delta = \mu_A - \mu_B$ is the random quantity; $\overline{X}_A - \overline{X}_B$ and s_d are fixed. Hence, one can estimate the probability that the difference in means falls within the equivalence limits, $P[d_l < \delta < d_u]$. This probability is given by

$$P[d_l < \mu_A - \mu_B < d_u] = F_{t,df}(t_u) - F_{t,df}(t_l)$$

where F_t is the cumulative distribution function of a t random variable with $n_A + n_B - 2$ degrees of freedom and

$$t_l = \frac{d_l - (\overline{X}_A - \overline{X}_B)}{s_d}$$

$$t_u = \frac{d_u - (\overline{X}_A - \overline{X}_B)}{s_d}$$

As before, s_d is the standard error of the difference, given by Equation 2.
For the nitrate concentration data, the statistics t_l and t_u are

$$t_l = \frac{-0.03 - 0.014}{0.010} = -4.33$$

$$t_u = \frac{0.03 - 0.014}{0.010} \quad 1.55$$

so the probability that the unknown difference in means lies within the equivalence interval is

$$P[-0.03 < \mu_A - \mu_B < 0.03] = F_{t,73}(1.55) - F_{t,73}(-4.33) = 0.937 - 0.00002 = 0.937$$

One possible decision rule would be to conclude that the two groups are equivalent if there is a sufficiently high probability (e.g., > 95%) that the true difference is inside the equivalence region. Following that rule, there is insufficient evidence to say that the treatments are equivalent because the observed probability is less than 95%. As with frequentist approaches, other choices of equivalence region or decision rule may lead to different conclusions.

In the Bayesian perspective, it is reasonable to compute and talk about the probability that the true difference falls within the equivalence region. In the frequentist approach, the confidence interval either includes the fixed true difference or it doesn't. Unfortunately, one never knows which is the case; all one can say is that the rule for calculating confidence intervals, repeated on many sets of data, produces intervals that include the unknown parameter with the specified frequency.

Nonparametric Methods

All the methods in the previous section assume, among other things, that the observations are normally distributed. This assumption can be relaxed slightly by using transformations. Many environmental data have skewed distributions with large positive tails. Such data can often—but not always—be transformed to normality with a log, square root, or some other power transformation (Gilbert, 1987; Stoline, 1991). There are some problems with the interpretation of transformed data. Most analyses assume that the correct transformation is known exactly and do not account for the uncertainty introduced if a transformation is estimated from the data. A more serious problem specific to equivalence tests is that a transformation changes the form of the equivalence hypothesis. For example, a log transformation changes a ratio hypothesis into a shift hypothesis. An alternative, favored by many (e.g., Seaman and Jaeger, 1990; Potvin and Roff, 1993; Johnson, 1995; and Stewart-Oaten, 1995), is to use a nonparametric test.

To test equivalence nonparametrically, the same two one-sided tests approach is used, but each subhypothesis is tested by a one-sided nonparametric test (Hauschke et al., 1990). For independent observations from two groups, an appropriate test is the Wilcoxon rank-sum (= Mann-Whitney) test (Gilbert, 1987), instead of the parametric t-test. This test has at least two different interpretations and associated sets of assumptions (Hollander and Wolfe, 1973, p. 73). The unknown shift interpretation is the one adopted here; the two groups of observations are assumed to have the same unknown distribution, except that one group is shifted in mean (or equivalently, shifted in median) by an unknown amount.

To test the subhypothesis H_{0a}: $\mu_A - \mu_B < d_l$ using a nonparametric test, one must first rewrite it in an equivalent form:

$$H_{0a}: \mu_A - (\mu_B + d_l) < 0$$

This is now in a form that can be tested by a one-sided nonparametric test of equality of means. If the observations are from independent groups, the appropriate test is the Wilcoxon rank-sum test. One group (A) is compared to a shifted version of the second group (B – d_l). The lower bound of the equivalence region is subtracted from each observation in the second group. The second subhypothesis, H_{0b}: $\mu_A - \mu_B > d_u$, is rewritten as:

$$H_{0b}: \mu_A - (\mu_B + d_u) > 0$$

The upper bound of the equivalence region is added to all observations in the second group (group B) before the test is computed. Details on the Wilcoxon test and other tests appropriate for paired data are available in introductory texts on nonparametric statistics (e.g., Hollander and Wolfe, 1973). Tables of exact critical values are widely available if each group has 10 or fewer observations. Tables for sample sizes up to 25 are available in Pearson and Hartley (1972). These tables are appropriate if there are no tied values in the data. Larger sample sizes or data with ties require the use of either special computer programs (e.g., StatXact, Cytel Software Corporation, 1991; or Testimate, idv Datenanalyse und Versuchsplanung, 1993) or a normal approximation to the distribution of the critical values (Hollander and Wolfe, 1973, pp. 68–69).

There is mild evidence of nonnormality in the nitrate concentration data, so a nonparametric test may be more appropriate than a t-test. A Wilcoxon rank-sum test with an equivalence region of (−0.03, 0.03) gives the following results for each subhypothesis. For H_{0b}, δ > 0.03, an exact analysis using the StatXact program gives p = 0.0080. A normal approximation gives a similar result: z = −2.36, p = 0.0092. For H_{0b}, δ < −0.03, both an exact analysis and a normal approximation (z = 4.173) give p < 0.0001. Both subhypotheses are rejected at α = 5% because both p-values are less than 0.05. The hypothesis of nonequivalence is rejected. There is reasonable statistical evidence that the mean nitrate concentrations are within ± 0.03 mg/L.

A nonparametric confidence interval for the median difference between the groups can be calculated and used to evaluate equivalence, just as parametric confidence intervals are analogous to parametric equivalence tests (Steinijans and Diletti, 1983). The objective is to find an interval, based on the data, such that the random interval includes the true difference with the specified confidence (Noether, 1972). This can be done nonparametrically by calculating all $n_A n_B$ pairwise differences between observations in group A and those in group B and ranking those differences from smallest to largest. The upper and lower endpoints of the nonparametric confidence interval are given by two of those pairwise differences; the problem is to decide which ones (Hollander and Wolfe, 1973). If we select differences close to the median difference, the confidence interval will be narrow, but it will include the unknown parameter relatively infrequently. If we select extreme pairwise differences, close to the largest and the smallest differences, then the interval will be wide, but it will include the unknown parameter too often. The appropriate endpoints for a 100(1−α)% confidence interval are given by the lth and u^{th} largest differences, where l and u are determined from tables (e.g., Hollander and Wolfe, 1973; or Gardner and Altman, 1991) if the sample sizes are small and the data do not contain ties, by a computer program, or by using a normal approximation. To use the normal approximation, calculate

$$l = n_A n_B / 2 - z_a \sqrt{n_A n_B (n_A + n_B + 1) / 12}$$

and

$$u = 1 + n_A n_B / 2 + z_a \sqrt{n_A n_B (n_A + n_B + 1) / 12}$$

z_α is the normal distribution critical value for a two-sided α-level test. This normal approximation does not account for ties in the data. If ties are present, the interval is conservative because l and u are farther from the median than they "should" be. Hence, the interval is wider than it should be.

In the nitrate data, there are 59 observations before and 16 after the station was moved, so there are 59 × 16 = 944 pairwise differences. Because of the ties, exact confidence intervals are not available; because of the large number of observations and the presence of ties, the problem is too large for StatXact to compute an exact confidence interval. I will use the normal approximation to calculate a 90% confidence interval. The appropriate critical value is $z_{0.9}$ = 1.645, so the locations of the lower and upper bounds are calculated as

$$l = 59 \times 16 / 2 - 1.645 \sqrt{59 \times 16 \times 76 / 12} = 472 - 127.2 = 345$$

$$u = 59 \times 16/2 + 1.645\sqrt{59 \times 16 \times 76/12} + 1 = 473 + 127.2 = 600$$

The lower location is rounded down, while the upper location is rounded up. The 345th largest difference is –0.01; the 600th largest difference is 0.02, so a 95% confidence interval on the median difference is (–0.01, 0.02). Since this interval lies entirely within the equivalence region (–0.03, 0.03), the nonparametric confidence interval procedure indicates equivalence, just as the nonparametric test did.

Implementation in S-Plus and SAS Packages

Most statistical packages include t-test calculations that can be adapted to test the nonequivalence hypothesis. Tables 12.2, 12.3, and 12.4 give code to test shift alternatives in two packages—S-plus and SAS. While the details of the code may not be useful to users of other statistical packages, the code does illustrate general principles that can be used to test the nonequivalence hypothesis in other packages. The S-plus package provides a general t-test function, t.test(), that computes p-values for each one-sided test (Table 12.2). The corresponding nonparametric test function is wilcox.test(). S-plus will compute exact nonparametric p-values if sample sizes are smaller than 50 and no ties are present. The output from t.test() includes one-sided 95% confidence bounds that can be combined into a two-sided 90% confidence interval. Testing the nonequivalence hypothesis for shift alternatives in S-plus is relatively easy because S-plus provides an option to test one-sided hypotheses.

If the statistical package doesn't test one-sided hypotheses, some extra programming will be needed. In the SAS package, the TTEST and NPAR1WAY procedures only test two-sided hypotheses (Table 12.3). The t-statistic (from TTEST) is the correct one, but a DATA step is needed to compute the one-sided p-values from the t-statistic (Table 12.4). SAS does not compute exact one-sided p-values for the Wilcoxon rank-sum test, so a DATA step function is used to compute the p-value from a normal approximation. The code in Table 12.4 assumes that sample sizes are sufficiently large for the normal approximation to be reasonable.

METHODS FOR RATIO ALTERNATIVES

In many cases, it is not possible to specify *a priori* the absolute bounds of an equivalence region. A difference of 10 mg/L nitrate may be considered a very large difference if the background concentration is only 20 mg/L and treated as essentially no change if the background concentration is 1000 mg/L. It may be more reasonable to specify the relative size of the equivalence region, rather than the absolute size. The appropriate equivalence hypothesis is a ratio alternative, rather than a shift alternative. Ratio alternatives are used to evaluate generic drugs, which must be within 20% of the name-brand drug.

Methods for Normally Distributed Data

Two slightly different methods based on one-sided tests have been proposed for ratio alternatives. Both methods assume normality, but they differ in their assumptions about the

Table 12.2. S-Plus Code to Read Nitrate Data and Test for Equivalence Using Shift Alternatives. Comments are Lines Starting with the # Symbol.

```
# read data file into an S-plus data frame,
#  first line (header) gives the variable names
water <- read.table('water.dat',header=T)

# create date variable from month and year
water$date <- water$year + (water$month-1)/12

# then split data into before and after groups
no3.a <- water$no3[water$date > 89.75]
no3.b <- water$no3[water$date < 89.75]

# syntax of t and wilcoxon tests is:
#   t.test(group 1, group 2, direction of alternative, shift)

# test H0a: after - before < –0.03, alternative is after - before > –0.03
t.test(no3.a,no3.b,alternative="greater",mu=–0.03)

# test H0a: after - before > 0.03, alternative is after - before < 0.03
t.test(no3.a,no3.b,alternative="less",mu=0.03)

# unequal-variance versions of t-test
t.test(no3.a,no3.b,alternative="greater",mu=–0.03,var.equal=F)
t.test(no3.a,no3.b,alternative="less",mu=0.03,var.equal=F)

# then do same nonparametrically using Wilcoxon rank-sum tests
wilcox.test(no3.a,no3.b,alternative="greater",mu=–0.03)
wilcox.test(no3.a,no3.b,alternative="less",mu=0.03)
```

variances. The method proposed by Erickson and McDonald (1995) assumes each group has the same variance. The ratio hypothesis

$$H_{0a}: \mu_A \leq b_1 \mu_B$$

is tested by calculating the t-statistic

$$T_c = \frac{\overline{X}_A - b_1\overline{X}_B}{s\sqrt{1/n_A + b_1^2/n_B}}$$

Here, s is the usual pooled estimate of the standard deviation

$$s = \sqrt{\frac{(n_A - 1)s_A^2 + (n_B - 1)s_B^2}{n_A + n_B - 2}}$$

where s_A^2 and s_B^2 are the sample variances in groups A and B. H_{0a} is rejected if T_c is larger than the $1 - \alpha$ quantile of the t-distribution with $(n_A + n_B - 2)$ degrees of freedom. The other subhypothesis

Table 12.3. SAS Code to Read Nitrate Data and Compute t-tests and Wilcoxon Rank-Sum Tests for Shift Alternatives. −0.03 and 0.03 are the Lower and Upper Bounds of the Equivalence Region for the NO$_3$ Example; Those Bounds Should be Replaced by the Appropriate Region. Text Enclosed Between /* and */ Symbols are Comments.

```
/* read in the data file and create shifted variables */
data water;
    infile 'water.dat';
    input month year no3;
    date = year + (month-1)/12;
/* use date of observation to decide whether after or before station move */
    if (date < 89.75) then do;
/* if before, create two new variables with shifted observations */
        move = 'Before';
        no3_sa = no3 – 0.03;    /* replace –0.03 with lower bound of equiv. */
        no3_sb = no3 + 0.03;    /* replace  0.03 with upper bound of equiv. */
        end;
    else do;
/* if after, don't shift the observations */
        move = 'After';
        no3_sa = no3;
        no3_sb = no3;
        end;

/* then do a two-sided t-test on the shifted observations, */
/*   comparing MOVE = before group to MOVE = after group   */
proc ttest;
    var no3_sa no3_sb;
    class move;

/* then do a two-sided Wilcoxon rank sum test on the shifted observations, */
/* comparing MOVE = before group to MOVE = after group   */
proc npar1way wilcoxon;
    var no3_sa no3_sb;
    class move;
    run;
```

$$H_{0b}: \mu_A \geq b_u\,\mu_B$$

is tested by calculating the t-statistic

$$T_d = \frac{\overline{X}_A - b_u\overline{X}_B}{s\sqrt{1/n_A + b_u^2/n_B}}$$

The null hypothesis H_{0b} is rejected if T_d is smaller than the α quantile of the t-distribution with $(n_A + n_B - 2)$ degrees of freedom. Just as with shift alternatives, the hypothesis of nonequivalence is rejected if both subhypotheses are rejected.

Figure 12.2 shows rejection regions for H_0 and each subhypothesis. The hypothesis of nonequivalence is rejected if the observed means fall in a wedge-shaped region inside the

Table 12.4. SAS Code to Compute One Sided p-Values for Shift Alternatives from t- or z-Statistics. Text Enclosed Between /* and */ Symbols are Comments.

```
data pvalues;
  input ta dfa tb dfb;

/* see if we have t-statistics (df > 0) or z score (df = 0)
  if (dfa > 0) then do;
/* parametric test, compute one-sided P[T > ta] and P[T < tb] */
    pa = 1-probt(ta,dfa);
    pb = probt(tb,dfb);
/* also compute Bayesian P[dl < delta < du] */
    pincl = probt(ta,dfa) - probt(tb,dfb);
    end;
  else do;
/* nonparametric test, compute one-sided P values            */
/*   N.B. this code assumes Na and Nb are large enough to assume */
/*   normality                                               */
    pa = 1-probnorm(ta);
    pb = probnorm(tb);
    end;

/* combine p values from subhypotheses into overall equivalence test */
  if (pa < 0.05) and (pb < 0.05) then
    decide = "Reject H0";
    else decide = "Accept H0";

/* Enter t-statistics and degrees of freedom from each test here */
/*   values are in the order:                              */
/*      t-statistic and d.f. for test of H0a, t-statistic and d.f. for H0b */
/*   df = 0 indicates output from nonparametric test */
cards;
  4.3273 73 -1.5532 73
  4.1736 0  -2.384 0
;

proc print;
  title 'Results of equivalence tests';
run;
```

lines defining each subhypothesis. The distance between the line defining the hypothesis and the boundary of the rejection region depends on the sample sizes (n_A and n_B) and the pooled variance estimate. If the sample sizes are large and the variance is small, boundaries of the rejection region are close to the lines defining the null hypotheses; if sample sizes are small and the variance is large, then the rejection region is much smaller. The rejection regions are not symmetrical around the line $\mu_A = \mu_B$ because the equivalence region is not centered around 0 on a log scale. If the equivalence region were specified symmetrically, e.g., as (0.8, 1.25) or (0.8333, 1.2), the rejection region would be symmetrical.

A slightly different test of the hypothesis H_{0b}: $\mu_A \geq b_u \, \mu_B$ was proposed by Stunkard (1994). He suggested multiplying all values in group B by the proportionality constant b_u,

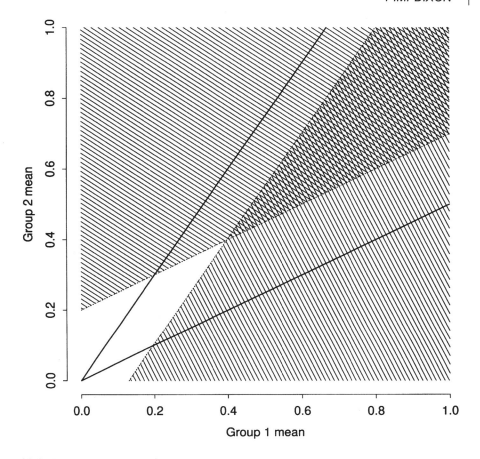

Figure 12.2. Rejection regions for equivalence ratio hypotheses with an equivalence region of (0.5,1.5). Thick solid lines indicate where the two ratio hypotheses ($\mu_A \leq 0.5 = \mu_B$ and $\mu_A \geq 1.5 = \mu_B$) are exactly equal. Each shaded polygon indicates the rejection region for that hypothesis. A rejection region is the set of observed $\overline{X}_A$ and $\overline{X}_B$ for which the hypothesis is rejected. The doubly shaded intersection is the rejection region for H_0, the hypothesis of nonequivalence.

then doing a one-sided t-test. This assumes that the variance among observations in group A equals the variance among the rescaled observations in group B. The t-statistic is

$$T^* = \frac{\overline{X}_A - b_u\overline{X}_B}{s^*\sqrt{1/n_A + 1/n_B}}$$

where

$$s^* = \sqrt{\frac{(n_A - 1)s_A^2 + (n_B - 1)b_u^2 s_B^2}{n_A + n_B - 2}}$$

Under Stunkard's assumptions, T* for each subhypothesis has a Student's t-distribution, so each subhypothesis is tested by comparing a value of T* to the same critical values used in Erickson and McDonald's test. The practical difference between the two tests can be seen clearly when the two sample sizes are the same ($n_A = n_B = n$). Then, the divisor in Erikson and McDonald's test statistic is

$$\sqrt{\left(\frac{s_A^2 + s_B^2}{2}\right)\left(\frac{1 + b^2}{n}\right)}$$

while the divisor for Stunkard's test statistic is

$$\sqrt{\frac{s_A^2 + s_B^2 b^2}{n}}$$

Erickson and McDonald's statistic assumes that the two groups have the same variances, so the sample estimates can be pooled. Stunkard's test assumes that the variance of the observed values in group B equals b^2 times the variance from group A, so it is appropriate to pool the quantities s_A^2 and $s_B^2 b^2$. While this assumption of proportional variances may be reasonable for a single one-sided test, it is not appropriate for both one-sided parts of an equivalence test. The test of H_{0a} assumes that $\sigma_A^2 = b_l^2 \sigma_B^2$; the test of H_{0b} assumes that $\sigma_A^2 = b_u^2 \sigma_B^2$. Both assumptions can not be true unless $b_l = b_u$, which is not a test of an equivalence region.

If data are normally distributed, it is possible to construct confidence intervals for the ratio of means, but they are either conservative (McDonald and Erickson, 1994), or complicated to compute (Chow and Shao, 1990; Chow and Liu, 1992).

Consider testing whether the nitrate concentration after the station was moved was within 20% of the pre-move concentration. The two subhypotheses to be tested are

$$H_{0a}: \mu_A \leq 0.8 \, \mu_B \text{ and } H_{0b}: \mu_A \geq 1.2 \, \mu_B$$

The Erickson and McDonald estimate of the pooled standard deviation is

$$s = \sqrt{(58 \times 0.00128 + 15 \times 0.00142) / 73} = 0.036$$

so the test statistics are

$$T_c = \frac{0.215 - 0.8 \times 0.201}{0.036\sqrt{1/16 + 0.64/59}} = 5.54, \text{ and } T_d = \frac{0.215 - 1.2 \times}{0.036\sqrt{1/16 +}}$$

If each null subhypothesis is true, each t-statistic has a t-distribution with 73 degrees of freedom. T_c is larger than 1.66, the 95th percentile of a t-distribution with 73 degrees of freedom, and T_d is smaller than −1.66, the 5th percentile of the same t-distribution. Alternatively, the p-values associated with each subhypothesis are (<0.0001 and 0.0086), so both subhypotheses are rejected. The mean concentration after the move is significantly within 20% of the pre-move level.

Nonparametric Tests Ratio Alternatives

Nonparametric tests of ratio alternatives can be computed by modifying one group of data in an appropriate fashion, then computing a nonparametric test. The hypothesis $\mu_A \leq b_1 \mu_B$ is tested by multiplying all observations in group B by b_1, then computing a one-sided nonparametric test of no difference. Strictly speaking, this procedure seems to violate the assumption that the two groups of data have the same distribution, except for perhaps a shift in location, i.e., they have the same variances. Multiplying one group of data by a constant changes its variance. However, this is not a problem for nonparametric tests based on ranks, because the ranks are unchanged by a monotonic transformation such as a log or exponential transformation. Shift alternatives could be tested by subtracting the boundary of the equivalence region from one group of data. If all observations are exp(x) transformed, that subtraction is transformed to a division without changing the ranking of the data points. The transformed hypothesis now specifies ratio alternatives.

The nonparametric test of ratio alternatives leads to the same conclusion as the parametric test of a ± 20% equivalence region. The first subhypothesis $\mu_A \leq 0.8 \, \mu_B$ is tested by multiplying all observations from before the move by 0.8 and using a Wilcoxon test of equality. The z score is large, 5.28, so the one-tailed p-value is very small and highly significant, <0.0001. The second subhypothesis $\mu_A \geq 1.2 \, \mu_B$ is tested in a similar fashion but multiplying by 1.2 rather than 0.8. The z score is very small, −2.80, so the one-tailed p-value is small and also significant, 0.0025. Since both subhypotheses are rejected, the hypothesis of nonequivalence is rejected.

Ratio Tests in S-Plus and SAS

Nonparametric tests and the Stunkard test of ratio alternatives are easy to compute in standard statistical packages. For either, one just multiplies all observations in one group by the proportionality constant, then uses a standard test of no difference (Table 12.6). In SAS, one-sided p-values are computed from the test statistics using the same approach used for test statistics for shift alternatives (Table 12.4).

The Erickson and McDonald test has to be programmed from first principles. An S-plus function to compute t-statistics and p-values is given in Table 12.5. A similar approach could be used to write a SAS DATA step, but a simpler approach is to use PROC TTEST to compute summary statistics for each group of data. Then, use a DATA step to compute t-statistics and one-sided p-values from those summary statistics (Table 12.7).

EXTENSIONS TO OTHER DATA AND QUESTIONS

This chapter has focused on equivalence tests for means. The methods illustrated here can be used to answer many different types of questions, but generalizations or extensions of the methods may need to be used for specific questions, such as paired observations

Table 12.5. S-Plus Function to Compute Erickson and McDonald t-Statistics and p-Values. Lines Starting with # Symbols are Comments.

```
ratio <- function(xa,xb,dl=0.8,du=1.2) {
# Erickson - McDonald t-test of equivalence between group A and group B
# bounds of equivalence region are (dl, du); default is (0.8, 1.2)

# remove missing values
xa <- xa[!is.na(xa)]
xb <- xb[!is.na(xb)]

# and compute summary statistics
n.a <- length(xa)
n.b <- length(xb)
m.a <- mean(xa)
m.b <- mean(xb)
sp <- sqrt((sum((xa-m.a)^2)+sum((xb-m.b)^2))/(n.a+n.b-2))

# compute lower and upper t-statistics
t.l <- (m.a - dl*m.b)/(sp*sqrt(1/n.a + dl*dl/n.b))
t.u <- (m.a - du*m.b)/(sp*sqrt(1/n.a + du*du/n.b))

# and p-values
p.l <- 1-pt(t.l,n.a+n.b-2)
p.u <- pt(t.u,n.a+n.b-2)

# return list with t-statistics and p-values
list(t.lower=t.l, t.upper = t.u, p.lower = p.l, p.upper=p.u)
}
```

(Steinijans and Diletti, 1983; 1985; Chow and Liu, 1992). The difference in means is a specific contrast in a linear model; parametric equivalence methods can be easily generalized to test equivalence in other linear models. Two simple examples are testing whether the slope in a regression model is equivalent to zero or whether two different slopes are equivalent. For example, the analysis of the nitrate data assumed that there was no temporal trend. The usual test of a regression slope equal to zero was nonsignificant. An alternative would be to define a range of slopes that are considered equivalent to zero and do an equivalence test for the slope (Dixon, 1997a).

It is more difficult to generalize nonparametric tests to complicated situations, e.g., non-parametric trend tests. Nonparametric tests rely on randomization, which is easy to do when the hypothesis is the usual one of no difference, but much harder with equivalence hypotheses. Nonparametric tests of equivalence hypothesis are possible because one sample of data can be shifted to convert the null hypothesis to one of no difference. If one has to test the equivalence of two trends using Kendall's τ_b or some other nonparametric trend statistic, it is less clear how to shift the data or otherwise convert the hypothesis to one of no difference. Some possible approaches for this problem are being developed (Dixon, 1997b).

Finally, environmental data frequently include below-detection-limit values. These are observations that are reported as "less than" some value (Gilbert, 1987). Many point estimates of the mean and the variance have been developed, but there are very few methods

Table 12.6. SAS Code to Read Nitrate Data and Compute t-tests and Wilcoxon Rank-Sum Tests for Ratio Alternatives. The SAS Code in Table 12.4 Can Be Used to Compute One-sided p-Values from the Wilcoxon Test Results. The SAS Code in Table 12.7 Should be Used to Compute the Erickson and McDonald t-Statistic and p-Value. Text Enclosed Between /* and */ Symbols are Comments.

```
data water;
  infile 'water2.dat';
  input month year no3;
  date = year + (month-1)/12;
  if (date < 89.75) then do;
    move = 'Before';
    no3_ra = no3 * 0.8;    /* replace 0.8 with lower bound of equiv. */
    no3_rb = no3 * 1.2;    /* replace 1.2 with upper bound of equiv. */
  end;
  else do;
    move = 'After';
    no3_ra = no3;
    no3_rb = no3;
  end;

proc ttest;
  var no3;
  class move;

proc npar1way wilcoxon;
  var no3_ra no3_rb;
  class move;
  run;
```

for testing hypotheses or constructing confidence intervals (Newman et al., 1989; Helsel, 1990). Akritas et al. (1994) generalize the nonparametric confidence interval procedure presented in the Nonparametric Methods section of this chapter to data with below-detection-limit observations. Another possible nonparametric method is to consider all observations smaller than the largest detection limit as tied with the largest detection limit (Gilbert, 1987). One complication of parametric approaches is that estimates of the mean and the variance are correlated if there are below-detection-limit values, so t-distributions are inappropriate. However, critical values for confidence intervals and tests can be estimated by a parametric bootstrap (Dixon, 1997c).

All the principles of good experimental design and data analysis (Fisher, 1960) are applicable to equivalence tests. Sample sizes and statistical power for various tests and hypotheses are considered by Metzler (1991), and Diletti et al. (1991; 1992), and Erickson and McDonald (1995). As with tests of no difference, statistical power depends on the sample size, the error variance, the size of the equivalence region, and the type of test.

DISCUSSION

Equivalence tests can be combined with the usual tests of equality of two groups to provide more information. For simplicity, consider shift alternatives. Ratio alternatives are

Table 12.7. SAS Code to Compute Erickson-McDonald t-Statistics and One-Sided p-Values for Ratio Alternatives from Group Summary Statistics. Text Enclosed Between /* and */ Symbols are Comments.

```
/* Enter results from PROC TTEST below the CARDS line */
/* input data are 8 values in the order:            */
/*    lower bound, upper bound to the equivalence region */
/*    group A: number of observations, mean, standard deviation, */
/*    group B: number of observations, mean, standard deviation  */

data ratio;
  input dl du na meana sda nb meanb sdb;

  s = sqrt((sda*sda*(na-1) + sdb*sdb*(nb-1))/(na+nb-2));
  sl= s*sqrt(1/na + dl*dl/nb);
  su = s*sqrt(1/na + du*du/nb);

  tl = (meana - dl*meanb)/sl;
  tu = (meana - du*meanb)/su;

  pl = 1-probt(tl,na+nb-2);
  pu = probt(tu,na+nb-2);

  if (pl < 0.05) and (pu < 0.05) then decide="Reject";
    else decide="Accept";
cards;
0.8 1.2 16 0.215 0.0377 59 0.201 0.0358
;

/* output is S: pooled s.d., Tl, Tu: lower & upper t-statistics */
/*    Pu, Pl: lower and upper p-values, Decide: test of H0 */
proc print;
  title 'Erickson-McDonald t-test of ratio equivalence';
  var s tl tu pl pu decide;
run;
```

conceptually similar—only the details are more complicated. The rejection regions for both the usual sort of test of no difference and a nonequivalence test can be graphed in terms of the observed difference and the observed standard error of that difference (Figure 12.3a,b,c). If both tests are performed, there are four possible outcomes (Figure 12.3d). The two tests give consistent results in two regions. In region 4, the difference in means is statistically significant in a test of no difference and the equivalence of the means is nonsignificant. The interpretation is that the difference is nonzero and may be larger than the equivalence bound. In region 2, the difference in means is nonsignificant, but the equivalence test is significant. The interpretation is that the difference between the groups is small compared to the *a priori* defined equivalence region.

The test of no difference and the test of nonequivalence give inconsistent results in the other two regions. In Region 1, neither test is significant. One likely interpretation is that the observations are too variable and/or the sample sizes are too small to make definitive conclusions about the difference between the groups. Such a conclusion depends on the

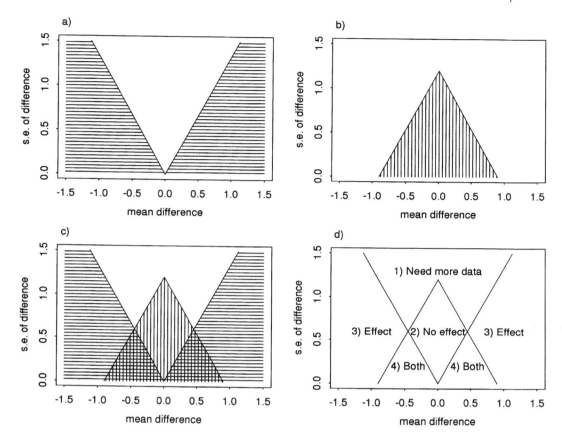

Figure 12.3. Rejection regions for (a) usual hypothesis of no difference and (b) nonequivalence hypothesis. In each panel, the shaded region is the set of observed difference and s.e. of the difference for which the null hypothesis is rejected. (c,d) the four possible pairs of outcomes from the superposition of rejection regions. If neither hypothesis is rejected (Region 1), more data are needed to evaluate the difference. If the nonequivalence hypothesis is rejected, but the hypothesis of no difference is accepted (Region 2), the tests consistently; the difference is equivalent to zero. If the nonequivalence hypothesis is accepted, but the hypothesis of no difference is rejected (Region 3), there is good evidence of a biologically important difference. If both hypotheses are rejected (Region 4), the difference is statistically different from zero, but small enough to be considered biologically equivalent to zero.

definition of the equivalence region. If a wider equivalence region were defined *a priori*, then the observed difference and standard error might lie inside Region 2 (no effect), instead of Region 1. In Region 3, both statistical tests are significant. The interpretation of this result is that the difference is not zero, but it is small compared to a difference that is equivalent to zero.

One caveat about my choice of nonequivalence methods is appropriate. Equivalence testing is an extremely active area of statistical research; new methods are being developed and older methods are being reevaluated. My opinions about methods have changed since Dixon and Garrett (1994), where we recommended a test developed by Anderson and Hauck (1983).

This test is always more powerful than the two one-sided tests, but the increase in power is associated with an unusual rejection region when the standard error of the difference is large. When the difference between the means is poorly estimated, the two one-sided tests method never rejects the nonequivalence hypothesis, but the Anderson and Hauck method can (Schuirmann, 1987). The choice between the two methods depends on which property of a statistical test is emphasized: maximum power or well-behaved rejection region. Another strong advantage for the two one-sided tests approach is that it can be used with many different types of one-sided tests (e.g., equal-variance t-test, unequal-variance t-test, Wilcoxon rank-sum test).

One alternative to the equivalence tests described in this chapter is to test the usual null hypothesis of no difference, but use the observed standard error to calculate statistical power (Schuirmann, 1987). Power is the probability that the test will reject the null hypothesis if the true difference equals some "interesting" difference. Two treatments are considered equivalent if the usual test of no difference is nonsignificant and the sample sizes are sufficiently large that the test has high enough power to detect the "interesting" difference. Like an equivalence test, one has to specify the size of an "interesting" difference, but the two approaches have very different characteristics (Schuirmann, 1987). Because the power approach is based on a test of no difference, it suffers from the confusion of biological and statistical significance in precise experiments. If the standard error is small, a small difference (e.g., within the equivalence region) will be statistically "significant" and the test will have high power.

The equivalence tests described here provide a statistically rigorous approach to demonstrating that two groups have equivalent means. This is possible only if the user is willing to define an equivalence region, within which two groups will be considered to be equal. All the statistical analysis depends on the specification of these bounds. However, the equivalence bounds provide a way to incorporate ecological and environmental knowledge directly into a statistical analysis, rather than using statistical analysis as a substitute for biological significance.

ACKNOWLEDGMENTS

This research was supported by Financial Assistance Award Number DE-FC09-96SR18546 from the U.S. Department of Energy to the University of Georgia Research Foundation. Many of the ideas were developed while on sabbatical leave at the Department of Mathematics and Statistics at the University of Otago, New Zealand, whose support is also greatly appreciated. Lyman McDonald, Wallace Erickson, two anonymous reviewers, and participants of this risk assessment symposium provided many useful comments that improved the manuscript. All errors remain my responsibility.

REFERENCES

Adler, H.I. and A.M. Weinberg. An approach to setting radiation standards. *Health Phys.* 34, pp. 719–720, 1978.

Akritas, M.G., T.F. Ruscitti, and G.P. Patil. 1994. Statistical analysis of censored environmental data, in *Handbook of Statistics, Environmental Statistics,* Vol. 12, Patil, G.P. and C.R. Rao, Eds., Elsevier, Amsterdam, 1994.

Anderson, S. and W.W. Hauck. A new procedure for testing equivalence in comparative bioavailability and other clinical trials. *Communications in Statistics—A: Theory and Methods* 12, pp. 2663–2692, 1983.

Bennett, C.S., T.W. Cooney, K.H. Jones, and P.A. Conrads. *Water Resources Data, South Carolina, Water Year 1990*. U.S. Geological Survey Water-Data Report SC-90-1, 1991.

Bross, I.D. Why proof of safety is much more difficult than proof of hazard. *Biometrics* 41, pp. 785–793, 1985.

Chow, S-C and J-P. Liu. *Design and Analysis of Bioavailability and Bioequivalence Studies.* Marcel Dekker, New York, 1992, p. 416.

Chow, S-C. and J. Shao. An alternative approach for the assessment of bioequivalence between two formulations of a drug. *Biom. J.* 32, pp. 969–976, 1990.

Cytel Software Corporation. *StatXact. Statistical Software for Exact Nonparametric Inference.* Cytel Software Corporation, Cambridge, MA, 1991.

Dannenberg, O., H. Dette, and A. Munk. An extension of Welch's approximate *t*-solution to comparative bioequivalence trials. *Biometrika* 81, pp. 91–101, 1994.

Diletti, E., D. Hauschke, and V.W. Steinijans. Sample size determination for bioequivalence assessment by means of confidence intervals. *Int. J. Clin. Pharmacol. Ther. Toxicol.* 29, pp. 1–8, 1991.

Diletti, E., D. Hauschke, and V.W. Steinijans. Sample size determination: extended tables for the multiplicative model and bioequivalence ranges of 0.9 to 1.11 and 0.7 to 1.43. *Int. J. Clin. Pharmacol. Ther. Toxicol.* 30, pp. S59–S62, 1992.

Dixon, P.M. Testing for no trend: are amphibian populations stable? Submitted to *Ecology*, 1997a.

Dixon, P.M. Nonparametric equivalence tests for trend. In preparation for *Biometrics*, 1997b.

Dixon, P.M. Testing for equivalence in data with below detection-limit values, in *Proceedings of the 2nd Statistics in Ecology and Environmental Monitoring Conference*. Dunedin, NZ, 1997c.

Dixon, P.M. and A.E. Ellison. Introduction: ecological applications of Bayesian inference. *Ecol. App.* 6, pp. 1034–1035, 1996.

Dixon, P.M. and K.A. Garrett. Statistical issues for field experimenters, in *Wildlife and Population Modelling: Integrated Studies of Agroecosystems*, Kendall, R.J. and T.E. Lacher, Jr., Eds., CRC Press, Boca Raton, FL, 1994.

EPA. *Data Quality Objectives Process for Superfund,* EPA-540-R-93-071. U.S. Environmental Protection Agency, Office of Emergency and Remedial Response, 1993.

EPA. *Guidance for the Data Quality Objectives Process*, EPA QA/G-4. U.S. Environmental Protection Agency, Quality Assurance Division, Washington DC, 1994.

Eberhardt, L.L. Appraising variability in population studies. *J. Wildl. Manage.* 42, pp. 207–238, 1978.

Erickson, W.P. and L.L. McDonald. Tests for bioequivalence of control media and test media in studies of toxicity. *Environ. Toxicol. Chem.* 14, pp. 1247–1256, 1995.

Fisher, R.A. *The Design of Experiments*, 7th ed. Oliver and Boyd, Edinburgh, 1960, p. 248.

Gardner, M.J. and D.G. Altman. *Statistics with Confidence*. British Medical Journal, London, 1991, p. 137.

Gilbert, R.O. *Statistical Methods for Environmental Pollution Monitoring*. Van Nostrand Reinhold, New York, 1987, p. 320.

Hauschke, D., V.W. Steinijans, and E. Diletti. A distribution-free procedure for the statistical analysis of bioequivalence studies. *Int. J. Clin. Pharmacol. Ther. Toxicol.* 28, pp. 72–78, 1990.

Helsel, D.R. Less than obvious. *Environ. Sci. Techn.* 24, pp. 1766–1774, 1990.

Hoekstra, J.A. and P.H. van Ewijk. Alternatives for the no-observed-effect level. *Environ. Toxicol. Chem.* 12, pp. 187–194, 1993.

Hollander, M. and D.A. Wolfe. *Nonparametric Statistical Methods.* Wiley and Sons, New York, 1973, p. 503.

idv Datenanalyse und Versuchsplanung. Testimate: Test and Estimation. Version 5.2. Munich, Germany, 1993.

Johnson, D.H. Statistical sirens: the allure of nonparametrics. *Ecology* 76, pp. 1998–2000, 1995.

McBride, G.B., J.C. Loftis, and N.C. Adkins. What do significance tests really tell us about the environment? *Environ. Manage.* 17, pp. 423–432, 1993. Errata, *Environ. Manage.* 18, pp. 317.

McDonald, L.L and W.P. Erickson. Testing for bioequivalence in field studies: Has a disturbed site been adequately reclaimed? in *Statistics in Ecology and Environmental Monitoring*, Fletcher, D. and B.F.M. Manly, Eds., Univ. of Otago Press, Dunedin, 1994.

Mandallaz, D. and J. Mau. Comparison of different methods for decision making in bioequivalence. *Biometrics* 37, pp. 213–222, 1981.

Metzler, C.M. Bioavailability. A problem in equivalence. *Biometrics* 30, pp. 309–317, 1974.

Metzler, C.M. Sample sizes for bioequivalence studies. *Stat. Med.* 10, pp. 961–970, 1991.

Millard, S.P. Proof of safety vs. proof of hazard. *Biometrics* 43, pp. 719–725, 1987.

Miller, R.G., Jr. *Beyond ANOVA, Basics of Applied Statistics.* Wiley, New York, 1986, p. 317.

Munk, A. An improvement on commonly used tests in bioequivalence assessment. *Biometrics* 49, pp. 1225–1230, 1993.

Newman, M.C., P.M. Dixon, B.B. Looney, and J.P. Pinder. Estimating mean and variance for environmental samples with below detection limit observations. *Water Resour. Bull.* 25, pp. 905–916, 1989.

Newman, M.C. *Quantitative Methods in Aquatic Ecotoxicology.* Lewis Publishers, Boca Raton, FL, 1995, p. 426.

Noether, G.E. Distribution-free confidence intervals. *Am. Stat.* 26, pp. 39–41, 1972.

Osenberg, C.W., R.J. Schmitt, S.J. Holbrook, K.E. Abu-Saba, and A.R. Flegal. Detection of environmental impacts: natural variability, effect size, and power analysis. *Ecol. App.* 4, pp. 16–30, 1994.

Pearson, E.S. and H.O. Hartley. *Biometrika Tables for Statisticians*, Vol. II. Cambridge University Press, Cambridge, 1972, p. 385.

Potvin, C. and D.A. Roff. Distribution-free and robust statistical methods: viable alternatives to parametric statistics? *Ecology* 74, pp. 1617–1628, 1993.

Rodda, B.E. and R.L. Davis. Determining the probability of an important difference in bioavailability. *Clin. Pharmacol. & Ther.* 28, pp. 252–257, 1980.

Schuirmann, D.J. A comparison of the two one-sided tests procedure and the power approach for assessing the equivalence of average bioavailability. *J. Pharmokinet. Biopharm.* 15, pp. 657–680, 1987.

Seaman, J.W. Jr. and R.G. Jaeger. Statisticae dogmaticae: a critical essay on statistical practice in ecology. *Herpetologica* 46, pp. 337–346, 1990.

Steinijans, V.W. and E. Diletti. Statistical analysis of bioavailability studies: parametric and nonparametric confidence intervals. *Eur. J. Clin. Pharmacol.* 23, pp. 127–136, 1983.

Steinijans, V.W. and E. Diletti. Generalization of distribution-free confidence intervals for bioavailability ratios. *Eur. J. Clin. Pharmacol* 25, pp. 85–88, 1985.

Stewart-Oaten, A. Rules and judgements in statistics: three examples. *Ecology* 76, pp. 2001–2009, 1995.

Stoline, M.R. An examination of the lognormal and Box and Cox family of transformations in fitting environmental data. *Environmetrics* 2, pp. 85–106, 1991.

Stunkard, C.L. Tests of proportional means for mesocosm studies, in *Aquatic Mesocosm Studies in Ecological Risk Assessment*, Graney, R.L., J.H. Kennedy, and J.H. Rogers, Jr., Eds., Lewis Publishers, Boca Raton, FL, 1994.

Weber, C.I. (Ed.). *Methods for Measuring the Acute Toxicity of Effluents and Receiving Waters to Freshwater and Marine Organisms.* 4th ed. EPA-600/4-90-027F. U.S. Environmental Protection Agency, Environmental Monitoring Systems Laboratory, Cincinnati, OH, 1993.

Westlake, W.J. Response to T.B.L. Kirkwood: Bioequivalence testing—a need to rethink. *Biometrics* 37, pp. 589–594, 1981.

13 ‖ Time, Space, Variability, and Uncertainty

Thomas B. Kirchner

INTRODUCTION

A fundamental methodology in science is to disaggregate complex systems into small components that can be studied individually to provide knowledge to explain the functioning of the larger system. Scientists usually assume that parts of the system can be more easily understood than can the whole system (Hauhs, 1990), although some scientists reject the idea that the functioning of complex systems can be totally explained by fully understanding the parts (Platt and Denman, 1975; Levins and Lewontin, 1980; Allen and Starr, 1982). The debate about the efficacy of reductionism versus holism is prevalent among many areas of environmental science. Regardless of the strengths and weaknesses of the holistic viewpoint in science, disaggregation of systems for the purpose of analysis and description is the dominant methodology in scientific research. Unfortunately, disaggregation of a system into smaller components for study may fail to collect data necessary for synthetic analyses of the larger system, e.g., understanding the interactions and correlations between components of the system.

In contrast to science, technology often seeks solutions to practical problems defined in terms of larger systems. Technology endeavors to integrate and extrapolate from the data provided by science to construct an adequate, if not fully truthful, model of the system. In general, technology assumes scientific knowledge is adequate (Hauhs, 1990). Environmental risk assessment is a relatively new field that encompasses both science and technology. Environmental risk assessment recognizes that scientific knowledge is not always adequate and attempts to deal with the uncertainties in knowledge that are bound to be present when making projections of risk. One of the challenges facing risk assessors is to properly handle the treatment of uncertainty so that assessments are judged credible by both scientists and nonscientists, and to clarify rather than obfuscate the information needed to make sound decisions. Properly aggregating and integrating data from experiments designed for scientific goals can be a significant challenge, particularly when uncertainties must be considered.

Estimating realistic levels of risk involving natural systems often requires simulation models. A simulation model is a mathematical representation of a system that is typically constructed by drawing on data and analyses from a variety of independent experiments or studies. Simulation models can also incorporate concepts and assumptions for which empirical evidence is lacking but which appear to be logical, e.g., the concentration of a contaminant in milk coming from a dairy can be calculated from the volume-weighted average of concentrations of the contaminant in milk from the various suppliers to the dairy. When constructing simulation models, scientists often must extrapolate from empirical studies in both the temporal and spatial domains: from individual studies to populations, population studies to individuals, from microcosm studies to fields, from controlled laboratory and field experiments to natural systems, from short-term experiments to questions of long-term consequences, and so forth. Unfortunately, the consequences of making such extrapolations in scale on the uncertainties inherent in the predictions are easy to overlook or misunderstand. This chapter reviews some of the problems to be considered when using empirical data to derive estimates of model parameters.

ERRORS, CONFIDENCE, UNCERTAINTY, AND VARIABILITY

When faced with uncertainty, scientists and risk managers often tend to emphasize different kinds of errors in their evaluation of a null hypothesis, such as "there is no significant impact." On the one hand, the goal of seeking the truth leads scientists to focus on minimizing Type I errors (rejecting a null hypothesis when it is true), often without regard to Type II errors (not rejecting a null hypothesis when it is false) (Eberhardt and Thomas, 1991). Scientists are also likely to focus on Type I errors because their probability can often be evaluated objectively, whereas Type II errors must often be evaluated subjectively (Zar, 1984). On the other hand, regulators are often concerned with minimizing the chance that an assessment will falsely conclude that there is no significant risk, i.e., with minimizing Type II errors (Eberhardt and Thomas, 1991). Although the consequences of making a decision based upon an erroneous conclusion may be different under Type I and Type II errors, the consequences are invariably detrimental to society or the environment (Mapstone, 1995). Therefore, risk assessors must often try to satisfy the proponents of both viewpoints when extrapolating from experimentally determined uncertainties to the environmental uncertainties required for projecting potential risks.

Most environmental risk assessments must deal with uncertainties due to natural variability and lack of knowledge about model parameters. In the past, uncertainty due to natural variability and lack of data were usually not distinguished in the methods used to propagate the uncertainties through models. Indeed, estimates of variability based on sparse data were often subjectively adjusted to account for the additional uncertainties associated with using the data in a different context. The goal in assigning subjectively derived distributions to parameters thought to be truly constant, and to subjectively adjusting distributions representing natural variability, was to increase confidence that decisions made using the model results would have a low likelihood of Type II errors. However, there is growing interest among scientists and desionmakers in producing realistic rather than conservative estimates of risk, in specifically identifying the components of confidence versus expected frequency in the results from risk assessments, and in partitioning the contributions from subjective and objective uncertainties (Helton, 1994; Hoffman and Hammonds, 1994).

The concept of uncertainty is usually framed in the context of probability. However, there are two dominant concepts for the meaning of probability, both of which are applicable to problems encountered in risk assessment. An objective definition of probability, favored by classical statistics, is the relative frequency with which an event occurs. A primary assumption of frequentist statistics is that there exists a distribution of values from which values can be sampled, and that the parameters of this distribution are constants. Inferences are made by drawing attention to a reference set of hypothetical data vectors that could have been generated by the probability model given the parameters for the distribution. Estimators for the parameters, which are functions of the data in a sample, are computed. Reference sets are generated for the parameters based on the hypothetical data vectors, and inferences are based upon the similarity between the sampled data and the reference data sets.

The second concept for the meaning of probability regards it as a mathematical expression of one's degree of belief or confidence with respect to a certain proposition (Savage, 1954; Box and Tiao, 1973). In classical statistics, confidence intervals are used to provide an idea of how far from the true values the calculated estimates might be, based on the empirical data. For example, one may test the hypothesis that the mean value of a quantity measured at one location is different from the mean value estimated at another. Although there are only two means, the test (a t-test) is based on the assumption that one would have observed variation in the two estimates of the means if one collected many sets of data at each location and computed a mean for each set. Confidence can thus be seen to represent a different concept than that of a measured frequency.

Otway (1985) defines probability, as typically used in risk assessment, succinctly: "...probability is no more than a degree of belief in an event or a proposition whose truth has not been ascertained." Risk assessors frequently incorporate subjective estimates of confidence into their analyses because adequate data for characterizing some parameters are unavailable and often unattainable. Subjectively derived distributions of confidence about constant but unknown parameters have been labeled Type B uncertainties, in contrast to objectively defined distributions of frequencies, which have been labeled Type A uncertainties (IAEA, 1989).

This viewpoint of probability as a measure of confidence lies at the heart of Bayesian statistics, and also tends to correspond to the viewpoint adopted by most people in their everyday consideration of probability (Walker, 1995). There may be no way to measure probability in this sense using frequency-based methods. For example, the probability of an individual getting cancer must be either 0 (the event will not occur), or 1 (the event will occur). The interpretation of a statement such as "one's chance of getting cancer from this activity is only 1 in 10,000" should be interpreted only as a statement of belief, as there is no way to treat the statement as a scientifically testable hypothesis. Bayesian methods enable an analyst to combine existing information, expressed as "prior" estimates of probabilities, with new data to estimate uncertainties from which inferences can be drawn (Ellison, 1996). The prior distributions can be derived from data or be entirely subjective in nature. Although Bayesian methods are being used in ecology for parameter estimation (e.g., Gazey and Staley, 1986), hypothesis testing (e.g. Crome et al., 1996), and decision analysis (Taylor et al., 1996; Wolfson et al., 1996), their use is controversial (Dennis, 1996; Edwards, 1996).

Unfortunately, it is not unusual for assessments of confidence to be ignored in the scientific literature when presenting estimates of statistics other than the mean, such as estimates of the population variance or the value associated with the upper 5% tail of a frequency distribution. Scientists are often satisfied knowing population means with some degree of confidence. Risk assessors often need to estimate the likelihood of low probability or extreme cases within the population, and thus need to be confident in estimates of both the means and the variances. Furthermore, objective estimates of confidence need to be combined with subjective levels of confidence to reflect the various assumptions and extrapolations used in making risk projections.

APPROPRIATENESS OF EXPERIMENTAL DATA FOR REPRESENTING NATURAL VARIABILITY

Scientific studies are frequently designed to meet the needs of the empirical models used for analysis. The variability observed in experimental data is sometimes used to represent the uncertainty in the parameters of simulation models. However, most experimental designs attempt to reduce variability to improve the detection of effects (Platt and Denman, 1975). Variation in unmeasured factors giving rise to natural variability in measurements is likely to increase with increasing spatial scale of population size. Thus, one should expect variability under natural conditions to exceed that observed under experimental conditions. Furthermore, experimental sampling programs are often designed to maximize independence of the samples because correlations between samples tend to violate assumptions of many statistical tests and reduce the inferential strength of analyses (Eberhardt and Thomas, 1991). The lack of knowledge about potential correlations between parameters of a system, or within a parameter across space or time, is problematic when attempting to incorporate related processes or to extrapolate information to different scales in a simulation model. Correlations among parameters can affect the predicted means as well as their uncertainties. For example, the mean of the product of two random variables will be larger or smaller than the product of the means of the variables, depending on whether the correlation between them was positive or negative, respectively.

Issues of scale tend to be less important in empirical models than for simulation models because most experimental studies are carried out at the scale of interest, and because many questions in experimental studies are concerned with relative magnitudes and differences rather than absolute descriptions. For example, physiologists may conduct experiments to better understand the effects and pharmacokinetics of a contaminant in individuals, using a laboratory control population to provide reference responses. These experiments can evaluate the impact of the contaminant under laboratory conditions, but are not directly applicable to evaluating questions about population level impacts under natural conditions. Experimental studies designed to test hypotheses rather than to estimate parameter values can often validly ignore potential bias in experiments. For example, experiments designed to evaluate whether an experimental treatment affects crop yield can produce valid results even if there is a consistent bias in estimating yield. Bias can be ignored because it is the differences in yields that are important, and the differences are expected to be unaffected by biases. Even when parameter estimation is the goal of an experiment, the potential for experimental bias may not always be reflected in the confidence intervals assigned to parameters (Henrion and Fischoff, 1986; Freudenburg, 1988; Schlyakhter, 1994).

An important aspect of assigning uncertainty to a parameter is to ensure that the uncertainty is consistent with the definition and use of the parameter. For example, suppose a model uses a transfer coefficient to estimate the concentration of a contaminant in cow's milk given the concentration of the contaminant in forage. In assigning uncertainty to the transfer coefficient one would need to consider whether the concentration in milk was expected to represent the concentration in milk coming from a dairy that obtains milk from several sources, the average concentration in a pool of milk obtained from a single herd of cows, or the concentration in milk from a single cow. Natural variability among cows and variability in the spatial distribution of the contaminant would be likely to cause greater uncertainty in the transfer coefficient used for estimating milk concentrations from a single cow than for estimating milk concentrations from a dairy. The problem of properly specifying uncertainty so that a meaningful probability distribution can be assigned led Howard and Matheson (1984) to suggest the clarity test as a conceptual means of ensuring proper specificity. The test involves the notion that, given sufficient resources, one could measure the parameter of interest. A properly specified parameter would thus give sufficient details to identify the conditions under which an experiment could be conducted to obtain the required data.

TRANSLATION AND EXTRAPOLATION OF MEANS AND VARIANCES

The translation of means and variances from one scale of measurement to another is a common task when adapting empirical data to provide estimates of model parameters. Such scaling typically involves a linear relationship and can be done exactly. Adding a constant, c, to a random variable, x, shifts the expected (mean) value, μ, of the distribution by the value of the constant and has no impact on the variance, σ^2:

$$\mu[x + c] = \mu[x] + c$$

$$\sigma^2[x + c] = \sigma^2[x]$$

Multiplying a random variable by a constant yields a distribution whose mean and variance are both dependent on the constant:

$$\mu[cx] = c\mu[x]$$

$$\sigma^2[cx] = c^2\sigma^2[x]$$

Extrapolation and interpolation allow one to estimate parameter values for one case based on data from another plus a mathematical function relating the two cases. In contrast to linear translations of scale, extrapolating or interpolating means and variances through non-linear functions is more difficult. In general, the function of a mean of random variables is not equal to the mean of the function of the random variables, i.e.,

$$f(\overline{X}) \neq \sum_i \frac{f(x_i)}{n}$$

nor is the function of the standard deviation equal to the standard deviation of the function (Table 13.1).

Table 13.1. Approximations of the Mean and Variance for Several Frequently used Mathematical Functions. The Equations for the Expressions Marked with a † or †† Assume That the Distribution for y is Normal or Lognormal, Respectively, and Give Exact Results for that Case. The Remaining Equations are First-Order Approximations Derived Using a Taylor's Series Expansion.

$x^\dagger = e^y$

$$m_x = e^{\mu_y + \frac{\sigma_y^2}{2}}$$

$$\sigma_x^2 = e^{2\mu_y + 2\sigma_y^2} - e^{2\mu_y + \sigma_y^2}$$

$x = \ln(y)$

$$\mu_x = \ln(\mu_x)\frac{\sigma_y^2}{2\mu_y^2}$$

$$\sigma_x^2 = \frac{\sigma_y^2}{\mu_y^2}$$

$x^{\dagger\dagger} = \ln(y)$

$$\mu_x = \ln(\mu_y) - \frac{1}{2}\ln\left(\frac{\sigma_y^2}{\mu_y^2} + 1\right)$$

$$\sigma_x^2 = \ln\left(\frac{\sigma_y^2}{\mu_y^2} + 1\right)$$

$x = \sin(y)$

$$\mu_x = \sin(\mu_y) \times \left(1 - \frac{\sigma_y^2}{2}\right)$$

$$\sigma_x^2 = \sigma_y^2 \times (\cos(\mu_x))^2$$

$x = \cos(y)$

$$\mu_x = \cos(\mu_y) \times \left(1 - \frac{\mu_y^2}{2}\right)$$

$$\sigma_x^2 = \sigma_y^2 \times (\sin(\mu_x))^2$$

$x = \sqrt{y}$

$$\mu_x = \sqrt{\mu_y} - \frac{1}{8} \times \mu_y^{-\frac{3}{2}} \times \sigma_y^2$$

$$\sigma_x^2 = \frac{\sigma_y^2}{4\mu_y}$$

$x = c^y$

$$\mu_x = c^{\mu_y} + \frac{1}{2}\sigma_y^2\ln(\mu_y)^2 c^{\mu_y}$$

$$\sigma_x^2 = \sigma_y^2 c^{2\mu_y}\ln(\mu_y)^2$$

$x^\dagger = c^y$

$$\mu_x = e^{\ln(c)\mu_y + \frac{\ln(c)^2\sigma_y^2}{2}}$$

$$\sigma_x^2 = e^{2\ln(c)\mu_y + 2\ln(c)^2\sigma_y^2} - e^{2\ln(c)\mu_y + \ln(c)^2\sigma_y^2}$$

$x = y^c$

$$\mu_x = \mu_y^c + \frac{1}{2}c(c-1)\mu_y^{c-2}\sigma_y^2$$

$$\sigma_x^2 = \sigma_y^2 c^2 \mu_y^{2(c-1)}$$

As an example, consider the case of extrapolating a dose response for a chemical contaminant from one species to another using an allometric relationship of body mass, r = am^c. It is well established that many physiological and life history parameters of organisms show an allometric relationship to body mass (Peters, 1983; Calder, 1984), including toxic doses (Travis and White, 1988; Goddard and Krewski, 1992; Travis and Morris, 1992; Watanabe et al., 1992; Andersen et al., 1995). Assuming that the uncertainty in c is represented by a normal distribution, the mean of r given body mass m is

$$\mu_r = ae^{\ln(m)\mu_c + \frac{\ln(m)^2\sigma_c^2}{2}}$$

and its variance is

$$\sigma_r^2 = a^2 \left[e^{2\ln(m)\mu_c + 2\ln(m)^2\sigma_c^2} - e^{2\ln(m)\mu_c + \ln(m)^2\sigma_c^2} \right]$$

Note that both the mean and variance of r depend upon the mean and variance of c. Although a was treated as a constant in this example, it too could have uncertainties that would need to be taken into account when estimating r.

AGGREGATION IN SPACE OR ACROSS INDIVIDUALS

Aggregation in modeling is the process of representing a collection of n states using fewer variables. There will always be error introduced by such aggregation (O'Neill and Rust, 1979; Gardner et al., 1982), but simplification in the representation of complex systems cannot be avoided. Even the order in which averaging is applied to computations in aggregated models can affect the results (Mosleh and Bier, 1992). Aggregation often involves creating variables to represent collections of states rather than all possible states, e.g., populations rather than individual organisms, or pastures rather than individual grass plants. As discussed earlier, scientific research often provides data at reduced scales, often for individuals or square meters, from which modelers must extrapolate to larger scales.

Hattis and Burmaster (1994) present an example that can be expanded upon to illustrate the effects and some of the problems of aggregation of states in a system. Their example involves potatoes contaminated with aldicarb, but it could be equally well applied to situations in which one must scale variability measured for square-meter plots to variability for entire fields. The distribution of concentrations of aldicarb in potatoes appears to be well characterized by a lognormal distribution. Hattis and Burmaster's example apparently assumes that potatoes are of uniform size, and so one can equate the amount of aldicarb consumed with the concentration. Assuming that an individual eats a potato at a meal, then the expected distribution for aldicarb intake per meal would be a lognormal distribution as well. However, if individuals consumed potatoes in meals which mixed two or more potatoes, such as mashed or fried potatoes prepared for a family or in a restaurant, then the variability for the concentration for the mixture would be expected to be lower because it would be represented by the average of the potatoes used to make the dish, assuming mixing is thorough. Likewise, the exposure of a single individual eating potatoes over a period of time could be based on the average concentration for the number of potatoes consumed. The expected distribution for the average concentration in n potatoes has a $\sqrt{n}$ distribution with a standard deviation, σ_a, equal to the standard deviation for the individual potatoes, σ_p, divided by $\sqrt{n}$. Note that this formulation applies to the arithmetic standard deviation, not to the geometric standard deviation. Additionally, it does not require the assumption of normality, as Hattis and Burmaster incorrectly stated. A consequence of averaging concentrations across n potatoes is that the expected distribution will not be lognormal, but will approach a normal distribution as n becomes large, as explained by the Central Limit Theorem. The rate—in terms of increasing sample size—at which the distribution of sample means approaches normal will depend on the magnitude of the geometric standard deviation of the population.

The reduced variance of the average concentration of aldicarb in potatoes eaten over an interval of time compared to the variance of concentrations in individual potatoes should not be interpreted as leading to lower variability in exposure as the time over which averaging takes place increases. Exposure is based upon the sum of the inventories in the potatoes eaten, not the average concentration. The total intake of aldicarb can be computed by multiplying the average concentration per potato by the mass of potatoes eaten to get the total amount of aldicarb ingested. The standard deviation for this product is $n\sigma_a$ (Mood et al., 1974). This is numerically equal to the standard deviation for the sum of n potatoes, i.e.,

$$n\sigma_a = n\frac{\sigma_p}{\sqrt{n}} = \sqrt{n}\sigma_p$$

Note that the uncertainty is not equal to the uncertainty that would result from eating n hypothetical potatoes that have the same mean value as those above and having a standard deviation of σ_a, i.e.,

$$n^2\sigma_a^2 = var[n\overline{X}] \neq var\left[\sum_n \overline{X}\right] = n\sigma_a^2$$

where var[] represents the variance of the function within the brackets and $\overline{X}$ is the mean concentration of n potatoes. Thus, averaging intakes across time does not reduce the uncertainty in the total intake even though the variability of the average concentration is less than the variability for the population of concentrations.

Hattis and Burmaster (1994) point out that this example ignores several other factors that might contribute to the uncertainty in the exposure, including the consumption rates for potatoes and correlations between concentrations in potatoes. For example, suppose that the way potatoes were collected and bagged introduced a positive correlation between concentrations of the potatoes in a meal. If the correlation coefficient between potatoes were r, then the standard deviation of the mean of n potatoes would be $\sigma_a = \sqrt{n\sigma_p^2(1 + p(n - 1))}$ (see Appendix).

One would also want to consider the distribution of sizes of potatoes. This would be important when an individual is eating several whole potatoes over time, and could be used as a weighting factor when computing the average concentration in a dish derived from two or more potatoes. The existence of correlations between the size of potatoes and the concentration of aldicarb would impact the mean of the resulting distribution as well as its uncertainty. If the mass of potatoes had a distribution with mean μ_m and standard deviation σ_m, and a correlation coefficient of r between potato size and concentration, then the inventory of aldicarb in potatoes would be distributed as

$$\mu_i = \mu_m \mu_p + \rho \sigma_m \sigma_p$$

$$\sigma_i^2 = \mu_p^2 \sigma_m^2 + \mu_m^2 \sigma_p^2 + 2\mu_m \mu_p \rho \sigma_m \sigma_p - (\rho \sigma_m \sigma_p)^2 +$$
$$E\left[(p - \mu_p)^2 (m - \mu_m)^2\right] + 2\mu_p E\left[1(p - \mu_p)^2 (m - \mu_m)\right] +$$
$$2\mu_m E\left[(p - \mu_p)(m - \mu_m)^2\right]$$

where μ_p is the mean concentration of aldicarb in the potatoes. Thus, correlations can have significant impacts on uncertainty of aggregate variables. In practice, failure to consider all possible correlations among inputs does not necessarily invalidate an uncertainty analysis. Smith et al. (1992) demonstrated that some correlations will have relatively low impact on the results of an uncertainty analysis. They suggest that correlations can often be neglected if (1) correlations are weak, (2) the correlations are between relatively well known variables, i.e., variables with low uncertainties, or (3) correlations are between variables with low influence on the results.

Assumptions about the correlation between, or the aggregation of, input variables can affect the assumptions that need to be made about the results as well. For the case of two or more people sharing a potato dish, as in a family or at a restaurant, the exposures of the people sharing the meal would be correlated. These correlations, in turn, can affect the uncertainty associated with across-person aggregate variables, such as collective doses. Thus, it would be inappropriate to estimate collective doses by simply summing the doses of individuals.

Aggregating states that show natural variability in their dynamics can be problematic. For example, consider the problem of predicting the average concentration in the levels of ^{131}I in milk for a herd of dairy cows starting with rate constants for individual cows. If the cows produce equal volumes of milk, then the average concentration would correspond to the concentration expected if the milk from the herd were pooled together. Like many processes in models of radionuclide transport, the decline in the concentration of ^{131}I in milk can be well described using first-order kinetics, i.e.,

$$\frac{dQ}{dt} = kQ$$

where Q is the level of ^{131}I in milk and k, the rate of loss, is negative (Sasser, 1965). The solution to this differential equation is $Q_t = Q_0 e^{kt}$, where Q_0 is the initial value of Q and Q_t is the value at time t. The rate of decline in the concentration of ^{131}I in milk is controlled by the combination of physical decay and biological elimination. Sasser (1965) measured half-lives for the rate of loss in 6 cows ranging from 0.60 to 0.85 days. A simulation model was constructed to represent the decline in concentration of ^{131}I in milk from each of the 6 cows, and from a hypothetical "herd" having an average rate of loss equal to the mean of the 6 observed rate constants. The dynamics of the herd were assumed to be identical to the dynamics for an individual cow, i.e.,

$$\frac{dQ_h}{dt} = k_h Q_h$$

The model also computed the average milk concentration across the 6 cows,

$$Q_a = \sum_{1}^{6} Q_i / 6$$

(Figure 13.1). Plotting the logarithm of Q_h yields a straight line, whereas the logarithm of Q_a is a concave curve. Using the average rate constant also results in Q_h being lower than Q_a. Given that the analytical solution for the model is an exponential function of the rate constant, an approximate solution for the curve of the mean milk concentration can be derived (see Table 13.1). The random variable being exponentiated is kt, which will have a mean of kt and a variance of σt^2. Hence the approximation for the function for the mean can be written

$$\overline{Q}_t = Q_0 e^{\overline{k}t + (t^2 \sigma^2)/2}$$

with the corresponding differential equation

$$\frac{d}{dt} \overline{Q} = \overline{Q}(\overline{k} + t\sigma^2)$$

Thus the rate of decline in mean ^{131}I milk concentration is expected to decrease as time increases. It should be noted that the mean milk concentration based on this approximation begins to show an increased rate of deviation from the simulated mean milk concentration after day 10 (Figure 13.2). This error is probably insignificant in the context of the example because it occurs after about 12 half-lives, at which time the concentration is about 0.02% of the original value.

This example demonstrates that predicting the average behavior for an aggregate variable may not be as simple as substituting average rate constants into the functions appropriate for the individual components of the aggregate. Even in this simple model it was necessary to substitute for the rate constant, k, a time-dependent function of the mean and variance of the rate constants, $k + t\sigma^2$, in order to approximate the dynamics of the mean.

EXTRAPOLATING POPULATION MEANS AND VARIANCES FROM SAMPLE MEANS AND VARIANCES

Hoffman and Hammonds (1994) discuss the advantages of providing estimates of the distribution of confidence for frequency distributions, but consider only the level of confidence in the predictions that is derived from subjective uncertainties assigned to constant but unknown parameters. However, there are objective measures of confidence as well that should be taken into account when describing predictions of frequency distributions. For

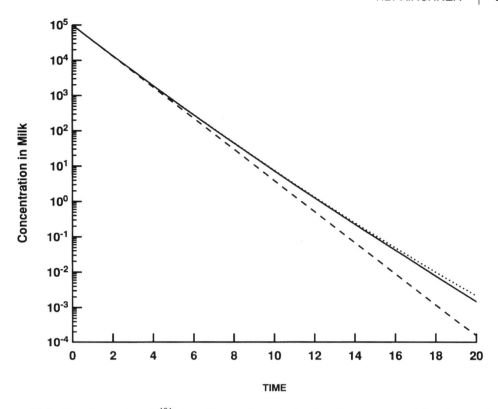

Figure 13.1. The dynamics of ^{131}I in milk are simulated using the average of 6 rate constants measured by Sasser (1965) (dashed line), and by simulating milk concentrations for each of the 6 cows and averaging the milk concentrations (solid line). The dotted line is an approximation derived for the average concentration, $\bar{Q}_t = Q_0 e^{\bar{k}t+(t^2\sigma^2)/2}$. The logarithm of the average concentration is not a linear function of time.

example, sample size introduces a predictable uncertainty to estimates of both the mean and variance of a distribution. The various calculations for the previous examples are expected to give the correct solutions only if the true means and variances of the input distributions are known. However, the analyses started with only estimates of the means and variances for the distribution; the true means and variances could be larger or smaller than the estimates, and hence affect the computed means and variances. For example, if one were to assume (falsely) that the distributions of the sample mean and sample variance were both symmetric, then there would be a 25% chance that the estimated mean and variance used as input to each assessment were both lower than the true mean and variance, hence underestimating the frequency of high exposures. There would also be an equal likelihood that the estimates of the mean and standard deviation were both greater than the true values, leading to overestimation of the highest exposures. In reality, the expected distribution for the sample mean approaches normal for large n, and the expected distribution for the sample variance is related to a chi-square. Specifically, if the distribution being sampled is normal, then $U = vs^2/\sigma^2$ has a chi-square distribution with $v = n-1$ degrees of freedom, where s^2 is the sample variance and σ^2 is the true variance (Zar, 1984). The confidence interval for σ^2 is

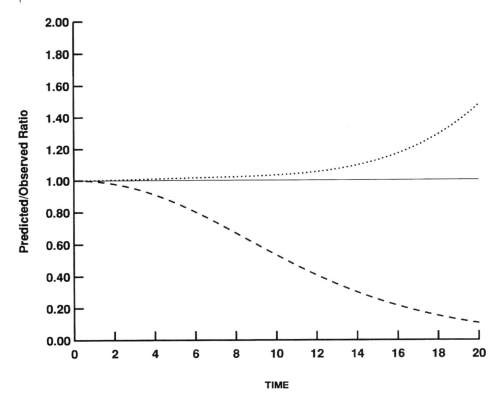

Figure 13.2. The predicted/observed ratios of the approximation for the average milk concentration to the concentration simulated using the average rate constant (dashed line) and the functional approximation derived, $\bar{Q}_t = Q_0 e^{\overline{kt+(t^2\sigma^2)/2}}$ (dotted line). Although the functional approximation represents the average milk concentration better than using the average rate constant, the error in the approximation increases rapidly after about day 10.

$$\frac{vs^2}{X^2_{\alpha/2,v}} \leq \sigma^2 \leq \frac{vs^2}{X^2_{(1-\alpha/2),v}}$$

where α is the confidence level (Zar, 1984). The confidence interval for the population mean, μ, is defined in terms of the t-distribution

$$\overline{X} - t_{a(2),v}s_x < \mu < \overline{X} + t_{a(2),v}s_x$$

where $\overline{X}$ is the sample mean and $t_{a(2),v}$ is the 2-tailed Student's t-value for a with n degrees of freedom. The sample mean and sample variance are independently distributed. The formulas can be applied to lognormal distributions as well by applying them to logarithmic transforms of the data.

There are two obvious ways to include the consideration of the impact of sample size on the propagation of uncertainty in a simulation model. The first is to use a nested Monte

Carlo sampling design, similar to those described by Hoffman and Hammonds (1994) and Price et al. (1996). In these designs, an inner loop was used to simulate natural variability in model parameters, and an outer loop was used to sample the parameters that were assumed to be constant but unknown. In their example, Hoffman and Hammonds assumed that the mean, variance, and type of distribution for variables in the inner loop were not known precisely and therefore were subjectively assigned distributions. To consider the case where sample means and variances are known, the outer loop would be used to sample the population means and variances for the distributions in the inner loop using distributions based on the sample size, sample mean, and sample variance of the original empirical distributions.

The second approach is to use "conservative" estimates for the population mean, μ, and standard deviation, σ, derived from the sample mean and variance. Deciding what is conservative requires one to decide whether to emphasize minimizing Type I or Type II errors. For example, one might consider using the upper 0.975 confidence level for the estimates of μ and σ for parameters that are positively correlated with risk. Because the sample mean and variance are independently distributed, using the 0.975 confidence intervals results in a joint probability of approximately 0.95 that neither the estimated mean nor the estimated standard deviation are less than μ and σ, respectively. For example, consider a distribution with a sample mean, $\overline{X}$, of 100 and a sample standard deviation, s, of 20 based on a sample size of 10. The 0.975 confidence interval for the mean is

$$\overline{X} \pm 2.685\frac{20}{\sqrt{10}}, \text{ or } \overline{X} \pm 16.98$$

The 0.975 confidence interval for the standard deviation is

$$\sqrt{\frac{9 \times 400}{19.02}} \leq \sigma \leq \sqrt{\frac{9 \times 400}{2.70}}, \text{ or } 13.76 \leq \sigma \leq 36.52$$

Thus, the uncertainty of the variance dominates the uncertainty of the distribution. Combining the uncertainty in μ with the upper estimate for σ would yield a conservative estimate for the 0.95 confidence interval of $\overline{X} \pm 88.5$. If the 0.95 confidence intervals for the mean and standard deviation are used instead of the 0.975 levels, the resulting confidence interval is $\overline{X} \pm 78.8$. To minimize the likelihood of a Type I error in drawing inferences about risks, one could use the lower confidence levels for μ and σ, whereas to minimize the likelihood of a Type II error, one could use the upper confidence level.

If the endpoint of an assessment can be computed using a fractile of an input distribution, such as the level associated with the 0.95 probability of the upper tail, then tables of tolerance coefficients can be used to compute the fractiles based on the sample mean, variance, and sample size, and given the fractile and the confidence level for the estimate (Hald, 1965). Using the table of tolerance coefficients from Hald (1965), the 0.95 tolerance level for the population distribution, with confidence of 0.95, is $\overline{X} \pm 3.39 \times 20$, or $\overline{X} \pm 67.8$. The comparison of the 0.95 interval based on the tolerance coefficients and the 0.95 interval based on the upper 0.975 levels for μ and σ indicates that using the 0.975 estimates for μ

and σ, suggested above, is likely to be quite conservative when the distribution is normal. Using the 0.95 levels for the estimated values of μ and σ appears to give reasonably conservative results. Given the interest in the tails of distributions among risk assessors, direct estimation of endpoints using extreme value statistics may be more appropriate and more efficiently estimated than other measures of variability in some cases (Gaines and Denny, 1993; Lambert et al., 1994; Lennox and Haimes, 1996).

The significance of this example is that it demonstrates that the impact of sample size on the estimators for the parameters of the input distributions should not be overlooked when adapting empirical data on variability for use in the uncertainty analysis of a model. If one had presumed that sample mean and standard deviation were accurate and precise estimators of μ and σ, then the 0.95 tolerance level for this example would have been computed to be $\overline{X} \pm 1.96 \times 20$, or $\overline{X} \pm 39.2$, rather than even the smallest interval of $\overline{X} \pm 67.8$. Given that environmental sample sizes are typically less than 8, and often as small as 4 (Eberhardt and Thomas, 1991), the importance of taking sample size into account is evident when scaling from empirical estimates of variability to uncertainty in model parameters.

TEMPORAL VARIABILITY

Treating model parameters as being stochastic through time rather than constant can have dramatic impacts on model results, including changes in the dynamics of the predictions through time (Kremer, 1983). However, both measuring and simulating variability in a rate through time are inherently problematic because of the continuous nature of time. Whereas variability among discrete objects, such as individual people or potatoes, is easy to visualize, time can only be divided into intervals of arbitrary length. Therefore, an important question is "What is the appropriate time interval for expressing variability in time?" There are two factors that affect the answer to this question: (1) the time interval associated with the empirical estimate of variability through time, and (2) the relevant time interval for evaluating impacts or risks.

Price et al. (1996) show that long-term averages tend to underpredict short-term variability. For example, consider a process such as weathering of a contaminant from the surface of vegetation through time. The rate of such weathering can be an important factor affecting exposure from foodchain pathways (Breshears et al., 1992). A model simulating the loss of a contaminant from plant surfaces was created to illustrate the relation between variability and the time interval. The loss from weathering was assumed to be a linear first-order process,

$$\frac{dC}{dt} = -rC$$

A graph of the logarithm of the contaminant inventory on plants versus time would yield a line whose slope is equal to the rate constant. A "random error" component was simulated by adding to the nominal rate a value drawn from a uniform distribution equal to ±50% of the nominal value of each time step. The error term is equivalent to a random walk imposed on top of the deterministic loss function, and is expected to yield a normal distribution in the "error" term when the model is run for many trials (Gard, 1988).

The model was run and data sampled over intervals from 1 to 10 time steps. Simulations were replicated to yield 400 estimates for the loss rate, which were determined from the inventory at the beginning and end of each time interval. The standard deviations for these estimates decrease as the length of the time interval increases (Figure 13.3). This example illustrates that the time interval associated with an estimate of uncertainty is an important dimension that must be taken into account when simulating such uncertainty in a dynamic simulation model. If the variability through time in a rate constant is based upon a 4-day measurement interval, then the corresponding model rate constant should be sampled every fourth simulated day as well. Applying the variability in a rate constant over an interval different than that on which it was measured can yield either higher or lower levels of uncertainty in the model predictions than are appropriate.

The time interval in which data were collected is not necessarily the appropriate time interval for assessing impacts in a risk assessment (Hattis and Burmaster, 1994). The dynamics of toxic effects should influence the time units in which variability of exposures are measured and expressed. For example, estimates of variability in long-term average exposures are likely to be inadequate when toxic impacts can occur from short-term, high-level exposures. Nonlinearities in response curves, including threshold levels, must be considered when determining the toxicologically relevant averaging time. Thus, the toxicologically relevant averaging time can be the main driver for determining the size of time step for some simulations. Toxicologically relevant averaging times for some contaminants are also likely to be species-specific, with metabolic rates and other correlates of body size contributing to the relationship. Therefore, assumptions about temporal variability need to be carefully examined when extrapolating dose effects from one species to another.

The time step used to solve the differential equations in a simulation model cannot be arbitrarily set equal to the time interval needed for sampling stochastic parameters. The size of the time step used for solving differential equations determines the magnitude of the local truncation error in the solution, with larger truncation errors associated with longer time steps. Adaptive time step methods for solving differential equations are often used to control the error within acceptable levels, but do so by adjusting the time step throughout a simulation. If the simulation is required to use a time step different than that from the interval on which variability was measured, one might be tempted to try to scale the variability to the size of the time step to yield the correct model behavior. For example, Figure 13.3 shows that the standard deviation of the estimated rate constant for the example varies approximately proportionally to the inverse of the square root of the interval, $1/\sqrt{\Delta t}$. Knowing this relationship, one could fit parameters to the curve and use it to predict variance for an interval within the range of measured values. However, this relationship arises in this example because of the consistency in the method used to vary the rate constant through time. One should not expect variance in rate constants always to scale with $\sqrt{\Delta t}$. In natural systems there could be many factors influencing the variability on different time scales, thus leading to an irregular relationship between interval length and variability. For example, adding additional random errors to the model rate constant every 3 and 5 days yields a completely different trend in variability (Figure 13.3). Therefore, scaling variability to the time step requires that sufficient data be collected to derive the relationship of variance as a function of time interval. An alternative approach is to design the code for the simulation model to enable parameters to be varied through time on intervals independent of those used to solve the differential equations. Although a design using nested loops, such as

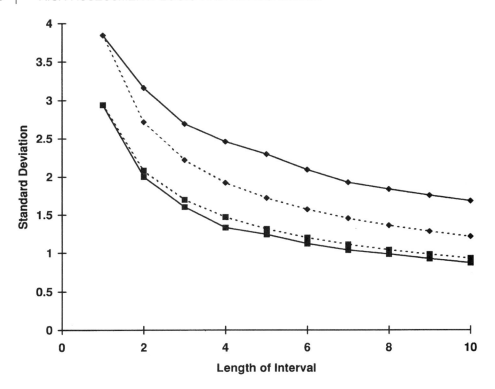

Figure 13.3. The estimate of variability in a rate constant depends on the length of the interval over which the measurements are made. Weathering of contamination, *C*, from plant surfaces was simulated as dC/dt = –rC. The lower solid line (squares) represents the standard deviation of the rate constant, *r*, that is estimated for the case of random error being assigned to *r* every day. The upper solid line (diamonds) is the estimated standard deviation of *r* when additional random error terms are added on intervals of 1, 3, and 5 days. The estimate of the variability in *r* declines as the length of the sampling interval, *t*, increases. The decline in the estimate for the first case is proportional to $t^{-1/2}$, where *t* is the length of the sampling interval (dotted line, squares). The decline of the estimate for the second case is not proportional to $t^{-1/2}$ (dotted line, triangles).

proposed by Price et al. (1996), might work in some cases, a more general approach would be to integrate a variable time step method for solving differential equations with a discrete event simulation strategy (Pooch and Wall, 1992).

Failing to measure time-dependent variability on the time interval at which that variability arises in the system being modeled will lead to errors, especially in the case of nonlinear dynamics. If the rate of a process is sensitive to temperature, and temperature has a random component, then the impact of temperature on the process can only be modeled well if the time interval for measuring variability in temperature is attuned to the time interval on which temperature varies. Instantaneous variability in continuous systems can be modeled, but often requires the development of analytical solutions specific to the rate functions and distributions of parameters (Gard, 1988).

Improvements in handling time-dependent variability in risk assessments can also be made by utilizing better sampling designs when collecting data for estimating rate constants. Sampling intervals that are well suited to estimating the mean value of a rate constant may be inadequate for describing variability in the rate through time. Under some circumstances it may be possible to derive estimates of long-term variability from short-term measurements (Wallace et al., 1994; Slob, 1996). However, it appears unlikely that short-term variability can be reliably predicted from long-term measurements. Underwood (1994) suggests that hierarchical designs be used in sampling to better detect and describe the pattern of temporal variability. In general, empirical studies designed to support environmental decisionmaking should emphasize designs that characterize the factors affecting variability in measurements at relevant time scales in addition to characterizing the factors affecting means.

SUMMARY

The disaggregation of systems into simpler components that can be studied individually has been a successful strategy in science, but tends to complicate the analysis of whole systems. On the one hand, scientists have provided much detailed data that can be used to help explain how various parts of a system function. On the other hand, this reductionist approach often fails to collect the information necessary to assemble a representation of the whole system that is as credible as the representation of the various components. Many experiments are specifically designed to reduce or eliminate correlations, interactions, and variability that must be taken into account when conducting environmental risk assessments. Furthermore, experimental designs used to collect data for testing hypotheses are often not the best designs for estimating parameters for use in simulation models. Thus, risk assessors must frequently interpolate and extrapolate model parameters from available data and, in turn, provide credible estimates of the uncertainties that result from such actions. Such extrapolations are likely to include both objective and subjective estimates of uncertainties in the parameters and inputs of a model. Risk assessment must try to balance the need for scientific credibility, which demands a low probability of Type I errors, with the need for credibility among regulators and the public, who tend to focus on achieving a low incidence of Type II errors. Risk assessors need to provide equitable and complete analyses of both types of errors to decisionmakers, who must balance the impacts of making Type I and Type II errors.

Mathematical manipulations involving random variables can be complex, and sometimes even counterintuitive. Often, the mean and variance of a distribution resulting from a nonlinear translation or extrapolation of a variable described by a distribution will both be dependent on the mean and variance of the original variable. Aggregate quantities, such as contaminant levels in a pool of milk from a herd of cows, can have dynamics different from those of the individual components. The difference in dynamics arises solely from variability among the components. Uncertainty arising from processes that show variability through time can be difficult to properly represent in a model because the magnitude of variability is likely to depend on the interval of measurement. Scaling temporal variability in a rate parameter to an arbitrary time step in a model may be possible, but requires scientists to use special sampling designs and to collect a number of samples far in excess of that needed to estimate the mean value of the parameter.

In general, sample sizes needed to provide adequate estimates of the mean for statistical analyses are often small, with the result that uncertainty in the estimate of the population variance of a parameter is often high. Given the recent interest in partitioning uncertainties into their components of variability and confidence, it is important that risk assessors adequately represent variability by explicitly accounting for the impact of sample size on the estimates of population means and variances. Representing confidence based on objective methods involving sample size and subjective methods involving uncertainties in knowledge will require risk assessors to explicitly consider whether confidence is being expressed in terms of the likelihood of making Type I or Type II errors.

Probabilistic risk assessors appear to be adopting methods that will permit better identification and analysis of the various dimensions of uncertainty in an assessment. It is going to be interesting to see whether the partitioning of uncertainties in risk estimates between subjective and objective components, or between confidence and variability, improves the credibility of risk assessment among decisionmakers, scientists, and the public.

REFERENCES

Allen, T.F.H. and T.B. Starr. *Hierarchy: Perspectives for Ecological Complexity*. Chicago Press, Chicago, 1982, p. 310.

Andersen, M.E., H. Clewell III., and K. Krishnan. Tissue dosimetry, pharmacokinetic modeling, and interspecies scaling factors. *Risk Anal.* 15, pp. 533–537, 1995.

Box, G.E.P and G.C. Tiao. *Bayesian Inference in Statistical Analysis*. Addison-Wesley, Menlo Park, CA, 1973, p. 588.

Breshears, D.D., T.B. Kirchner, and F.W. Whicker. Contaminant transport through agroecosystems: assessing the relative importance of environmental, physiological and management factors. *Ecol. Appl.* 2, pp. 285–297, 1992.

Calder, W.A. *Size, Function, and Life History*. Harvard University Press, Cambridge, MA, 1984, p. 431.

Crome, F.H.J., M.R. Thomas, and L.A. Moore. A novel Bayesian approach to assessing impacts of rain forest logging. *Ecol. Appl.* 6, pp. 1104–1123, 1996.

Dennis, B. Discussion: Should ecologists become Bayesians? *Ecol. Appl.* 6, pp. 1095–1103, 1996.

Eberhardt, L.L. and J.M. Thomas. Designing environmental field studies. *Ecol. Monogr.* 61, pp. 53–73, 1991.

Edwards, D. Comment: the first data analysis should be journalistic. *Ecol. Appl.* 6, pp. 1090–1094, 1996.

Ellison, A.M. An introduction to Bayesian inference for ecological research and environmental decision making. *Ecol. Appl.* 6, pp. 1036–1046, 1996.

Freudenburg, W.R. Perceived risk, real risk: Social science and the art of probabilistic risk assessment. *Science* 242, pp. 240–245, 1988.

Gaines, S.D. and M.W. Denny. Extremes in ecology. *Ecology* 74, pp. 1693–1699, 1993.

Gard, T.C. *Introduction to Stochastic Differential Equations*. Marcel Dekker, New York, 1988, p. 234.

Gardner, R.H., W.G. Cale, and R.V. O'Neill. Robust analysis of aggregation error. *Ecology* 63, pp. 1771–1779, 1982.

Gazey, W. and M.J. Staley. Population estimation from mark-recapture experiments using a sequential Bayes algorithm. *Ecology* 67, pp. 941–951, 1986

Goddard, M.J. and D. Krewski. Interspecies extrapolation of toxicity data. *Risk Anal.* 12, pp. 315–317, 1992.

Hald, A. *Statistical Theory with Engineering Applications*. John Wiley and Sons, New York, 1965, p. 783.

Hattis, D. and D.E. Burmaster. Assessment of variability and uncertainty distributions for practical risk analyses. *Risk Anal.* 14, pp. 713–730, 1994.

Hauhs, M. Ecosystem modeling: Science or technology? *J. Hydro.* 116, pp. 25–33, 1990.

Helton, J.C. Treatment of uncertainty in performance assessment for complex systems. *Risk Anal.* 14, pp. 483–511, 1994.

Henrion, M. and B. Fischoff. Assessing uncertainty in physical constants. *Am. J. Phys.* 54, pp. 791–798, 1986.

Hoffman, F.O. and J.S. Hammonds. Propagation of uncertainty in risk assessments: the need to distinguish between uncertainty due to lack of knowledge and uncertainty due to variability. *Risk Anal.* 14, pp. 707–712, 1994.

Howard, R.A. and J.E. Matheson. *Readings in the Principles and Practices of Decision Analysis*. Strategic Decision Systems, Menlo Park, CA, 1984, p. 955.

I.A.E.A. *Evaluating the Reliability of Predictions Made Using Environmental Transport Models*. Safety Series No. 100. International Atomic Energy Agency, Vienna, Austria, 1989, p. 106.

Kremer, J.N. Ecological implications of parameter uncertainty in stochastic simulation. *Ecol. Modell.* 18, pp. 187–207, 1983.

Lambert, J.H., N.C. Matalas, C.W. Ling, Y.Y. Haimes, and D. Li. Selection of probability distributions in characterizing risk of extreme events. *Risk Anal.* 14, pp. 731–742, 1994.

Lennox, M.J. and Y.Y. Haimes. The constrained extremal distribution selection method. *Risk Anal.* 16, pp. 161–176, 1996.

Levins, R. and R. Lewontin. Dialectics and reductionism in ecology. *Synthese* 43, pp. 47–78, 1980.

Mapstone, B.D. Scalable decision rules for environmental impact studies: effects size, Type I, and Type II errors. *Ecol. Appl.* 5, pp. 401–410, 1995.

Mosleh, A. and V. Bier. On decomposition and aggregation error in estimation: Some basic principles and examples. *Risk Anal.* 12, pp. 203–214, 1992.

Mood, A.M., F.A. Graybill, and D.C. Boes. *Introduction to the Theory of Statistics*. McGraw-Hill, New York, 1974, p. 480.

O'Neill, R.V. and B. Rust. Aggregation error in ecological models. *Ecol. Modell.* 7, pp. 91–105, 1979.

Otway, H. Multidimensional criteria for technology acceptability: a response to Bernard Cohen. *Risk Anal.* 5, pp. 271–273, 1985.

Peters, R.H. *The Ecological Implications of Body Size*. Cambridge University Press, New York, 1983.

Platt, T. and K.L. Denman. Spectral analysis in ecology. *Ann. Rev. Ecol. Syst.* 6, pp. 189–210, 1975.

Pooch, U.W. and J.A. Wall. *Discrete Event Simulation: a Practical Approach*. CRC Press, Boca Raton, FL, 1992, p. 432.

Price, P.S., C.L. Curry, P.E. Goodrum, M.N. Gray, J.I. McCrodden, N.W. Harrington, H. Carlson-Lynch, and R.E. Keenan. Monte Carlo modeling of time-dependent exposures using a microexposure event approach. *Risk Anal.* 16, pp. 339–348, 1996.

Sasser, L.B. "The uptake and secretion of iodine-131 by lactating dairy cows resulting from a contaminated pasture," MS. Thesis, Colorado State University, Ft. Collins, CO, 1965.

Savage, L.J. *The Foundation of Statistics*. Wiley, New York, 1954, p. 294.

Schlyakhter, A.I. An improved framework for uncertainty analysis: Accounting for unsuspected errors. *Risk Anal.* 14, pp. 441–447, 1994.

Slob, W. A comparison of two statistical approaches to estimate long-term exposure distributions from short-term measurements. *Risk Anal.* 16, pp. 195–200, 1996.

Smith, A.E., P.B. Ryan, and J.S. Evans. The effect of neglecting correlations when propagating uncertainties and estimating the population distribution of risks. *Risk Anal.* 12, pp. 467–474, 1992.

Taylor, B.L., P.R. Wade, R.A. Stehn, and J.F. Cochrane. A Bayesian approach to classification criteria for spectacled eiders. *Ecol. Appl.* 6, pp. 1077–1089, 1996.

Travis, C.C. and R.K. White. Interspecific scaling of toxicity data. *Risk Anal.* 8, pp. 119–125, 1988.

Travis, C.C. and J.M. Morris. On the use of 0.75 as an interspecies scaling factor. *Risk Anal.* 12, pp. 311–313, 1992.

Underwood, A.J. On beyond BACI: sampling designs that might reliably detect environmental disturbances. *Ecol. Appl.* 4, pp. 3–15, 1994.

Walker, V.R. Direct inference, probability and a conceptual gulf in risk communication. *Risk Anal.* 15, pp. 603–609, 1995.

Wallace L.A., N. Duan, and R. Ziegenfus. Can long-term exposure distributions be predicted from short-term measurements? *Risk Anal.* 14, pp. 75–85, 1994.

Watanabe, K., F.Y. Bois, and L. Zeise. Interspecies extrapolation: a reexamination of acute toxicity data. *Risk Anal.* 12, pp. 301–310, 1992.

Wolfson, L.J., J.B. Kadane, and M.J. Small. Bayesian environmental policy decisions: two case studies. *Ecol. Appl.* 6, pp. 1056–1066, 1996.

Zar, J.H. *Biostatistical Analysis*. Prentice-Hall, Englewood Cliffs, NJ, 1984, p. 718.

APPENDIX

The variance for a sum of random variables is

$$\sigma^2_{sum} = \sum \sigma^2_{x_1} + 2\sum_{i>j}\sum_j cov[x_i, x_j]$$

where $cov[x_i, x_j] = \rho\sigma_{x_1} s_{xj}$. Assuming that $\sigma_{xi} = \sigma$ and ρ is positive, then

$$\sigma^2_{sum} = n\sigma^2 + \frac{2(n^2 - n)\rho\sigma^2}{2} = n\sigma^2(1 + \rho(n - 1))$$

14 ‖‖‖ Deconvolution Can Reduce Uncertainty in Risk Assessment

Scott Ferson and Thomas F. Long

OVERVIEW

The operations implemented in familiar software packages such as @Risk, Crystal Ball, and Analytica routinely estimate convolutions of random variables. In comprehensive risk analyses, however, deconvolutions should be used whenever justified because they will often yield narrower distributions containing less overall uncertainty. Furthermore, in some circumstances, failing to use deconvolutions when appropriate can lead to conclusions that are insufficiently protective. Deconvolutions reduce uncertainty by taking into account knowledge about the underlying relationship among the variables. Although they are almost never used in current risk analyses, situations that justify their use arise frequently in practical situations. Both probabilistic and traditional worst-case or bounding-estimate risk analyses can benefit from appropriate use of deconvolution. We outline when deconvolutions can be used and how they are computed in both analytical settings.

INTRODUCTION

When combining random variables in a probabilistic risk analysis, whether additively, multiplicatively, or otherwise, one is performing a convolution on their probability distributions (Springer, 1979). Recently available software tools such as Crystal Ball[®1] (Burmaster and Udell, 1990), @Risk[2] (Barton, 1989; Salmento et al., 1989) and Analytica[3] (the successor to Demos; see Morgan and Henrion, 1990) implement convolutions with Monte Carlo methods. These software packages have made convolution of probability distributions almost as simple as doing spreadsheet arithmetic on ordinary numbers and, as a consequence, the technique is now routinely used in many risk analyses.

---

[1] Crystal Ball is a registered trademark of Decisioneering, Denver, Colorado.
[2] @Risk is a product of Palisade Corporation, Newfield, New York. (RISK is a registered trademark of Parker Brothers, Division of Tonka Corporation, which Palisade Corporation uses under license.)
[3] Analytica is a registered trademark, and Demos is a trademark, of Lumina Decision Systems.

Deconvolution is the inverse operation of convolution. In this chapter, we give two examples that illustrate the benefits of using deconvolution in terms of reduced uncertainty, as well as the hazards of ignoring the technique in circumstances where it is indicated. The first is in the context of a probabilistic risk assessment. The second is in the context of a traditional risk analysis using the bounding estimates currently favored by the U.S. Environmental Protection Agency (EPA, 1987; 1992; Norton et al., 1988). Finally, we summarize the optimal use of convolutions and deconvolutions in different situations.

DECONVOLUTION IN PROBABILISTIC ANALYSES

We introduce the need for deconvolution with a simple numerical example. Consider the relationship between the statistical distributions of dose, potency, and risk for a hypothetical carcinogen. Suppose that dose has a uniform distribution on the range [1000, 2000] and the potency factor has a uniform distribution on the range [0.02, 0.04]. These distributions represent the heterogeneity among individuals in the underlying population. If risk is defined by

$$\text{dose} \times \text{potency} = \text{risk}$$

then the implied distribution for risk is nonuniform and ranges on [20, 80]. We simulated the distribution of risk using ordinary Monte Carlo techniques assuming independence, i.e., we randomly selected a dose and a potency factor from their respective ranges and multiplied them together. The resulting products were tallied in a histogram. Figure 14.1 depicts these distributions for dose, potency, and risk. For the purpose of this example, assume perfect knowledge regarding the depicted relationship. Now suppose that one can directly observe the statistical distribution of dose (perhaps by concentrations of urinary metabolites) and the statistical distribution of risk (from, say, published mortality data). It is not unreasonable to assume, because of the nature of mortality data, that the measurement of risk cannot be made on a particular individual, so we cannot generate pairs of dose and risk for individuals. Therefore we only know the marginal distributions for each variable.

Consequently, the defining equation and histograms for risk and dose are the only information that would be available empirically. The potency factor distribution would be unknown to an observer. How would an analyst go about estimating the distribution for the potency factor? A natural way might be simply to solve the equation for the unknown and then use the Monte Carlo techniques available in @Risk, Crystal Ball, or Analytica to estimate it. Figure 14.2 shows the result of using the expression

$$\text{risk} / \text{dose} = \text{potency}$$

and Monte Carlo convolution to estimate potency from the statistical distributions for risk and dose.

Figure 14.3 depicts both the original PDF for potency (solid line) and the reconstructed estimate of potency (dashed line). It is clear that this approach does *not* recover the original distribution. This qualitative result does not depend on the particular forms of the underlying distributions, nor the details of the equation relating the variables involved. In general, in fact, convolutions using an inverse mathematical operation cannot be used to solve a

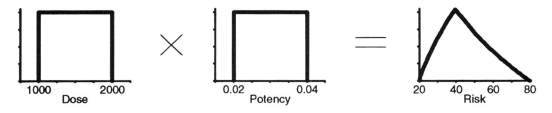

Figure 14.1. The (hypothesized) relationship among probability density functions for risk, dose, and potency. The ordinates are probability.

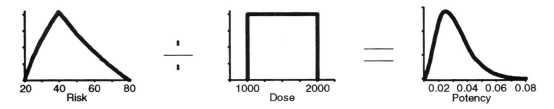

Figure 14.2. Attempt to reconstruct the distribution of potency from risk and dose using ordinary Monte Carlo methods (i.e., convolution). The ordinates are probability.

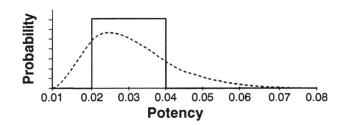

Figure 14.3. Discrepancy between the original distribution of potency (solid line) and the estimate computed by division convolution (dashed line).

problem like this. What is needed is some method of *de*convolution that undoes a convolution. A method for probabilistic deconvolution exists and, as we will see below, does correctly recover the original hypothesized distribution of potency.

Note that if we had *paired* data on risk and dose for exposed individuals, we could have correctly estimated the distribution of the potency factor, simply by forming a histogram of the quotients of the pairs. It is the ignorance about which risk comes from which dose that invalidates the use of division to estimate the potency factor. Unfortunately, in predictive risk assessments, such individual-based data are often expensive, or even impossible, to collect. The circumstance of having only information on a marginal distribution is probably the usual case in risk analyses. Indeed, the case for which sufficient data have been collected to estimate the joint probability distribution is unusual—even in comparatively well studied problems (Barraj, 1993). In this case, a form of convolution, albeit of a very much more complex kind, is the appropriate method to propagate uncertainty through calculations (Springer, 1979). However, whenever one has only marginal distributions, the use of deconvolution operations should be considered.

In general, as in this example, the appropriate deconvolution can produce a tighter distribution (and a steeper CDF) than does convolution. When a variable for which deconvolution is the appropriate method of estimation is nevertheless estimated by the convolution, the dispersion of the variable can be overestimated. In other words, the tails of the estimated distribution can contain more of the probability than is justified. Using deconvolution when appropriate can therefore *reduce the uncertainty in the results of an analysis* because it yields a distribution with the correct, smaller weight in the tails.

DECONVOLUTION FOR BOUNDING ESTIMATES

The need for deconvolution also occurs in any analyses making bounding estimates, even if they do not involve probabilistic statements of any kind. Such analyses include worst-case analysis, maximally exposed individual (MEI) analysis, and theoretically upper bounding estimate (TUBE) analysis (EPA, 1992), all of which seek upper or lower bounds of one kind or another on exposures, doses, or other variables of concern. In all such analyses, interval arithmetic (Moore, 1966; Alefeld and Herzberger, 1983; Neumaier, 1990) provides the necessary formal quantitative methods for use whenever the problem exceeds the trivially simple.

Let us consider another numerical example. Suppose that we have estimated an acceptable dose of some toxicant and now want to compute the environmental concentration that should not be exceeded in order to ensure that the acceptable dose is not exceeded. To do this, we can use the expression

$$\text{dose} = \frac{\text{concentration} \times \text{intake}}{\text{body weight}}$$

and solve for concentration. Suppose the intake ranges between 1 and 3 L/day, and that we are assuming body weight ranges between 60 and 80 kg. If our estimate of the maximal acceptable dose we want is, say, 3 mg/(kg·day), then what is the range of concentrations we could permit? The answer obtained by straightforward interval analysis

$$\text{concentration} = \frac{\text{dose} \times \text{body weight}}{\text{intake}} = \frac{[0,3] \times [60,80]}{[1,3]} = [0,240]$$

seems reasonable. However, when we check this answer by putting it back into the original formula, we see that

$$\text{dose} = \frac{\text{concentration} \times \text{intake}}{\text{body weight}} = \frac{[0,240] \times [1,3]}{[60,80]} = [0,12]$$

which means we would be allowing doses *four times* larger than we intended. Obviously, something is seriously wrong with this approach.

The error arises because we are computing bounding estimates. The algebraic manipulation we used to solve for concentration is invalid when the underlying variables are bounding estimates rather than precisely known real numbers. Interval arithmetic offers a formula for computing the correct bounding estimate for concentration that would lead to the prescribed dose. (The answer, as we shall see below, is [0, 60] mg/L.) Clearly, the solution is analogous to probabilistic deconvolution because it untangles the relationship that is encoded in the original definition of dose. It is a property of bounding estimates (like probability distributions and unlike point estimates) that special deconvolution operations are required to solve these kinds of problems. Furthermore, any general approach based on a worst-case analysis will encounter this complexity. Ignoring it can lead to either overestimation or underestimation of a particular quantity, depending on the details of the analysis; it would therefore be a mistake to ignore opportunities to use deconvolution whenever it is justified to do so. This is true whether the bound is interpreted as the "worst case," the dose of the "maximally exposed individual," or the "theoretical upper bounding estimate."

HOW TO COMPUTE DECONVOLUTIONS FOR BOUNDING ESTIMATES

In this section, we review the basic arithmetic operations and their associated deconvolutions on intervals of the positive real numbers. This discussion relates directly to bounding analyses such as worst-case, MEI, or TUBE analyses. Because these computations are so simple, they can also be used for hand-calculation checks on the probabilistic deconvolutions generated by computer algorithms.

Consider the intervals A, B, and C of the positive real numbers $\Re^+$,

A = $[a_1, a_2]$,
B = $[b_1, b_2]$ and
C = $[c_1, c_2]$,

such that $0 \le a_1 \le a_2$, $0 \le b_1 \le b_2$, and $0 \le c_1 \le c_2$. Interval analysis (Moore, 1966) allows us to define the basic arithmetic operations on such intervals.

C = A + B	is defined by $[c_1, c_2] = [a_1+b_1, a_2+b_2]$;
C = A − B	is defined by $[c_1, c_2] = [a_1 - b_2, a_2 - b_1]$;
C = A × B	is defined by $[c_1, c_2] = [a_1 \times b_1, a_2 \times b_2]$;
C = A / B	is defined by $[c_1, c_2] = [a_1/b_2, a_2/b_1]$; and
C = A^B	is defined by $[c_1, c_2] = [a_1^{b1}, a_2^{b2}]$.

These are analogous to the probabilistic convolutions implemented with Monte Carlo methods in @Risk and Crystal Ball, by which distributions of random numbers are combined together in arithmetic operation. Indeed, interval analysis on the supports of the random distributions will give the same support that probabilistic convolution yields.

When A and C are known and B is to be estimated by appeal to knowledge about the underlying arithmetic relationship among the variables, the procedure necessary to make the estimate correctly is not what one might assume from experience with the arithmetic of real numbers. In particular, if one knows that A + B = C, one cannot simply subtract A from C to find B. Consider a numerical example. Suppose A is [2, 3] and B is [4, 5]. C, the sum of A and B, is therefore [6, 8]. If we know only A and C and try to estimate B as the difference C − A, we obtain [3, 6], the wrong answer. The correct answer can, however, be recovered with a deconvolution operation

$$B = decon(A, C) = [c_1 - a_1, c_2 - a_2]$$

for which a solution exists whenever $(c_1 - a_1) \le (c_2 - a_2)$.

Trying to estimate B by computing the difference C − A will generally yield the wrong answer, since $[c_1 - a_2, c_2 - a_1] \ne [c_1 - a_1, c_2 - a_2]$ except in special circumstances. In general, the deconvolution will have less uncertainty than the difference, and the improvement can be measured by how much narrower the interval estimating B from deconvolution is compared to that from the difference convolution. This is

$$width (C - A) - width (decon(A,C)) = (c_2-a_1) - (c_1-a_2) - [(c_2-a_2) - (c_1-a_1)]$$

which simplifies to

$$2(a_2 - a_1)$$

which is nonnegative, since $a_1 \le a_2$. Thus, the deconvolution is narrower than the difference by twice the width of A. Indeed, it is a general fact about all forms of uncertainty propagation that deconvolution will yield narrower results than convolution. Ordinarily, arithmetic with uncertain numbers—whether they are intervals or probability distributions—tends only to increase uncertainty in the cascade of mathematical operations. By construction, deconvolution is the inverse operation, so it must tend to decrease overall uncertainty in the reverse way.

There are two deconvolutions for the subtraction operation, depending on whether A or B is the unknown. But both can be expressed in terms of the additive deconvolution function because the subtraction A − B = C can be expressed as A + (−B) = C. Therefore, if A and C are known and B is unknown, it can be estimated as −decon(A, C). If A is the unknown, it can be estimated as decon(−B, C).

For intervals on the positive real numbers, multiplicative deconvolution is the operation

$$factor(A, C) = [c_1/a_1, c_2/a_2]$$

that provides the solution B to

$$A \times B = C$$

when A and C are known. A solution will exist whenever $c_1/a_1 \le c_2/a_2$. The deconvolutions for a division operation are related to the multiplicative deconvolution as the deconvolutions for subtraction are related to additive deconvolution. The solutions are outlined in Table 14.1.

The discussion in this section has so far been limited to intervals taken from the positive real numbers $\Re^+$. While instructive and often useful in many practical cases involving biological variables, the more general situation in which intervals can include both positive and negative numbers should also be addressed. Although addition and subtraction and

Table 14.1. Estimation Formulas for Minimizing Uncertainty. For Suitably Defined Operations and Functions, this Table is Valid Both for All Approaches Based on Interval Arithmetic (Worst-Case Analysis, MEI, TUBE) as Well for a Full-Blown Probabilistic Risk Analysis.

When you know that	and have empirical estimates of	use this formula to find the unknown
$A + B = C$	A, B A, C B, C	$C = A + B$ $B = \text{decon}(A, C)$ $A = \text{decon}(B, C)$
$A - B = C$	A, B A, C B, C	$C = A - B$ $B = -\text{decon}(A, C)$ $A = \text{decon}(-B, C)$
$A \times B = C$	A, B A, C B, C	$C = A \times B$ $B = \text{factor}(A, C)$ $A = \text{factor}(B, C)$
$A / B = C$	A, B A, C B, C	$C = A / B$ $B = 1 / \text{factor}(A, C)$ $A = \text{factor}(1/B, C)$
$A^B = C$	A, B A, C B, C	$C = A^B$ $B = \text{factor}(\log A, \log C)$ $A = \exp(\text{factor}(B, \log C))$

their deconvolutions extend naturally to the general case, multiplication and division for intervals in $\Re$ are made more complex by possible changes in sign. Division for intervals in $\Re$ is also constrained by the necessity that the divisor not include the number zero. Multiplicative deconvolution is considerably more complex if the intervals can include both positive and negative numbers. Kaufmann and Gupta (1988) review this situation. Although extremely cumbersome in hand-calculation, the operations for intervals in $\Re$ have been implemented in computer software (Kuhn and Ferson, 1994).

HOW TO COMPUTE DECONVOLUTIONS FOR PROBABILITY DISTRIBUTIONS

In this section we review the definition of probabilistic convolution and offer a straightforward formula for computing deconvolutions from specified marginal probability distributions.

If A and B represent two probability distributions whose discrete PDFs are given by the functions A(x) and B(y), then the discrete convolution of these two distributions is given by

$$C(z) = \sum_{z=x+y} A(x) \times B(y)$$

C(z) describes that distribution C formed by the sum of random variate pairs from A and B. Strictly speaking, the term "convolution" applies only when we are talking about the sum. However, since there are no special terms that denote the analogous operations for other simple arithmetic functions, when we have need to refer to

$$C(z) = \sum_{z=x-y} A(x) \times B(y)$$

$$C(z) = \sum_{z=x\times y} A(x) \times B(y)$$

$$C(z) = \sum_{z=x/y} A(x) \times B(y), \; y \neq 0$$

$$C(z) = \sum_{z=x^y} A(x) \times B(x)$$

—which, respectively, give the distributions for differences, products, quotients, and powers of pairs of random variates—we will use the terms subtractive, multiplicative, divisive, and exponential convolutions. This usage is justified by the fact that these operations can be reduced to strict, additive convolution by using appropriate transformations involving negation and logarithms.

Although the expressions above could be used to implement convolutions in a computer program, the most commonly used algorithm employs a Monte Carlo method in which random deviates from the A distribution are paired with random deviates from the B distribution, arithmetically combined, and tallied in a histogram to estimate the C distribution. Iman and coworkers (Iman and Conover, 1980; Iman et al., 1981a,b; Iman and Shortencarier, 1984) have described very efficient methods for doing this, which have been implemented in several software packages for doing Monte Carlo analysis.

From the expression that defines additive convolution, we can derive a formula to recover B from given distributions for A and C. Suppose that the C distribution is discretized into n bins each having some fixed width and that the A distribution is discretized into m bins of the same width. Let the notation D_i denote the probability mass in the ith bin of the distribution D. Then the recursive expression

$$\text{decon }(A,C) \approx B_i = \frac{C_i - \sum_{j=1}^{i-1} A_{i-j+1} \times B_j}{A_1}, \; i = 1, \ldots, n - m + 1$$

provides a discrete estimate of B with nm+1 bins. This formulation assumes that the discretization of A can be chosen so that $A_1 \neq 0$.

After B is estimated, the putative solution should be tested by checking that A+B does in fact equal C within the error of approximation. This is necessary because in some cases—depending on the distributions A and C—a solution may not exist.

This estimation method should only be considered as exemplary. Note, for instance, that only the first nm+1 bins of C are used in the estimation. Techniques to make more efficient use of the information contained in C—as well as several other possible algorithmic improvements to minimize the effects of noise—will be obvious to those inclined to implement deconvolution for themselves. Deconvolution is widely used in many diverse fields ranging from spectroscopy (Jansson, 1984) to seismology (Bernabini et al., 1987). Jansson (1984) reviews the literature on deconvolution algorithms. An iterative method (Ferson, 1995) may be useful in some circumstances to produce approximate deconvolutions.

A formulation for multiplicative deconvolution is somewhat more complex because the discretization bins cannot be assumed to have constant width. However, it will often be convenient to estimate multiplicative deconvolution using a simple transformation such as

$$factor\ (A,C) = \exp\ (decon\ (\log A, \log C))$$

which will provide an excellent approximation in many practical circumstances.

We know of no method using purely Monte Carlo techniques to deconvolve probability distributions in the general case. This means it would be awkward (although still not infeasible) to implement deconvolutions in extant software packages such as Crystal Ball or @Risk. However, we note that performing a deconvolution needs no more empirical information than that which is already required by a Monte Carlo convolution. Therefore, insufficient data should never be a reason to use convolution over deconvolution. Since the two operations will generally yield different results, using the correct one in a given circumstance will be worth some additional effort.

DISCUSSION AND CONCLUSIONS

Deconvolution is an underutilized technique that can often reduce uncertainty both in probabilistic and traditional worst-case risk analyses. For this reason, the use of deconvolution should be considered in any comprehensive risk assessment. In addition to the benefits of uncertainty reduction, there are also serious hazards that can arise from failing to use deconvolution in situations when it is warranted. Although convolution yields a wider distribution than deconvolution, this alone does not guarantee the final answer will enclose the true value(s). In some situations, using convolution instead of deconvolution will yield erroneous answers that are too narrow and therefore represent wishful thinking.

Table 14.1 spells out which formula will give the best (narrowest) results justified given the nature of the underlying relationship among the variables and estimates for two out of three of them. The formulas in the table apply both to probabilistic and traditional (bounding estimate) risk analyses. In a probabilistic analysis, the arithmetic operators (+, −, ×, /, superscripting) represent the respective probabilistic convolutions, and the functions "decon" and "factor" represent the associated deconvolution functions. For a traditional analysis, the symbols +, −, etc., represent interval addition, subtraction, etc. (Moore, 1966), and "decon" and "factor" represent the interval-valued functions defined above for positive intervals (or by Kaufmann and Gupta (1988) for the general case).

In probabilistic assessments, the choice between convolution and deconvolution turns on a question about independence assumptions. It is logically impossible for the three variables in these equations all to be mutually independent of one another. Because it makes a difference in the calculation which two variables are assumed independent, the primary question to ask is where the independence lies. Under the equation A+B=C, for example, if A and B are independent, then convolution should be used. But if C and one of the other variables are independent, then deconvolution must be used to get the correct answer. The table specifies what operations to use when the variables on the left side of the equal sign in the first column are assumed to be independent. In probabilistic risk assessment as it is currently practiced, analysts commonly assume that the elements of every pair of variables entering the calculation are independent. They ignore the fact that the result of an assessment depends, sometimes very strongly, on exactly which variables they make this assumption about. Failing to explicitly consider whether and how an assumption of independence should be used in calculations will inevitably lead to erroneous results. The magnitude of the error can range from minor to severe.

In a traditional, nonprobabilistic risk assessment, there is no such notion of independence. Nevertheless, methods analogous to deconvolution are still needed to untangle the equations. In this setting, the question is about which mathematical expressions are considered to be defining. If C is defined to be A+B, and empirical estimates are available for A and B, then convolution should be used. But if C and either A or B are empirically estimated, then deconvolution is required to compute an estimate for the unknown. One cannot escape the necessity of having both convolution and deconvolution, because which variables can be empirically estimated in practice and how the relationship among the variables is defined in theory are separate considerations.

Of course, deconvolution is no panacea in either methodological setting. Nor can it be used in all circumstances in which an analyst might like to use it. A deconvolution does not necessarily exist for every possible pair of variables. In general, if there is more uncertainty in A than in C then one cannot deconvolve A out of C. For instance, if A is the interval [1,4] and C is [1,2], there is no interval B such that A + B = C. In this case, and whenever the best estimate for A is wider than the estimate for C, deconvolution will not be useful in computing an estimate for B. Similar considerations apply in the probabilistic case to the supports of the respective distributions. In practice, some empirical information obviously counterindicates the use of deconvolution.

Assumptions that amount to using convolution are often hidden within the derivation of the model supplied to a Monte Carlo analysis. In the original deterministic model, which uses real numbers, these assumptions are perfectly valid, and are commonly made at many points in the derivation of a model. Within a Monte Carlo or worst-case analysis, however, the assumptions are sometimes invalid and can lead to incorrect estimation of the results. Whether one should use convolution or deconvolution to combine two variables in a particular case is a somewhat subtle consideration that depends on the defining form of the equation that relates them, or upon which variables in this relationship are assumed to be independent. The risk analyst must determine which estimation technique to use in any particular case.

Beyond simple convolution or deconvolution, there are methods that can be used in cases in which independence cannot be assumed between *any* two variables in an equation. When one has empirical information about the observed correlation between two variables, this information can be incorporated into a Monte Carlo convolution by generating deviates from marginal distributions with a specified product-moment (Scheuer and Stoller, 1962) or rank correlation (Iman and Conover, 1982). Of course, having such information is rare in real-world problems. When empirical information is so sparse that one cannot assume (or is unwilling to assume) anything about the correlation or dependency among two variables, then dependency bounds analysis (Williamson and Downs, 1990; Ferson and Long, 1994) can be used to compute bounds on the distribution of their arithmetic combination (sum, etc.) and thereby make appropriate estimations.

ACKNOWLEDGMENTS

We thank Lev Ginzburg (State University of New York at Stony Brook), Rüdiger Kuhn (Universität Bielefeld), David Burmaster (Alceon Corporation), Mark Burgman (University of Melbourne) and Lorenz Rhomberg (Harvard University) for helpful discussions. Publication of this manuscript was made possible by a Small Business Innovation Research (SBIR) grant (Number 1 R43 ES0 6857-01A1) to Applied Biomathematics from the National Institute of Environmental Health Sciences (NIEHS) of the National Institutes of Health.

REFERENCES

Alefeld, G. and J. Herzberger. *Introduction to Interval Computations.* Academic Press, New York, New York, 1983, p. 333.

Barraj, L.M. Evaluating total multiple pesticide exposure residues in food: methods and implications, paper presented at the 1993 Annual Meeting of the Society for Risk Analysis, 1993.

Barton, W.T. Response from Palisade Corporation. *Risk Anal.* 9, pp. 259–260, 1989.

Bernabini M., P. Carrion, G. Jacovitti, F. Rocca, and S. Treitel. *Deconvolution and Inversion.* Blackwell Scientific Publications, Oxford, U.K., 1987, p. 355.

Burmaster, D.E. and E.C. Udell. A review of Crystal Ball. *Risk Anal.* 10, pp. 343–345, 1990.

EPA. Risk assessment guidelines of 1986. *Federal Register* 51, pp. 33992–34054, 1987.

EPA (Environmental Protection Agency). Guidelines for exposure assessment of 1986. *Federal Register* 57, pp. 22888–22938, 1992.

Ferson, S. Using approximate deconvolution to estimate cleanup targets in probabilistic risk analyses, submitted for the proceedings of the Ninth Annual Conference on Contaminated Soils, P.T. Kostecki, Ed., Amherst Scientific Publishers, Amherst, MA, 1995.

Ferson, S. and T.F. Long. Conservative uncertainty propagation in environmental risk assessments, in *Environmental Toxicology and Risk Assessment,* Third Volume, ASTM STP 1218, Hughes, J.S., G.R. Biddinger, and E. Mones, Eds., American Society for Testing and Materials, Philadelphia, PA, 1994.

Gabel, R.A. and R.A. Roberts. *Signals and Linear Systems.* John Wiley & Sons, New York, 1973, p. 415.

Iman, R.L. and W.J. Conover. Small sample sensitivity analysis techniques for computer models, with an application to risk assessment. *Commun. Stat.* A9, pp. 1749–1842, 1980.

Iman R.L. and W.J. Conover. A distribution-free approach to inducing rank correlation among input variables. *Commun. Stat.* B11, pp. 311–334, 1982.

Iman, R.L. and M.J. Shortencarier. *A Fortran 77 Program and User's Guide for the Generation of Latin Hypercube and Random Samples for Use with Computer Models* (NUREG/CR-3624, SAND83-2365). Sandia National Laboratories, Albuquerque, New Mexico, 1984.

Iman, R.L., J.C. Helton, and J.E. Campbell. An approach to sensitivity analysis of computer models, Part 1. Introduction, input variable selection and preliminary variable assessment. *J. Qual. Technol.* 13, pp. 174–183, 1981a.

Iman, R.L., J.C. Helton, and J.E. Campbell. An approach to sensitivity analysis of computer models, Part 2. Ranking of input variables, response surface validation, distribution effect and technique synopsis. *J. Qual. Technol.* 13, pp. 232–240, 1981b.

Jansson, P.A., Ed. *Deconvolution with Applications in Spectroscopy*. Academic Press, Orlando, FL, 1984, p. 342.

Kaufmann, A. and M.M. Gupta. *Fuzzy Mathematical Models in Engineering and Management Science*. North-Holland, Amsterdam, The Netherlands, 1988, pp. 70f.

Kuhn, R. and S. Ferson. *Risk Calc*. Applied Biomathematics, Setauket, New York, 1994.

Lloyd, K.J., M.M. Thompson, and D.E. Burmaster. Probabilistic techniques for backcalculating soil cleanup targets, in *Superfund Risk Assessment in Soil Contamination Studies* (ASTM STP 1158, Hoddinott, K.B. and G.D. Knowles, Eds., American Society for Testing and Materials, Philadelphia, PA, 1992.

Morgan, M.G. and M. Henrion. Demos: a case study in computer aids for modeling uncertainty, in *Uncertainty: A Guide to Dealing with Uncertainty in Quantitative Risk and Policy Analysis*, Cambridge University Press, Cambridge, UK, 1990.

Moore, R.E. *Interval Analysis*. Prentice-Hall, Englewood Cliffs, NJ, 1966, p. 145.

Neumaier, A. *Interval Methods for Systems of Equations*. Cambridge University Press, Cambridge, UK, 1990, p. 255.

Norton, S., M. McVey, J. Colt, J. Durda, and R. Hegner. *Review of Ecological Risk Assessment Methods*, EPA/230-10-88-041. Environmental Protection Agency, Office of Policy Analysis, Washington, DC, 1988.

Salmento, J.S., E.S. Rubin, and A.M. Finkel. A review of @Risk. *Risk Anal.* 9, pp. 255–257, 1989.

Scheuer, E.M. and D.S. Stoller. On the generation of normal random vectors. *Technometrics* 4, pp. 278–281, 1962.

Springer, M.D. *The Algebra of Random Variables*. Wiley, New York, 1979, p. 470.

Williamson, R.C. and T. Downs. Probabilistic arithmetic I: Numerical methods for calculating convolutions and dependency bounds. *Int. J. Approx. Reason.* 4, pp. 89–158, 1990.

Part 4
Conclusion

15 ||| Summary

Carl L. Strojan and Michael C. Newman

OTHER NEEDFUL THINGS

Toward the end of the Dickens' novel, *Hard Times*, Mr. Gradgrind realizes that the "Facts" quoted in the introductory chapter of this book are not at all the one needful thing in life. Such a realization is also crucial to scientific progress and moving beyond the boundaries of "normal science" to use the terminology of Kuhn (1970). Facts are certainly critical, but they must be used in the context of a conceptual or quantitative framework to generate understanding. Furthermore, our ability to know facts may be limited by the methods we use to generate them. As pointed out in Chapter 1, four things needed to improve risk assessment are relevant data, a logical framework, quantitative measurement techniques, and appropriate models.

Implicit in the title of this book is an assumption that risk assessment is a finite process, that there is logic to the process, and that measurement is an important part of the process. The development and evolution of risk assessment into its current form indicate that this indeed is the case. Each of the foregoing chapters provides insight into some aspect of the logic and measurement inherent in risk assessment, while also pointing out current limitations and suggesting ways to improve the risk assessment process.

Risk Assessment

Risk assessment is based on the principle that if sufficient information is known about a system and the various factors that affect it, then the probability of certain adverse events happening can be estimated with some degree of confidence. This principle is the foundation upon which the entire insurance industry is built, and it is the same principle that forms the basis for human and ecological risk assessment.

While relatively simple in concept, risk assessments can be difficult to carry out, particularly for complex systems where adequate information may not exist. This is frequently the case for ecological systems, which are usually characterized by many interacting species that vary over time and space, and which may be subject to multiple, competing risks at the same time. Furthermore, simply defining risk to ecological systems may not be totally straightforward. Suter (1993) defined risk as "the probability of a prescribed undesired effect." Competing definitions are given in several of the preceding chapters (cf., Chapters 4, 5, 6, and 7). In part these various definitions reflect the relatively young and changing nature of the field, as well as individual interpretations of the authors. Fortunately, the definitions are variations of a similar theme rather than major differences of opinion. All of the definitions have in common that risk is characterized by the *probabilistic* occurrence of a specified adverse effect. The expression of risk in terms of its probability of occurrence is what most distinguishes modern risk assessment from more qualitative forms of impact assessment. As is shown throughout the various chapters, however, the application of probabilistic ecological risk assessment is frequently still an ideal rather than a reality.

Logic

Bascietto (Chapter 2) describes some of the historical events associated with the development of ecological risk assessment, and the trend away from traditional reliance on qualitative approaches and descriptive ecology to more quantitative methods. As described in Chapter 2 and other chapters (viz., 3, 4, and 6), the logic underlying the risk assessment process is spelled out in a series of documents published in recent years. For example, in 1983 the National Research Council published a report titled *Risk Assessment in the Federal Government: Managing the Process* (NRC, 1983), which built the foundation for modern risk assessment. The report proposed a conceptual basis for managing risks to human health through a process involving research, risk assessment, and risk management. The NRC risk assessment process was divided into four components:

- hazard identification
- dose-response assessment
- exposure assessment
- risk characterization

Having focused just on risks to human health, the NRC approach had certain limitations when applied to ecological risk assessments. For example, in human health risk assessments, individual humans or populations are usually the focus of concern and harmful effects may be illness, injury, or death. In contrast, ecological risk assessments may need to focus on populations or communities of many species, and harmful effects may be more wide-ranging. Such concerns led to the development of a complementary model for determining ecological risks that was published by the Environmental Protection Agency as *A Framework for Ecological Risk Assessment* (EPA, 1992). The EPA framework, in its simplest form, has the following three components:

- problem formulation
- analysis, including characterization of exposure and characterization of ecological effects
- risk characterization

These components are illustrated in Figure 4.2, and they are elaborated upon in Figures 4.3 to 4.7. It should be noted that neither the NRC process nor the EPA framework is a prescriptive guideline for conducting risk assessments.

Conceptual and quantitative models are useful ways to approach a topic in a logical manner. Chapters 4, 5, and 6 review various models and their application in risk assessment. Voit and Schubauer-Berigan propose the use of certain types of canonical models as a unifying framework for risk assessments. Mauriello et al. provide an example of how data from field studies can be used in simulation models to estimate risk from a chemical— 4-nonyphenol—in aquatic ecosystems. In Chapter 5, Campbell and Bartell point out that the real issue in determining whether models can contribute to ecological risk assessment is not validity, but credibility. Models for use in risk assessment may come into play at many levels of biological organization, including molecular, cellular, individual, population, community, ecosystem, and landscape levels. Models have certain appealing aspects for use in risk assessments, including the potential to project over space and through time. Their greatest contribution may be still to come, however, if they can be used to solve the very difficult problem of determining risk from multiple stressors acting simultaneously.

Although the logic behind risk assessments may be relatively simple and compelling, the actual application of risk assessments may not be. In Chapter 3, for example, Ryan discusses the difficulty and complexity associated with doing just the exposure assessment component in human risk assessments. The historical development of exposure assessment presented in this chapter also documents an increased sophistication over time in conceptual approaches and experimental methodology that has clearly led to a better understanding of exposure.

As difficult and complex as accurate human risk assessments are to conduct, ecological risk assessments are usually even more so. Chapters 3 and 6 discuss some of the similarities and differences in human and ecological risk assessments. Chapter 7 uses a case study approach to compare and contrast human and ecological risks from exposure to radiation. In Chapter 7, Hinton identifies the following factors as being responsible for making it difficult to accurately determine probabilistic risk for ecological systems:

- Measuring the dose received by most nonhuman species is a difficult task.
- Tables of dose conversion factors that transform a unit concentration of contaminant to a dose received by the organism exist for only a few nonhuman species.
- It appears there are no risk factors established for nonhumans, i.e., the ability to equate the body burden or dose of a contaminant to a probabilistic risk is limited.
- There are difficulties in interpreting the effects of sublethal endpoints.

In spite of the difficulties in trying to quantify human and ecological risk, the logic of doing so remains sound. Limitations to the process usually occur as a result of insufficient data or lack of appropriate methodology.

Measurement

Quantitative risk assessments may frequently be limited by insufficient or inappropriate data. This makes the measurement and analysis portion of risk assessments more important than it might otherwise have to be. The foregoing chapters contain a wealth of information on specific methodological approaches and their application in various case studies.

In Chapter 7, Hinton uses a case-study approach to estimate risk from exposure to radiation. This is an area where enormous efforts and expenditures have been made over the years to quantify dose rates, determine exposure pathways, derive risk factors, and develop databases—at least for humans. As a result, this is an area where the probabilistic determination of risk is best developed and its usefulness most clearly seen. On the ecological side, the absence of similar information means that conservative assumptions must be made to account for extreme possibilities. This in turn very likely inflates the true risk that exists.

Having insufficient data is frequently a major problem in risk assessments. Data that do exist may also need to be used for purposes that are not ideal, e.g., extrapolating between acute and chronic exposures, or across different levels of biological organization. This requires an understanding of the relationship between what is actually being measured and the effect that is being assessed. In Chapter 8, Suter reviews a variety of effects extrapolation models that can be used for exactly these purposes. These begin with the relatively simple technique of classification, which assumes that observed responses are representative of some broader class of responses. Another approach is to multiply a response by some factor to account for uncertainties or biases in the relationship between test and assessment endpoints. Distribution techniques, such as the species sensitivity distribution, assume that an assessment endpoint is a random variable represented by the distribution of test endpoints. Regression analysis is a commonly used extrapolation technique. Finally, allometric scaling models can be used for extrapolation based on the fact that responses to a contaminant may vary among organisms or species as a function of their size or weight.

Historically, most toxicity testing has been done on individual test organisms, under controlled conditions, and for specific toxicants. Kooijman et al. (Chapter 9) review the use of so-called static methods, such as analysis of variance and the logit model, that quantify effects at a standardized exposure time. They point out limitations of these methods, including the fact that information about the rate at which effects build up during exposure is not used to predict effects after prolonged exposure. This is usually done simply by using extrapolation factors. The authors contrast these static approaches with a dynamic method that quantifies effects as functions of contaminant concentration and exposure time. The dynamic model has three essential components: (1) a kinetics component that links internal concentrations to external concentrations, (2) an effects component that links effects on a target parameter to the internal concentration, and (3) a physiological component that links output variables, such as body length, number of offspring, etc. The Dynamic Energy Budget (DEB) model is proposed and illustrated as the simplest choice that links all essential processes: feeding, digestion, respiration, maintenance, growth, development, reproduction, and aging.

Chapters 5, 10, and 11 cover models and measurement techniques that assist with understanding environmental risk at higher levels of ecological organization, such as communities and landscapes. Matthews et al. suggest that three major tasks are needed to apply community-level toxicity testing to environmental risk assessment. These are: (1) creating an appropriate community-level data set that is based on measurements likely to show a direct or indirect effect from exposure to a contaminant; (2) extracting information from the data set through appropriate analytical methods to find both expected and unexpected patterns; and (3) converting the information from the data set into a set of expectations or recommendations that address risk. Creating an appropriate data set obviously begins with defining a study's objectives and then determining an appropriate experimental design. In terms of extracting information, Matthews et al. discuss many specific examples involving four general techniques for extracting information from multivariate data sets: indices, metrics, classifiers, and clustering. Incorporating this information into the risk assessment and risk management processes then involves determining the kinds of community-level changes that are acceptable, and those that are not. This means not just statistically acceptable changes, but also those that are politically, economically, and socially acceptable.

In Chapter 11, Richards and Johnson point out the importance of considering spatial and temporal scales when trying to detect change at the landscape level. Measurements at scales that are smaller than appropriate may be too sensitive, while measurements at scales that are larger than appropriate may be too insensitive, and thus miss detecting patterns of change. These considerations obviously apply to all levels of environmental analysis, but they may be more apparent at the landscape level because most sampling and analysis methods have been developed for use at the individual, population, or community levels. Another important factor in assessing risk at the landscape level is the selection of appropriate endpoints for estimating effects. A number of these are discussed, such as physical and chemical endpoints, as well as population, community, ecosystem, and regional endpoints, and a case study is presented that demonstrates their applicability. Still, there are factors that currently limit the integration of landscape-level processes with risk assessment, such as extrapolation across time and space, quantification of uncertainty, validation of predictive tools, and appropriate databases. The development of new approaches and regional databases, along with analytical tools such as Geographic Information Systems, suggest this is an area where considerable progress may be made in the future.

Some quantitative methods can be used at different scales and across many levels of biological organization. For example, risk assessment, along with many other scientific fields, typically involves measurement techniques and statistical methods that seek to determine whether two or more categories or treatments are statistically different from one another. In such situations, statistical analyses of data are used to evaluate a null hypothesis of no significant difference. If the data are statistically different, they are usually assumed to be biologically different. Dixon, in Chapter 12, argues that this approach is inappropriate when the intent is to show that categories or treatments are equivalent, e.g., such as whether two sampling sites are equivalent. The reason is that the classical null hypothesis, i.e., no significant difference, may be accepted if the true difference is close to zero, if the number of replicates is too small, or if random variation is too large. As an alternative, Dixon discusses and illustrates the use of equivalence testing, in which scientifically relevant boundaries are chosen *a priori* for an equivalence region within which two groups are considered to be equal. This is an innovative approach that involves not only a different way of doing statistical analyses, but also a fundamentally different way of looking at an issue.

One attraction of quantitative risk assessments is that they seem to confer a degree of certainty to any results that are obtained. It should be remembered, however, that the data used in such assessments, as well as the measurements used to generate the data, have some degree of variability and uncertainty associated with them. The final two chapters discuss the important topic of uncertainty and ways to deal with it. In Chapter 13, Kirchner points out that most environmental risk assessments must deal with uncertainties due to natural variability and a lack of knowledge about model parameters. As this uncertainty is propagated through models, there has been a tendency to estimate risk conservatively as a way to compensate for this uncertainty. There is a developing trend, however, to produce realistic estimates of risk, rather than conservative estimates, to identify components along with risk estimates, and to partition the contributions from subjective and objective uncertainties.

In the final chapter of the book, Ferson and Long describe the importance of considering convolution and deconvolution procedures as ways to reduce uncertainty, both in probabilistic and traditional worst-case risk analyses. They illustrate that, when dealing with probabilistic distributions, one cannot employ simple algebraic computations. Instead, methods outlined in the chapter must be used to do probabilistic analyses of risk.

CONCLUSION

It is our hope that the ideas and methods set forth in this book will be useful now and in the future to students and professionals in the field of risk assessment. More importantly, we hope they stimulate others to continue developing innovative ideas and better quantitative methods. As pointed out in Chapter 1 and repeated at the beginning of this chapter, four things needed to improve risk assessment are relevant data, a logical framework, quantitative measurement techniques, and appropriate models. By continuing to make improvements in these areas, we can enhance the scientific foundation of risk assessment and the credibility of the overall process.

ACKNOWLEDGMENT

This work was supported by Financial Assistance Award Number DE-FC09-96SR18546 from the U.S. Department of Energy to the University of Georgia Research Foundation.

REFERENCES

EPA. *Framework for Ecological Risk Assessment*. EPA/630/R-92-001. Washington, DC, 1992.

Kuhn, T.S. *The Structure of Scientific Revolutions*. 2nd ed. University of Chicago Press, Chicago, IL, 1970, p. 210.

NRC (National Research Council). *Risk Assessment in the Federal Government: Managing the Process*. National Research Council, National Academy Press, Washington, DC, 1983, p. 191.

Suter, G.W., II. *Ecological Risk Assessment*. Lewis Publishers, Boca Raton, FL, 1993, p. 538.

Index